Real Property in Australia

T0295740

Real property in the form of investment, ownership and use pervades almost every aspect of daily lives and represents over 40% of Australia's wealth. Such assets do not exist in isolation – they are dynamic and forever evolving, impacted by a range of physical, economic, demographic, legal and other forces.

Consequently, a true appreciation of individual assets and of the property sector as a whole demands an understanding of both the assets themselves and the context and markets in which they exist. The sector is complex and, on the face of it, confusing. It is however, not without logic and underlying themes and principles.

This book provides a wider understanding of how the real property sector works. It covers topics such as the nature of real property and its functions, economic drivers, valuation principles, legal and tenure parameters, property taxation, land development and subdivision, asset and property management and sustainability – all critical components in this complex and critically important sector. It provides a wide and balanced perspective for experienced practitioners, investors, students and anyone involved in property decision-making or wishing to secure a deeper understanding of these areas.

The book integrates research-based theory with practical application and first-hand insights into a sector that underpins the Australian economy, its communities and its sustainability.

Michael J. Hefferan is the recently retired Pro Vice-Chancellor (Engagement) and Professor of Property and Development at the University of the Sunshine Coast.

He has a PhD, Masters of Applied Science and postgraduate management qualifications. He is a Registered Valuer, a fellow of the Australian Property Institute and was a long-term fellow of the Royal Institution of Chartered Surveyors, Urban Development Institute of Australia and of the Australian Institute of Company Directors. He is a past president of the Australian Property Institute, Queensland Division, and past Chair of that organisation's National Education Board.

He was previously Director of the Institute for Sustainable Resources at the Queensland University of Technology, Brisbane – a major cross-faculty research organisation with particular interest in resources management, sustainable built environments and social dimensions of change.

For almost 20 years previously, he was an executive director within the Queensland State Government, with direct responsibility for industry and regional development. In those areas, he managed some of the largest development projects and commercial and industrial property portfolios in the State.

His particular research interests include issues relating to statutory valuation and property taxation, building use and adaptation, master planned communities, sustainability within the built environment and regional development.

He was, for many years, co-editor of the *Australasian Journal of Regional Studies* and continues with academic and research involvement with the University of the Sunshine Coast and Queensland University of Technology and, in North America, with Purdue University, Indiana and Memorial University, Newfoundland. He consults widely and is involved in numerous programme and project reviews for governments and international bodies.

He has now been awarded Emeritus status by his home university.

Real Property in Australia

Foundations and Application

Michael J. Hefferan

THIRD EDITION

Routledge
Taylor & Francis Group

LONDON AND NEW YORK

Third edition published 2021
by Routledge
2 Park Square, Milton Park, Abingdon, Oxon, OX14 4RN

and by Routledge
52 Vanderbilt Avenue, New York, NY 10017

Routledge is an imprint of the Taylor & Francis Group, an informa business

© 2021 Michael J. Hefferan

First edition published by Mackenzie Green Research and Publishing 2012
Second edition published by Mackenzie Green Research and Publishing 2016

British Library Cataloguing-in-Publication Data
A catalogue record for this book is available from the British Library

Library of Congress Cataloging-in-Publication Data
A catalog record has been requested for this book

ISBN: 978-0-367-48589-4 (hbk)
ISBN: 978-0-367-48588-7 (pbk)
ISBN: 978-1-003-04178-8 (ebk)

Typeset in Bembo
by codeMantra

for
Zoë, Ruby, Elsa and Jarvis

Contents

Figures

Tables

Preface

Real property (land, buildings and infrastructure) is integral to human existence – providing sustenance, shelter, a store of wealth and 'space and place' for practically all human interactions and economic activity.

It is, however, far more than a simple backdrop to events and its characteristics and markets are never homogenous and rarely synchronous. It is a dynamic asset class positioned within a dynamic environment. Each of the literally millions of individual parcels is unique, different from all others in ways that are sometimes subtle but often striking and dramatic. Further, numerous economic/financial, political, legal and social and community forces continually influence decision-making and use. The COVID 19 pandemic provides a stark example of the potential impact of those forces. Time lags in property development and dealings further distort the normal 'stimulus–market reaction' relationship.

For all of that, it is incorrect to consider that the market is chaotic or without rules or rationale.

This text makes two fundamental observations. The first is that, despite all of that diversity and seemingly disordered activity, there are fundamental principles that determine how the sector behaves. As with any capitalist market, basic drivers are economic but here nuanced to reflect the unique character and wider role that real property fulfils within the social order. These principles or 'general rules' create a framework by which the overall property sector and its numerous sub-sectors and variations can be reasonably understood.

The second, fundamental observation is related to that. If (as is patently obvious) property assets do not exist in isolation, then studies in this area cannot simply be 'the study of real property'. Before any of those parcels can be truly understood, managed and predicted, the physical, economic, legal, community, social and political 'context' of that property must be appreciated.

In a capitalist country such as Australia, real property is typically identified in economic terms – combined with labour, capital and entrepreneurial skill and knowhow – as underlying factors of production. As a functional asset, real property's effective and efficient use is critical to economic activity, prosperity and income and wealth creations for individuals, households, firms and communities.

In the wider context too, real property represents a key determinant of liveability, identity and association, culture and custom. It might be reasonably argued that the dominance of economic parameters in property activities needs to be tempered to some extent to accommodate contemporary issues including urbanisation, globalisation, sustainability and climate change and growing inequality in wealth distribution.

Real property is quite dissimilar from those others factors; its basic component (land) is not man-made and its use or misuse has repercussions far beyond the normal primacy of self-interest that underpins fundamental economic rationale. Consequently, to consider real property simply in economic terms (as important as that is) is to grossly underestimate its importance and overall impact. A more holistic view is required.

To make the situation worse, real property is typically controlled and managed through its component parts – design, finance, construction, sales, management and servicing, et cetera – and, perhaps, with insufficient consideration of how all these elements combine to effect function and outcome.

A comprehensive appreciation also needs to recognise the long-time frames typical of real property investment, development, use and sustainability (in the wider and correct use of that word).

This is the third edition of this text. The support, feedback and input by both academic and industry colleagues leading to its publication have been of great value and are received with thanks. Based on that feedback, several additions have been made including new chapters on compulsory acquisition and summary observations of a range of urban and rural land uses. Additional sections have been added to existing chapters, particularly as regards Native Title and advances in carbon capture and related initiatives. Updated statistics and a range of references to published research that have emerged since previous editions have also been included.

In this edition, Chapters 1 to 7 identify the philosophy, academic theory, history, law and public policy in Australia upon which real property activities and dealings are based. The latter chapters, 8 to 15, describe in more detail various specialist areas pertaining to the property sector, individual properties and the interface with the wider environment.

The overall approach and rationale of the text remains the same as previous editions. It aims to provide an overview of the ownership, use and development of real property in the Australian context to a readership that includes students, practitioners, other professionals and members of the wider community seeking a greater understanding of the sector beyond sales and transactions alone. The book does not presume detailed prior knowledge.

As reflected in the various chapters, the topics addressed in the text are diverse. Many such study areas are worthy of an entire text in their own right and, indeed, in a number of cases, such specialist books exist. For each of these topic areas, this work offers an insight and provides a sound working understanding but, in the interests of conciseness, must confine the depth of analysis of each of those areas. In all cases, however, comprehensive referencing is provided to facilitate further study.

Chapter themes reasonably align with various knowledge fields in tertiary property programmes. The book can be read in its entirety, but each chapter is discrete in its own right. Due to this approach, there are some minor duplications between chapters.

References are presented in chapter order to assist further study on specific areas addressed within each chapter – and again, for completeness, there are minor duplications in those schedules.

In this contemporary area of study, a problem frequently arises where terms in common use do not have a precise or agreed definition (even the whole area of discipline is variously called 'real property', 'property', 'real estate', 'property assets', 'land' and a range of others). As much as possible, definitions of key terms and concepts are included in the text. A wider Glossary of Terms has recently been completed by the Property Council of Australia https://www.propertycouncil.com.au/Web/EventsServices/Research_Services/Glossary_of_Term.aspx and provides a valuable additional reference here.

Most of the legislative and administrative structure for real property and real property dealings in Australia are state based. While there is now a high degree of commonality across all states, only general observations about these arrangements can be made in this text. More detailed investigation and research will be required to appreciate local variations and parameters within each jurisdiction.

Emeritus Michael J. Hefferan, PhD
Sunshine Coast
Queensland, Australia

Abbreviations

ABARE	Australian Bureau of Agricultural and Regional Economics
ABN	Australian Business Number
ABS	Australian Bureau of Statistics
ACCC	Australian Consumer and Competition Commission
ACT	Australian Capital Territory
ADB	Asian Development Bank
ADR	Alternative Dispute Resolution
AI	Artificial Intelligence
AIHW	Australian Institute of Health and Welfare
API	Australian Property Institute
APRA	Australian Prudential Regulation Authority
A-REIT	Australian Real Estate Investment Trust
AS	Australian Standards
ASFA	Australian Superannuation Funds Association
ASIC	Australian Securities and Investment Commission
ASX	Australian Stock Exchange
ATIC	Australian Trade and Investment Commission
ATO	Australian Taxation Office
BA	Building approval
BAS	Business Activity Statement
BIM	Building Information Modelling
BREEAM	Building Research Establishment Environmental Assessment Method
BRI	Building-Related Illness
CAD	Computer-Assisted Design
CAPM	Capital Asset Pricing Model
CBD	Central Business District
CCS	Carbon Capture and Storage
CFC	Chlorofluorocarbon
CGT	Capital Gains Tax
COAG	Council of Australian Governments
CPI	Consumer Price Index
DA	Development Approval
DOGIT	Deed of Grant in Trust
EBIT	Earnings before Interest and Tax
EFT	Electronic Funds Transfer

EOI	Expression of Interest
ESD	Ecologically Sustainable Development
ESP	Environmentally Sustainable Practice
EU	European Union
FIRB	Foreign Investment Review Board
GDP	Gross Domestic Product
GFC	Global Financial Crisis
GHG	Green House Gas
GIS	Geographic Information System
GLA	Gross Lease Area
GOC	Government-Owned Corporation
GOE	Government-Owned Enterprises
GST	Goods and Services Tax
HWM	High Water Mark
IAQ	Indoor Air Quality
ICOMOS	International Council on Monuments and Sites
ICT	Information and Communication Technology
IFRS	International Financial Reporting Standards
ILUA	Indigenous Land Use Agreement
ISO	International Standards Association
IMF	International Monetary Fund
IRR	Internal Rate of Return
IVSC	International Valuation Standards Council
KPI	Key Performance Indicator
LEED	Leadership in Energy and Environmental Design
MAT	Moving Annual Turnover
MCA	Material Change of Use
MPC	Master Planned Community
NABER	National Australian Building Environment Rating Scheme
NBFI	Non-Bank Financial Intermediary
NBN	National Broadband Network
NGO	Non-government organisation
NIR	Notice of Intention to Resume
nla	Net Lettable Area
NPV	Net Present Value
NNTT	National Native Title Tribunal
NTA	Native Title Act
NTBC	Native Title Body Corporate
OECD	Organisation for Economic Cooperation and Development
OTA	On-line Travel Agency
PAYE	Pay As You Earn
PCA	Property Council of Australia
p:e	Price to Earnings (ratio)
PPP	Public Private Partnership
RBA	Reserve Bank of Australia
REIA	Real Estate Institute of Australia
REIT	Real Estate Investment Trust

RICS	Royal Institute of Chartered Surveyors
RNTBC	Registered Native Title Body Corporate
ROI	Return on Investment
SBS	Sick Building Syndrome
SCCA	Shopping Centre Council of Australia
SME	Small-to-Medium-Enterprise
SV	Site Value
TAFE	Technical and Further Education
TBL	Triple Bottom Line
TDR	Transferable Development Rights
TICA	Tenancy Information Centre of Australia
TOD	Transport-Orientated Development
TVM	Time Value of Money
UCL	Unallocated/Unalienated Crown Land
UDIA	Urban Development Institute Australia
UN	United Nations
UV	Unimproved Value
VAV	Variable Air Volume
vcl	Vacant Crown Land
WALE	Weighted Average Lease Expiry
WTO	World Trade Organisation

Chapter 1

A contemporary approach

1.1 Introduction

It seems everyone has an opinion on real property (or 'real estate' as it is commonly called); and their interest is understandable and justified. For most Australians, real estate (in the form of the home they may own) represents their only large-scale financial asset. Furthermore, any investment they might have in superannuation would likely derive part of that fund's income from property in one form or another.

Some opinions regarding real property are based on experience. However, even at corporate and government levels, many opinions and decisions on property investment and utilisation are only based on only a generalised understanding of how property markets really work.

There is, in fact, a large body of theoretical and practical knowledge that can provide a framework for sound decisions. That is based on longstanding economic and legal principles; industry and community standards and government regulation.

This chapter provides an overview of that framework and, in general terms, the study of property. It reflects on how, in a comparatively short period of time, real property has grown into a sophisticated and discrete discipline. The chapter also puts forward some fundamental principles and structures that will assist in contemporary study and research.

1.2 The evolution of studies in real property

The comprehensive and cohesive study of real property is a relatively new discipline and remains an area that is demonstrably under-represented in formal, public-domain research. This lack of attention is remarkable when one considers that the whole of the physical world has just three, underlying components: the natural environment, human activity and the built form (i.e. real property in its various forms).

In a wider context, considerable scientific endeavour had been applied to the related areas of the 'natural sciences' since the Age of Enlightenment (the late seventeenth and early eighteenth centuries). These advances were particularly prevalent in Britain and, subsequently, in the US (Himmelfarb 2005). Much of this knowledge is relevant to real property and its links with the wider physical, economic, social and political realms. However, this largely science-based knowledge could not in itself address fundamental questions regarding property allocation and ownership, potential use or effective and efficient management of resources. Those questions required extensive, detailed and specific studies only recently recognised as a freestanding, organised branch of knowledge.

Real property fulfils numerous roles. In the form of shelter and food production, it is fundamental to human existence and represents a basic factor of production within the economy. Depending on how it is defined, the built form represents between 44 per cent and, perhaps, up to 60 per cent of the nation's wealth. It provides the collateral for the most significant financial dealings (Fiorilla et al. 2012; Stein 2019). Global built asset is estimated to have an almost incomprehensible value of $217 trillion (Stein 2019).

The familiarity with and pervasiveness of the built form often lead to an approach based on personal experience rather than structured research and analysis.

Recent decades however have seen a significant increase in the profile now ascribed to real property assets and the built environment more generally. This has been based, at least in part, on:

- their ability to earn income and to accumulate wealth within the more sophisticated, but potentially more volatile environment post the Global Financial Crisis (GFC)
- their importance as a key factor in the efficient and effective performance of the economy as a whole, and of the firms and commercial operations within it
- their contribution to the liveability of a location or region for individuals, households and communities
- the essential role that the built environment plays in securing a sustainable future for the physical environment, the economy and communities.

A range of influences contribute to this new profile, but two important catalysts are rapidly increasing urbanisation and, related to that, the increasingly integrated and complex functions required of built assets.

Urbanisation

In the study of the many facets of economics, there is an often quoted adage: 'demography is destiny' – that is, the size, age and key characteristics of any population will largely pre-determine the direction, activity and future prospects of that population, its region or country.

In a remarkably short period of time, world population has grown exponentially and, at the same time, has concentrated itself into urban environments – typically in coastal or estuarine areas. The overall impact (both positive and negative) on prosperity, liveability and overall sustainability is profound. Population growth encapsulates the legacy of human activity over the past 200 years (Brugmann 2009; Glaeser 2011). It would be naïve and simplistic, however, to believe that these significant changes could be effected without attendant detrimental side effects.

World population now stands at over seven billion and will continue to grow at about 1.1 per cent per annum, or about 83 million net increase per year, for some decades (UN 2017, 2019). However, there are now sound predictions that world population growth will slow and potentially begin to decline beyond mid-century – the first since the Black Death 700 years ago (Pearce 2010). An ageing population and falling birth rates are also creating their own challenges across many OECD[1] countries. Nevertheless, the pressures created by overall population growth on the sustainability of the planet are critical issues for all.

This population trend is reflected in how property, and particularly urban land, is now allocated, held, developed, used, managed and valued. Further, those resources form part of what are now much more complex 'urban systems'. The political, economic and social fabric of a country or region is now largely held within urban environments, and the operations of any component, including real property, can only really be understood in the context of the wider system or environment.

These fundamental changes will enhance the functionality and value of certain properties and property subsectors while detracting from others. Typically, long-term fixed assets such as property do not accommodate radical change well – leading to obsolescence and deterioration in value. Their ability to adapt varies from property to property, influenced by a range of attributes including location, overall design, available services and the regulatory regime in place.

This more holistic approach, now widely recognised by a range of disciplines, sees the city as a living system rather than its unrelated component parts. The approach establishes, above all, that cities are about people and their relationships with each other and the physical environment (Giradet 2004).

In countries like Australia, urban environments also need to accommodate more heterogeneous communities, impacted by large-scale immigration, ageing populations and changes in household composition.

Further, in a single generation, economies and communities have had to address radical change including the downsizing of manufacturing and the emergence of knowledge-based and service sectors that integrate more readily into the urban environment. Information and communication technologies (ICT) are critical, but, contrary to earlier predictions, these have, in fact, encouraged closer settlement and interrelationships in clusters and precincts (Mitchell 1996).

An important milestone occurred in 2006–2007 where, for the first time in history, more of the world's population lived in urban rather than rural areas (UN 2017). This simple milestone, significant as it may be, represents only part of a critically important story. A little more than a generation ago, in 1950, only 30 per cent of the population lived in an urban area; even between the '50–50 point' of 2006–2007 and the last recorded statistics in this area in 2019, that figure had risen again to 55 per cent – or about 55 million additional residents per year living in urban locations. By 2050, 68 per cent of the world population will be urban dwellers (UN statistics quoted in Phys.org 2019).

This population shift is not simply to 'non-rural' areas but represents a move to cities where increased concentrations of economic activity, jobs, prosperity, lifestyle, education, services, social interactions, amenities and opportunities will further entrench urban dominance (Brugmann 2009; Glaeser 2011).

In 1990, there were ten cities of over ten million inhabitants; by 2014 that the number had increased to 28 (UN 2017). By way of contrast, in 1950, there were 83 cities worldwide with a population of one million. That number has now risen to over 400 (Population Reference Bureau 2015).

These matters are discussed in detail in Chapter 4.

These quite remarkable statistics now define the global, human environment as 'predominantly urban', where the driving forces – economically, politically, culturally and socially – are underpinned by a network of very large, global cities. Around the world, the city that is the hub of economic activity, and that aggregation of power, wealth and urban population continues to grow disproportionally compared with rural regions (Short 1996; Reader 2004; Montgomery 2007; Glaeser 2011).

Rural production and rural regions have always been important in Australia, as perpetuated by notions of a 'bush' heritage and culture. That importance, however, has often been overrated. Unlike any other country, Australia had towns even before it was fully explored and widely settled. With a few exceptions, such as the discovery of gold (and later mineral development), Australia was always politically, economically and socially dominated by its cities, towns and the concentrated urban environment they created. While physically, culturally and politically important, the agriculture sector in June 2015 represented only about 2.7 per cent of Australia's Gross Domestic Product (Australian Government [ABARE] 2019) and the rural population, that is, outside urban areas and provincial cities, accounted for approximately 10.7 per cent of the total population in 2014 (ABS 2015a).

Although Australia's physical environment, huge land mass and very low population density (3.1 people per square kilometre [ABS 2015b]), may appear unique, it is in fact one of the most urbanised countries in the world. Nearly two-thirds of the population live in capital cities, and over three-quarters in cities with populations above 85,000 (ABS 2015b; Department of Infrastructure and Transport 2015). Practically all of Australia's major cities are located near the coast, with 90 per cent of the population residing within 80 kilometres of the ocean, mostly along the south-eastern seaboard. Thus, the pressures of urbanisation and the competition to develop land in prime locations are comparable to those of North America and Europe.

These issues, which have obvious links to optimum population size and density, are matters of ongoing political and community debate (O'Connor & Lines 2008; Smith 2011).

Note that change is neither linear nor sequential, but rather, for better and for worse, the physical manifestation of complex and diverse forces. These include increasing wealth, investment and consumption, demographic changes, investment patterns, economic transformations and ever improving health and educational standards, technological changes and innovation.

As noted above, real property does not often adapt quickly or easily, and changes made to either the natural or built environment are often irreversible. Unsurprisingly, therefore, sustainability issues are now integral to a full appreciation of the role and long-term use of real property. As the debate on sustainability, energy efficiency and global-climate change intensifies, the study and analysis of the built environment will inevitably gain prominence, given that buildings worldwide account for about 32 per cent of all energy consumed (International Energy Agency 2015).

These extensive changes demand a far more sophisticated approach to critical resources, including real property.

Complexity and integration

Not unrelated to the above observations, the density, diversity and complexity of current urban land use and development contrast with the relatively simple issues of ownership, finance and the regulatory environment of real property in earlier waves of urban development, especially those that characterised the second-half of the twentieth century.

As also noted above, the physical manifestation in question represents a confluence of changing demography, new waves of non-manufacturing, service-based industries and a financial sector increasing in size, complexity and dominance. This more concentrated and increasingly hybrid built form will continue to change rapidly, driven also by new imperatives such as the management of climate change, congestion, energy usage and future transportation options.

Real property assets are immovable and the land component is non-renewable. The asset life cycle is significantly longer than normal commercial time frames and even human life spans. All of these characteristics mean that these assets exhibit significant inertia to change and present challenges to adaptation.

As discussed in Chapter 2, the matter is further complicated in that most property ownership is held, and more investment decisions undertaken, by individuals and firms who (understandably) adopt a microeconomic (property-by-property) decision-making approach. While individuals respond to the wider demands of the market, their decisions will be based primarily on self-interest with a lower priority given to macro issues including urbanisation, 'place building' and long-term liveability and sustainability challenges.

Wider control and integration are currently applied through statutory-based town planning and development legislation. However, this prescriptive framework, which attempts to provide strategic direction and overall management in a complex environment, has limitations. Moreover, it is increasingly obvious that a more responsive, adaptive and innovative approach to planning and land-use management is required.

Probably more than any other discipline, the study of property combines knowledge from many diverse fields: the natural sciences, particularly geography; law; town planning; engineering; building design and construction; finance and economics; business studies and management; demography and social sciences; ethics; and even physiology and psychology (important in issues of workplace environment and environmental health). Relevant parts from all these disciplines must be considered and integrated in order to derive full benefit from property assets for all stakeholders.

Of all the related disciplines identified above, economics and management are arguably most linked to property and its use. Economics is critical because the study of property is essentially concerned with the allocation and effective use of a scarce resource ('land'). Similarly, to manage this valuable and long-term resource, effective systems must be established to ensure that economic imperatives are translated into operations and outcomes over time. Both economics and management issues of real property dominate contemporary politics, business and community behaviour; they also provide the underlying approach and structures underpinning the ownership and use of the physical asset base.

In Australia, company property assets typically represent a significant proportion of the total asset base. In the past, these assets, described as 'non-current' on balance sheets, were often seen as passive, 'book value' investments. However, in today's highly competitive, globalised business environment, most firms cannot afford the luxury of having significant parts of their asset base underperform or be mismanaged.

Further, property can provide much more than simply a 'place, space and time' for contemporary businesses. The 'value add' involves the provision of a physical environment that is cost effective and creates a working environment conducive to the operations and production of a firm and supports its image, identity and culture.

All these characteristics demand a more sophisticated, integrated and structured approach to property, including its ownership, analysis and management.

1.3 An holistic approach

Given the characteristics of the contemporary property environment, some realignment of the approach taken to its ownership, use and management is timely, particularly emphasising the following tenets:

View particular situations from a broad, long-term perspective

Property is a high-value, long-term strategic asset, influenced by many economic, non-economic, systemic and non-systemic factors. The ability to appreciate all these influences, risks and opportunities, to prioritise their importance and to understand their long-term effects, is fundamental to success.

In property dealings, management and operations, certain decisions and actions that may appear quite minor can, in fact, have longer-term impacts on function, costs, income and value. Consequently, the overall impact of any decision – short-, medium- and long-term – must always be considered. (These matters are discussed in detail in Chapter 4.)

An up-to-date, sound working knowledge in a range of disciplines, together with an understanding of the current economic, social and political events likely to affect property, is also important and should instil a philosophy of 'continuous learning' for property professionals.

Because of the unique characters of real property, the application of theory differs in each case. Consequently, an innovative, but practical and realistic approach is needed for each asset – subscribing to the overall priorities of the owner but adapting them to accommodate the particular characteristics of each property.

Take a structured, analytical approach

As noted above, making logical, rational observations about property is difficult because of the intimate links with personal experience and preferences through daily interactions and associations. This familiarity has risks for property professionals. For example, all will have subjective opinions on what represents a 'desirable' neighbourhood or property; what type of shopping centre best meets a particular or personal need, and so on. Property professionals must guard against allowing personal bias to affect their judgement.

Rather, a well-researched, structured, analytical approach is required to establish the wider market prospects of a particular property scenario.[2]

The task of the property professional – analyst, financier, developer, owner or manager – is unlike many who may be quite willing to advance unsubstantiated or generalised opinions regarding particular property matters. The property professional is trained to comprehensively analyse the circumstances of a case and deduce defendable, balanced conclusions from that analysis. A tentative or negative approach, which might reflect a personal rather than an analytical stance, needs to be avoided. Nevertheless, given the value of the assets and the potentially negative consequences of a poor property decision, a measured, moderately conservative approach is justified.

In particular, valuation and property practice employ a range of methods and structures to assist with property investigations based on education, technical competency and experience. The most relevant methods of assessment are chosen and applied to the unique physical, environmental and economic characteristics of an individual property. The approach should be, in essence, analytical – collecting and interpreting information to present a full and logical report with recommendations for the particular case (refer Chapter 10).

Data, models and human experience

The function and therefore value of real property is largely expressed in numeric/monetary equivalents – a point recognised in the much-referenced 'Spencer' legal principle Spencer vs Commonwealth [1907 5 CLR 418] and discussed in detail in Chapter 10 of this text. On that premise, mathematical and economic modelling and statistics are fundamental tools in establishing cost and income cash flows, sinking funds, time value of money assessments, sensitivity and comparative analysis and many other techniques to assist in property decision-making.

Over recent decades, two, largely unrelated, evolutions have occurred that have had a profound influence on these methodologies. In the first instance, the real property and development sector and the financial sector that supports it have grown exponentially in scale and complexity and, with that, the need to employ much more rigorous and multifaceted research and analysis. Second and through about the same period – now often referred to as the start of the 'second machine age' – the internet and new communication and other technologies were securing a central role in business processes and across the wider community. This was aided by huge increases in computing capacity, speed and applications (Brynolfsson & McAfee 2014).

Many of these advances were applicable to the assembly and analysis of property information. That trend continues to gain momentum, not only in financial analysis but also in incorporating geographic information systems (GIS), building information modelling (BIM) and high-resolution graphics.

In property sectors, these new capacities have been of particular benefit in areas where very large data sets ('big data'), derived from various sources, were involved. These include mass appraisal for property taxation and related purposes and improvements in hedonistic modelling comparing property value determinants, again often involving larger scale markets such as residential subsectors. Important applications also lie within the development and investment sectors. There, specialist software and hardware capacity can now accurately address a range of scenarios and 'what if' options in complex cases that would be near to impossible using manual calculations.

Given the exponential rise in computing power and artificial intelligence more generally over recent decades, it would be naïve to think that that trend will not continue into the future – and indeed, with proper quality control, will lead to more comprehensive and accurate assessment into the future. As in many other areas of business and professional activity, the question may now be posed as to the extent to which the 'automation' of real property analysis and decision-making will advance into the future, perhaps effectively taking over the decision-making process altogether. This, of course, is currently a matter of wide debate surrounding driverless transportation, robotics, drone technology and many others (Alder 2017).

Walsh (2018) would suggest that, without under-estimating such advances, the limitations on artificial intelligence are profound and will remain so long into the future. He notes that much human imagination, innovation and creativity and the ability to put, otherwise random, ideas and experiences together are innately human abilities and, even those involved cannot really articulate how that process happens. Consequently, because those processes cannot be translated to instructions (i.e. a formula or algorithm), it is not possible, on the face of it, to give those instructions to a machine. The challenges are even greater when criteria such as ethics, morals and professional standards and non-financial priority setting are involved.

Analysis and assessment of real property provide an important case in point here. The financial and other models for some property-related undertaking have become increasingly accessible and reliable. Computers however are literal and do not adapt well to multiple variants interacting in complex ways – as is typically the case within property markets. Invariably, therefore, the assessment will require the direction of an experienced professional to guide the process, to ensure that all nuances and inter-relationships pertaining to that property and its physical and economic contexts are appropriately taken into account in arriving at the final assessment.

Even though in aggregate or simple cases, property analysis and assessment of value will be increasingly available through computer-based algorithms, it is perhaps a little ironic that the larger, more complex, multifaceted property cases will typically require more human conceptualisation and input, not less. Overall, therefore, it is difficult to conceive that basic partnership – processed data with experienced human oversight – will change significantly even in the longer term.

1.4 An education and research base

In Australia, universities and Technical and Further Education (TAFE) institutions provide a range of property study programmes. Many are also externally accredited by government and professional bodies.

TAFE courses largely address competency and skill development in areas including real estate licencing, residential property management and building management and operations.

University undergraduate degrees and postgraduate programmes typically require a longer investment of time and focus on analysis, decision-making and more specialised areas. Many programmes have professional accreditation with one or both of the Australian Property Institute (API) and the Royal Institution of Chartered Surveyors (RICS). These programmes may provide an academic pathway for registration as a property valuer as well as access to a range of other property-related careers. Universities are well equipped to address many property issues beyond technical requirements and to foster skills in complex analysis and professional judgement.

Like most other professional disciplines, studies in property (often using names as 'property economics', 'valuation', et cetera) are best undertaken through blended learning – a mix of theoretical and practical ('in service') experiential studies. This approach also reflects the nature of the professional tasks and the diverse careers that graduates will later undertake. Given the dynamic and changing environment of the property sector, as summarised in Sections 1.2 and 1.3, graduation must be seen as simply the end of an initial phase of the ongoing learning required for professional competency and success. This implies that further specialist (post graduate) skill development and targeted, professional development programmes are required.

It is counterproductive to criticise any stakeholder for not providing all the required skills, competencies and knowledge. That full suite of skills can only eventuate with the commitment and involvement of education institutions, employees, professional associations, government and, most importantly, from the individual student.

The recent rise in the number and variety of tertiary courses in property studies reflects the growing scale and capacity of the sector. Historically universities have:

- drawn together existing knowledge and disciplines
- disseminated that knowledge through teaching and learning
- analysed, interpreted and applied theory to particular circumstances
- generated new knowledge through research
- provided leadership and direction, particularly in times of change and uncertainty
- provided a hub for business, social and community interaction
- facilitated the general pursuit of the arts and sciences
- fostered innovation and lateral thinking, which generate ideas and constructively challenge the status quo for the advancement of communities and individuals (Harding et al. 2007; Kealey 2008; Cole 2009).

Practically all of these attributes and activities are relevant to property studies. Most tertiary property programmes in Australia involve a number of key property fields – a sectoral overview, law and tenure, urban economics, valuation and analysis, investment, finance and taxation, property management and property development. These fields serve largely as the themes for the chapters within this text.

The transformation of data (i.e. information) through research, teaching and learning into knowledge represents the underlying process of education. Data is essentially based on observations, statements or other information collected for analysis or reference. Often, a challenge for contemporary students and researchers is the sheer volume of data and data sources now available – some biased, some thinly disguised advertising and a proportion that is misleading or simply wrong. This 'information overload' can confuse rather than assist. Raw (unprocessed) data is of limited use in learning and research and requires verification, collation, analysis and testing in order to draw reliable conclusions (see Sections 1.5 and 1.6).

These processes – the questioning, consolidation and subsequent advancement of knowledge and understanding – are central to university teaching and research activities. Combined with pre-existing experience and observations, they provide accurate knowledge of the subject, which can then be applied to specific cases or problems.

This investigation and research can take a number of forms. A literature review may be performed, and previously published material can thus be collated, verified, critiqued and re-presented. Original or primary research involves activities such as surveys, fieldwork, trials, experiments and market studies to establish new knowledge. Secondary research will typically involve the collation, further review, and combination and analyses of earlier research or published data. Both forms are legitimate provided that the broader research parameters outlined in Section 1.5 are applied. Research may also be categorised as 'foundational studies' or 'applied research'. Typically, foundational studies, sometimes called 'discovery' or 'pure' research, investigate abstract or theoretical issues. As the name suggests, applied research solves particular, often practical, problems and then, perhaps, generalises the outcomes for application to other cases.

Well-recognised analytical structures and the use of appropriate research methodologies ensure that research outcomes are properly verified, that results are sound and that the data sources are suitably acknowledged. Other research protocols address issues of ethics and confidentiality and provide guidance on the best forms of investigation and valid statistical analysis. This research management and quality control are the fundamental activities of universities.

1.5 Critiquing information

A thorough, critical approach to the review of existing information sources and publications is essential as it will assist in providing the evidence base for the work undertaken. These are not simply matters of research interest for academics, but are fundamental to commercial, legal or governmental decision-making. Today, the access to vast amounts of information through the internet and other data platforms makes 'information dumping' very simple. However, these are only random facts, data or information – not knowledge or true research.

When reviewing existing literature and publications, the following issues need to be taken into account (Cooper & Schindler 2001; Maylor & Blackmon 2005; Hair et al. 2007; Saunders et al. 2009).

Currency of information

Given the rapid and compounding rate of contemporary change, currency of information to be relied upon is critical. With the exception of foundational theory and principle, publications more than five to eight years old may not reflect current knowledge. The 'shelf life' of some knowledge may be even shorter, depending on the subject material. While some seminal works may be decades old, generally the older a work, the more careful and critical the researcher must be regarding its contemporary value.

It needs to be remembered, too, that published material is usually based on research performed well before the date of publication. Thus, the time taken to publish hardcopy makes obsolescence a real threat in today's research environment. Further, a check needs to be made of the list of references of a research paper or the bibliography of a book to confirm the quality and depth of the research, and the age of the wider research upon which the publication is based.

Quality

Clearly, not everything published represents research, nor should it be relied upon as such. Before accepting material, critical regard must be paid to the qualifications and experience of the author, the research methodology employed, its currency and its research and statistical base. In this context, 'critical' does not mean being negative or dismissive, rather, it warns to be careful. Prior findings require confirmation and the opinions given must be accepted as sound before incorporating them into future work.

Newspapers (and their advertising components) and various commercial and industry magazines, newsletters and so forth fulfil a valuable and important role in recording current events and in presenting opinion. The property professional must realise, however, that they normally do not provide the level of research required to provide an evidence base for complex property projects and tasks. Caution must always be exercised in relying on these resources, which are often little more than single-source observations or opinions.

For more substantial published articles, the purpose and quality of the publication, the approach taken, the reputation and known stance of the author and, particularly, whether the article is well referenced, are all important indicators as to the value, or otherwise, of that work.

Breadth and depth

Sound research is normally broadly based and, quality notwithstanding, the over-reliance on a particular source can limit the overall perspective.

Even reputable journals will often exhibit a particular culture or commonality of opinion. For example, articles on 'buildings' published in an architecture journal will typically approach the topic differently from, say, an article on buildings in journals that relate to economics, published for valuers and property economists. Neither publication is wrong or misleading; nor, however, will either provide a comprehensive research base.

Also, care must be taken not to accept any single paper as definitive in a particular area. Verification, or at least the use of more than one reliable source, is essential before making deductions.

Published versus unpublished data

Some material uncovered during a search will be published (i.e. printed and circulated as a text, journal article or similar); others will be unpublished (e.g. an individual's private notes, or discussion or conference papers and reports that have not been widely circulated).

Though both sources may provide quality information, publication usually signifies that the work has been previously peer-reviewed and then exposed to wide professional or public critique. This process normally confers a higher level of intellectual rigor and validity on published rather than unpublished material.

The Australian environment

The volume and breadth of available Australian studies into real property and related issues, while improving, are small compared with international research, particularly from the US or the UK.

Although the internet provides relatively easy access to a range of overseas studies, researchers need to be aware of various pertinent differences – legal, governance, scale, and so on – between Australia and other countries, particularly the US. Further, the research emphasis varies across various countries. In Australia, and the UK, real property tends to be often recognised as a distinct asset grouping, addressing both physical ('built environment') and financial/economic components; however, US 'real estate' research typically focusses heavily on financial issues and modelling.

Obviously, quality research into Australian issues and cases is preferred. Overseas works are useful, provided that the inherent differences are recognised. Research on law and legal structures, government structures, economic variables and scale are all matters of major difference that require careful interpretation.

Finally, it should be noted that the recommended rigorous approach to research may sometimes need to be tempered depending on the particular topic or field involved. Some topics are well researched, and they yield a breadth and depth of quality information. However, quality prior research on lesser known topics may be limited, which makes the researcher's task more difficult. In the latter case, less comprehensive or reliable sources may be necessary to build an evidence base. Nevertheless, the attendant risks need to be recognised and managed and suspect sources discarded. If not, the basis of further research will be prejudiced.

1.6 About the numbers

As noted previously in this chapter, mathematics and statistics represent a critical component in understanding and interpreting property assets and markets. This aspect takes several forms.

In the first instance, as will be elaborated in the following chapters, there are mathematical underpinnings to the operations of property markets. These are, in essence, case studies in applied economics. They include marginal analysis, various types of modelling and the assessment of both micro and macro impacts of investments and other activities. Second, much of the work undertaken by valuers and other analysts into particular properties and property projects involves detailed investigations of complex cash flows and the financial comparisons between projects and other investment opportunities. Finally, success in property and property investment will often be based on the careful analysis of a range of statistical trend data. This can be drawn from within the property sector and from wider economic and demographic sources. If this data is correctly interpreted, it will greatly improve the level of certainty and reliability of predictions and estimates.

All of these investigations require a good working knowledge of mathematics and of financial systems and require the ability to test and interpret statistical data. Although this research may well become complex in large commercial properties and projects, the key operations and logic of these markets are consistent and relatively easy to understand. These issues are discussed in Chapters 10 and 12. Quantitative analysis should aim to provide a better understanding of the fundamental components of investment cash flows, timing, marketability, relativity and risk profiling. Care needs to be taken, however, to ensure that the market realities of these key issues do not become clouded or confused in the complexity of the analysis undertaken. The use of charts and graphs is now commonplace in the popular press and elsewhere to help in explaining often complex issues. While that may be a valid objective, it is essential to ensure that a simple two-dimensional graph adequately reflects the range of issues that may be relevant to the case. The scales used on both the X and Y axes on any graph need to be carefully examined to ensure that the graphic representation produced truly reflects the relationship between the variables (Cairo, 2019). This reinforces earlier observations regarding the need to always take a holistic view in property investigations and projects (Fung 2010).

Statisticians are the first to recognise that while figures are by nature neat and precise, the real world is simply not like that. Rather it is dynamic, continually changing and 'messy'. The undertaking of any statistical analysis involves, obviously enough, counting. But counting implies the ability to adequately define those things being counted – some things will be included, and some things excluded – and that is where the 'messiness' of real life becomes important (Blastland & Dilnot 2009).

A property economist may, for example, wish to investigate housing demand and supply in a particular area. That, on the face of it, and with access to past data, may seem a reasonably straightforward task. But definitional issues soon emerge: what is 'a house'; is it the same as 'a residence'; should a demountable structure, caravan or tent be included if currently occupied; should the definition include a guesthouse, hotel or serviced apartment; how is a 'household' to be defined?

This is not to imply that statistical analysis is somehow worthless – far from it – in fact, its worth in the verification of observations is often underestimated. Problems arise, however, where the intrinsic limitations of statistics are forgotten, where definitions (i.e. of what is being counted and why) are unclear or where numbers are taken out of context. Numbers help enormously in building knowledge of the real world, but they are not, and do not purport to be, the real world.

It is only through a combination of a sound statistical analytical base and wider qualitative observations of the particular case (particularly by those with expert, professional experience) that reliable scenarios can be developed.

There is a continual need for property professionals to analytically and critically assess the huge array of statistical and trend data readily available through the press, internet and other formal and informal sources. Some of this data will prove correct and of great value, but much will be shown to be incorrect, incomplete, misleading and, in some cases, simply deceptive.

Very high-quality statistical information is available in Australia through the Australia Bureau of Statistics (ABS), state statisticians and treasuries and other reputable bodies. Even in these cases, however, 'raw figures' should never be accepted without establishing exactly what was being counted, the timing of the survey, the context of the research and, where possible, comparisons with other available data sets.

Some simple rules which can help in such assessment are as follows:

- No statistical analysis can ever explain everything. Researchers must recognise the critical role that statistics and calculations play but must not expect them to provide a full resolution or definitive recommendations. This is particularly true in a sector as large and as complex as real property, where numbers alone cannot explain the qualitative characteristics encountered. Further, it needs to be understood that, within practically any event or investigation or sampling, there will be a natural and inherent level of randomness often based on the interaction of complex forces or influences occurring at a particular point in time. The larger the sample, the more probable that random clusters and patterns and coincidences may be encountered and be identified (Blastland & Dilnot 2009; Mazur 2016).

Randomness creates surprise, excellent anecdotes and stories of interest for the popular press. Difficulties and confusion emerge in research, analysis and policy development, where (as increasingly occurs) precise reasoning and absolute answers are demanded for practically all things when the event or relationship is the result of simple concurrence, chance or fluke (Mazur 2016).

- Consequently, care needs to be taken not to attempt to deduce conclusions or to attempt to force 'rational explanations' onto every piece of statistical information that, in fact, may be little more than an interesting aberration. Large and relatively slow-moving sectors such as property require longer-term analysis to establish real and sustainable trends (Kaplan & Kaplan 2006).
- Many research works or statistical analysis will, quite reasonably, include projections or predictions of future events or outcomes. Such estimates need to be carefully considered before being relied upon. Sometimes they will be simple extensions of existing observed trends. There are inherent dangers because any number of new influences could significantly change these simple estimates. Other estimates, however, may be based on more sophisticated predictive modelling and sensitivity analysis and, therefore, are probably of greater reliability.
- Before accepting any statistical information, it must be confirmed that the information is accurate, current and has been properly collected and competently analysed. Be particularly wary of arguments and analysis put forward by those with vested interests. This is not to imply that this information be dismissed, but simply that care needs to be taken to ensure that preconceived ideas did not prejudice the research direction.

Note also, that even very high-quality data, such as that provided by the ABS, will become dated and less reliable over time. The perceptions and predispositions of those receiving the information/figures are also important considerations. Depending on the prevailing economic and political mood, as well as the confidence levels within the community or target group, the reaction to certain information (good or bad) can result in quite varied outcomes (Ropeik 2010).

Such pre-established views are now further challenged by volatile environments created by rapidly advancing technologies, innovations and scientific breakthroughs, combined with new business models and overall social change. As a result, emerging ideas and concepts can move, over a very short period of time, from the 'highly improbable' options to fully operational courses of action. It is often very difficult for existing organisations, analysists and policy makers to appropriately accommodate such change, balancing the need to quickly move forward in a sustainable way while not compromising fundamental economic reality, community, social and political standards nor legal compliance (Taleb 2010).

- Raw or abstract figures, particularly percentages, are meaningless unless placed in context. It is therefore essential to always ask: 'Is the figure or percentage being presented a big and significant number?' Where a percentage is provided, or some overall increase or decrease is claimed, the question must immediately be asked: 'A percentage of what?' A particular quoted figure should not be considered as either 'large' or 'small' until it is placed in the scale or context of the whole matter under consideration, including the time frames and historical trends involved. For example, a figure of one million dollars would be considered an extraordinarily large amount in the context of a household budget; but that same amount, considered as part of the study of the national economy, would be seen as of little consequence.

- It is common that many summaries of data include the use of averages (the sum of all of the results divided by the number of responses) and medians (the responses provided by the respondent exactly half way along the responses, ranked in order). Both of those are of value, but it needs to be recognised that such simple analyses can hide great variations and a spread of results. Common examples include data referring to 'average household incomes', 'average house price' or even 'average age'. These figures are of some value and are concepts to which the public can readily relate. However, in reality, they present a less than comprehensive picture and can, in practice, prove to be misleading (Fung 2010).
- Important for sectors such as property is the statistical truism that, simply because one event occurs simultaneously or at about the same time as another event (even on a number of occasions), it does not follow that the two events depend on each other or that one represents a cause of the other. While that may be the case, those conclusions depend on further investigation.
- Finally, surprising or sensational figures, such as those that sometimes provide headlines in newspapers, are a reason for caution rather than concern or action. The reports may well be valid, but, before further consideration, it would need to be established that the findings were correct, in context, statistically meaningful and properly interpreted. Again, the possibility of naturally occurring random, short-term aberrations should not be overlooked.

The emphasis placed on statistics in the media and by contemporary business interests is a good thing. Political and economic environments are complex and it is natural to attempt to interpret and translate that complexity into something that makes sense and gives comfort in providing predictability. Issues arise, however, when very basic analyses, such as averages of raw data, are accepted as definitive and as 'the full story'. This can lead to an anticipated, but often unconfirmed, 'bell' curve around an average that, in turn, becomes a basis for policy and/or decision-making. That is a potentially dangerous, but not an uncommon event (Kaplan & Kaplan 2006; Ropeik 2010). Again, such uniformity is extremely rare across any statistical sample and a presumption of that type will almost certainly result in important characteristics of the study being hidden by the apparent consistency of results.

The inclusion in reports of relatively simple, additional statistical interpretation techniques, such as standard deviations, would help greatly in providing a more realistic appraisal. Unfortunately, the typical understanding of, or tolerance for, statistical analysis by the media, and probably the general public, does not allow this. Worse still, are clichéd expressions such as 'middle Australia' or 'average households' that are so lacking in definition they are of little or no use in analytical assessment.

Notes

1 The Organisation for Economic Cooperation and Development (OECD); a United Nations-based group made up of the 34 most developed countries.
2 Property case law often refers to this detached, professional analysis as the actions of 'the hypothetical prudent purchaser who possesses all the relevant facts and is willing, but not forced, to transact'.

Chapter 2

Economic foundations

2.1 Introduction

Over recent decades, there has been a strong tenancy across the professions and in tertiary education and training to specialise, often into quite narrow and esoteric areas. There are sound reasons for this and the trend is obvious across most disciplines including law, medicine, science and engineering and almost all business undertakings – including primary and secondary production and service and 'knowledge-based' sectors. In real property, that trend manifests itself through the large number of specialist professionals now typically involved in development, construction and asset management.

In reality, however, the environment in which individuals, businesses and communities operate is rarely so precise. While high-level, specialist skills will be required, it would also be observed that complex, contemporary demands and problems require more integrated and multi-disciplinary solutions.

Given this scenario, the real value of economics becomes evident by providing an understanding of systems and a framework for decisions to ensure the most efficient allocation of scarce resources – problems common to practically all disciplines and activities.

This chapter discusses, from first principles, the theoretical underpinning of economics and, in particular, its relevance to the property sector.

Changing economic, political and social priorities have also focussed increasing attention on the impact and cost of decisions, and the very basis and criteria for making decisions is being questioned across many disciplines.

Many individuals involved in (real) property dealings have a fairly general understanding of economics, typically based on theory provided in secondary education. Therefore, many may not fully appreciate that the laws of economics and the rationale behind the property market are, in fact, one and the same.

On that basis, this chapter:

- defines and explains the philosophy of economics and how current theories developed
- summarises the primary elements necessary for an understanding of economic theory and practice
- identifies the relevance and application of economics to resources and real property.

This chapter necessarily presents only a summary of these complex topics but should provide a basic understanding and a platform for further study. References included later in this text will facilitate further enquiry.

2.2 The nature of economics: some definitions

Economics appears to suffer from a mixture of bad publicity, generalisations and mis-understandings. Variously called the 'science of scarcity' and the 'dismal science', the discipline is trivialised by such hackneyed expressions as 'well, it's just about supply and demand' (the latter comparable to reducing Newton's theory of gravitation attraction to collecting apples).

Difficulty also arises from the fact that economics affects, and is often critical to, nearly all forms of human activity. However, when examined in isolation, the subject can, like mathematics, appear abstract or obtuse. In defence of economic theory, problems often arise because politicians and makers of public policy have a long history of applying only that part of an economic approach or theory that happens to meet their immediate purposes, with little or no regard for the entirety of the proposal. When the (partially implemented) policy then fails to deliver the desired result or creates other difficulties, flawed theory is blamed.

Economics would never claim to have definitive models or processes that can be universally applied or somehow accurately predict the future; no discipline can offer that. It would confidently claim, however, that there are common, observable, 'rules of the game' that provide a sound basis for rational decision-making across a range of activities. This includes, of course, structures for decisions in the use of resources such as property.

Economics can be defined, in both theory and practice, as the science of how and why individuals and groups make decisions and choices and why they undertake certain actions. Traditionally, this analysis was based almost exclusively on financial considerations. In today's more complex contemporary environment, however, a broader understanding of resource, and political, environmental and social considerations form part of that process. Thus, economics can be described as the study of options and the study of incentives used to encourage desirable outcomes (Wheelan 2010).

Economics can also be defined as a social science that studies the production, distribution and consumption of goods and services. The word comes from the Greek 'oikos' meaning house and '*vouoc*' meaning custom or law; in other words, the 'rules of the household' or 'how things are organised and how they run'. More specifically, Robbins (1998) defines economics as the science which studies human behaviour as a relationship between ends and scarce means, which have alternative uses. It can involve the analysis of what currently exists or, alternatively, a normative study; that is, proposing what ought to be, projecting trends and making recommendations for improvements in future decisions, choices and distribution.

Robbins' definition raises the issue of scarcity and the negative connotations referred to earlier. Certainly, given the almost endless list of human demands and wants, and the finite quantities of resources available for production (e.g. land, labour, capital and management), scarcity and shortages are inevitable. The positive aim of economics, however, is to provide the context (settings) and methods to maximise the efficient and effective use of those limited resources to meet, as far as possible, the demands of individuals and communities. Hence, the various branches of economics emphasise decision-making, the setting of priorities and the distribution of resources. 'Property economics' is one of those branches.

Economists correctly rely on mathematical models and the analysis of statistical data to help explain such decisions, but they recognise that, at the end of the day, decisions are made by individuals or groups. Consequently, the study of human behaviour and psychology are also relevant. Issues such as demographic profile (age, family structures, education levels, cultural and ethnic background), perceptions of risk, time preferences and community sentiment will all potentially influence final behaviours, to a greater or lesser extent away from pure economic reasoning or mathematical models (Häring & Storbeck 2009).

Hence, the earlier description of economics as a 'social science' – it is about the incentives or disincentives to act and make certain choices, about how individuals and groups send, interpret and respond to information and, from that, how decisions and prioritisation of actions are made The understanding of the behaviour of buyers and sellers are even more important in markets such as real property where the scale of transactions is significant and financial consequences, positive or negative, serious.

Economics can be divided into two distinct but interrelated parts, 'microeconomics' and 'macroeconomics' (Conway 2009).

Microeconomics

Microeconomics examines the economic behaviour, actions and choices of individual units within the economy. The 'units' include individuals, households and businesses that interact within the 'markets' and communities in which they operate. A market is a transaction point for economic interactions or dealings. In a particular market, those involved can act as producers (suppliers) or as consumers, depending on whether they buy or sell. Some markets are obvious; for example, going to the local shop to exchange money for groceries. Others are less obvious; for example, an employer exchanging money (wages) for an employee's time and skills. Nevertheless, a clear transaction point also exists in those cases.

At any point in time, huge numbers of markets exist. One basic objective of economics is to examine the operation of these markets to determine the general 'rules' that seem to regulate behaviour. Establishing these rules allows for better management, stabilisation and prediction of the markets – all laudable objectives in a mature, sophisticated economy such as that in Australia.

In this context, the prefix 'micro' does not necessarily imply small. For example, the largest firm in the country makes 'microeconomic' decisions. Individuals, single households and small and large firms are all driven by what is known in economics as 'the primacy of self-interest' (Akerlof & Schiller 2009) That is, they can be expected to take the actions and make the decisions they believe will maximise their own security, income and wealth. For example, in this case, large firms act for the benefit of their shareholders. This primary objective is balanced by the risks taken and managed and the effective, efficient use of available resources to maximise benefit – 'satisfaction' in economic language. Statutory requirements also place constraints on this pursuit of perceived interest.

On the face of it, these activities appear simple and straightforward and are frequently encountered in the dealings of individuals and households and in business transactions. In practice and as certain decisions have greater consequences, matters become more complex with an increased number of influences weighing on the decision-maker and a range of options for action emerging.

Consumers, for example, are influenced by their hierarchy (i.e. priority) of needs and wants and their level of income, which affects their inclination (called 'marginal propensity') to direct their income into either consumption or savings. US psychologist, Abraham Maslow (1908–1970), attempted to explain these thought processes in his seminal work, *The Theory of Motivation* published in 1943. Therein he described human needs in terms of a pyramid (Maslow 1943, 1998, 2000; McLeod 2014). More primitive (basic) needs form the base of an ascending hierarchy of needs that become progressively more sophisticated, complex, subjective and often more difficult to define (Figure 2.1).

Although this classification is somewhat generalised, it is reasonable nevertheless. Maslow also made several important observations pertinent to economics, and particularly to consumer behaviour. First, he observed that, if a lower (more basic) need is unmet, then satisfying this need will dominate or take priority in the individual's behaviour and consumer preferences. However, once that need is satisfied, or addressed to an acceptable level for that individual, the demand may become inconsequential and other (higher level) needs now demand priority. These needs also change over time.[1]

The decisions of producers are also influenced by many factors – from their understanding of that market to the scale of production, available capital, the risk profile of the proposed activity, competitors' actions, time lags and input costs (fixed and variable). Each producer has available a unique combination of the key elements of production – capital, land, labour and management skills (entrepreneurship) – that dictate the optimal level of production for maximum profit. In other words, maximum production does not necessarily maximise profits. Rather, seeking the optimum level of production to maximise profit over time is the producer's primary goal (Akerlof & Schiller 2009). That essential task is certainly not an easy one and will never be exact. However, an estimate based on the best information available is essential if sound and sustainable profits are to be achieved.

The mantra of optimum rather than (necessarily) maximum output to secure a higher sustainable profit is fundamental to production decision-making. The cost at which suppliers can actually produce the goods and the number to be produced are the key components of that analysis.

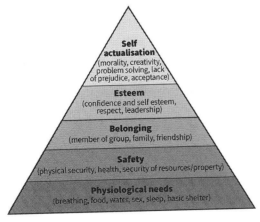

Figure 2.1 Maslow's hierarchy of needs (adapted from Maslow 1943).

The production cost-per-unit will almost invariably initially fall as the quantity produced increases. This fall occurs because fixed business costs can be amortised over a larger number of production units, input materials can be purchased more cheaply and more sophisticated methods of production and specialised labour can be employed. This trend, common to most production activities, is known as 'economies of scale'. It, however, does not continue indefinitely, and an optimum point is reached at a certain level of production. After this point, the 'marginal cost', which is the cost of producing the next unit of the commodity, begins to rise again.

This is called the 'Law of Diminishing Returns'. The law is activated for several reasons, including limitations (at least in the short term) imposed by one or more production factors such as land or capital, by increasing overheads or by production issues often associated with large businesses (e.g. production bottlenecks, risks of stockpiling and high levels of inventory held). Again, this analysis suggests that optimal, rather than maximum, production will be a long-term objective.

Though explained here in theory with only simple examples as illustrations, these principles are profoundly important across almost any form of production, including property development and other components of that sector (Bowles 2004; Baumol & Blinder 2006).

Such analysis has short-term and long-term aspects. In the short term at least, supply functions (i.e. the operations and responses to changing market situations of producers) tend to be less volatile than demand functions for the same commodity. This is because producers often have existing investment in current production that cannot easily be changed. Consumers, however, typically have a much greater ability to withdraw from a particular market and take up other options, often in a very short time frame. These observations are particularly true of property and development markets, given the scale of prior commitments required of those who produce (i.e. supply and build) property assets.

Macroeconomics

Macroeconomics is the study of aggregates of the economy. It attempts to explain the broad components of the national economy: government; regions, industry sectors and economic variables, such as national income and output; interest and exchange rates; employment levels; inflation; global markets; international trade and so on (Feigenbaum & Hafer 2012). Further, it investigates the links between these elements and thereby provides direction for better management and policy into the future.

Clearly, both microeconomics and macroeconomics are related. However, they deal with different aspects of the economic environment. The following simple examples may illustrate the difference. Consider a wheat farmer who (as an individual or household) produces grain. The farmer has to make microeconomic decisions about what crop to plant (e.g. what wheat variety) and the timing and nature of the inputs (labour, fertiliser, irrigation, et cetera) that will produce the best, most profitable crop for that season. The correct choices and the right combination of production factors (land, capital, labour and the farmer's skills) will produce the optimum return (i.e. the greatest net profit). However, 'externalities' such as weather can also affect returns. The farmer is also obviously interested in, and concerned about, wider factors such as interest rates, world grain prices and inflation, but realises these matters are effectively beyond his / her individual control – especially since the cultivation decisions particularly were made and committed months earlier. That is, the farmer has already irreversibly committed the investment (i.e. planted the crop) for the production period. In this illustration, the farmer is making 'microeconomic' decisions.

However, that farmer is one of many thousands of farmers in Australia who together (i.e. in aggregate) represent the producers within the 'rural sector' (where wheat production is a sub-sector). The economic fundamentals/rules remain the same but, once aggregates are considered, the forces at work and the nature of decision-making vary considerably. Matters such as world prices and analysis, national marketing strategies, the exchange rate and interest rates then become the critical issues for this 'aggregate sector' in its relationship with the government and the other components of the economy. All these factors comprise an 'aggregate' view of the economy: macroeconomics.

2.3 The historical development of mainstream economic theories

Issues related to the production of required goods and the challenges of scarcity are fundamental to human existence, as is starkly shown in subsistence economies in many parts of the world today.

However, when an individual is able to produce more of a certain commodity than that person requires for their own needs (i.e. 'a production surplus'), then trade with others begins. Trade facilitates the exchange of these surpluses for other goods in short supply. Producers or regions with comparative advantages over others in producing a particular good or service would, over time, purposely specialise to increasing surpluses for trade with others. Money was then introduced as a unit of exchange to facilitate trade and, over centuries, markets grew and became far more sophisticated and complex. In parallel, theories emerged as to the nature and operations of markets and how they could be managed, stabilised and made more efficient (Robbins 1998; Mills 2003; Vaggi & Groenewegen 2003).

The first of these major theories was put forward towards the end of the eighteenth century. Since then, many theories have been proposed as economic, political and social environments have changed. Some of these theories addressed more minor economic components. Others, however, provided commentary on and presented comprehensive principles that significantly influenced economic activity during that era and, often, well into the future (Ekelund & Hebert 1990; Screpanti & Zamagni 2005).

Several of these mainstream theories are outlined below. However, economic history should be viewed overall as a complex and continually evolving process, with one or more of these important theories underlying particular periods or heralding significant change.[2]

Classical Theory

The first mainstream economic theory, later known as Classical Theory, was proposed by Scottish political economist and philosopher Adam Smith (1723–1790), principally through his 1776 work, An Inquiry into the Nature and Causes of the Wealth of Nations.

The nature of this work needs to be understood in its historical context. The late eighteenth century was a period known as The Age of Enlightenment – a time of major discoveries, theories and upheavals in many fields of scientific and philosophical endeavour. It was accompanied by great ventures in world exploration and dramatic political change. The general movement and change manifested itself differently in the leading countries of the day. In France, it took the form of resistance to, and later revolution against, the established power of the monarchy and the church. In England, it was largely based on growing scientific knowledge and emerging technologies that would form the basis for the urbanised Industrial Revolution. In the US, it coincided with the early establishment of the nation and provided many of its civic, political, philosophical and educational foundations.

In each case, the impact across practically all sections of society including economic activity was profound and, in effect, marked the beginning of the modern era (Himmelfarb 2005). Key aspects of Western philosophy and culture – based on the premise that reason, intelligence and individual freedom were the primary sources of legitimate authority – developed during this period.

A generation before, Isaac Newton (1643–1727) had undertaken his seminal work. Adam Smith was also a contemporary of Benjamin Franklin (1709–1790), J.S. Bach (1685–1750), Thomas Jefferson (1743–1826), Voltaire (1692–1778), Mozart (1756–1791) and the explorer James Cook (1728–1779).

The American and French revolutions also took place in this period. Philosophers of the day, including Smith, were inspired by great discoveries about the natural environment and scientific truths which, for them, demonstrated mankind's earthly supremacy (often with theological overtones). Influenced by these scientific discoveries, many philosophers perceived an order and balance in the natural world.

For Smith, who was particularly interested in political thought and the distribution of wealth, it seemed reasonable that natural principles should apply to matters of trade, capital and enterprise. Smith's contributions to history and philosophy are important but can only be summarised here as follows.

Few other disciplines have a single founder as exemplified by Smith in economics. Contextually, The Wealth of Nations (1776) followed his philosophical work on social relations, The Theory of Moral Sentiment (1759). A third text on government and jurisprudence was unfinished at the time of his death. These contributions to philosophy in general, and economics in particular, are complex and unfortunately, have often been oversimplified and misunderstood (O'Rourke 2007).

The economic component of The Wealth of Nations was, of course, conceptual as no data had previously been collected – even the concept and use of graphs only appeared some years later. On its merits alone, the approach adopted in Smith's text may seem overly pragmatic and promote a 'survival of the fittest' mantra within the context of a completely unregulated environment. However, this view is incorrect. Smith's earlier work and recommendations had set a social premise of a fair and equitable society. In this new society, individuals were free to pursue the development of their skills without the artificial hurdles of restrictive trade groups and, as far as possible, without the costs of 'mercantile middle men' who, in Smith's opinion, added little value; he considered middle men produced little other than extra costs and windfall profits for themselves.

To summarise, in contemporary terms, Smith took a microeconomic approach and regarded the supply of goods and services as the more important determining factor of economic activity, rather than demand. He writes:

> ... It is not from benevolence of the butcher, the brewer or the baker that we can expect our dinner but from their regard to their own self-interest. We address

ourselves not to their humanity but to their self-love, and never talk to them of our
own necessities but of their advantages.

(Smith 1776)

In later economic theories and writings, this concept would be referred to as 'the animal spirit' (Akerlof & Shiller 2009). In other words, Smith had an unrelenting faith in
the ability of the market mechanism, the 'invisible hand' as he called it, to regulate and
control economic activity. He correctly suggests that individual producers, such as the
butcher, brewer and baker, will act in their own self-interest by supplying certain goods
and services. This supply, he said, would be balanced, as measured by market price, the
consumers' demand for those goods and services, if the system were left alone – what is
known as a 'laissez-faire' approach (Haakonssen 2006).

Over time, producers, again driven by self-interest, will produce the optimum
amount of that product and the market will reach equilibrium ('at rest'), with no net
gains or losses in supply or demand (Figure 2.2).

Smith contended that all individuals in an economy or market best know their own best
microeconomic requirements, and that the aggregation of the opinions and actions of all
the individuals concerned will therefore represent the best macroeconomic position and
outcomes for the entire community. Consequently, any external action that improves the
economic position of one individual or group at the expense of another individual or group
will be detrimental to the free workings of the market mechanism and, therefore, to the
economy and the welfare of the community as a whole. Government actions, which will
almost certainly favour one group over another, are an example.

Smith's Classical Theory is based on a fundamentally correct, if simplistic, premise that the market mechanism does exist and individuals do act with the primacy of
self-interest. However, in a medium- to long-term analysis, particularly in the contemporary environment, the approach has obvious flaws. Markets are, by nature, complex and volatile and not based solely on price-quantity equilibriums. In most markets,
knowledge is imperfect, as both producers and consumers rapidly enter and leave.
Moreover, markets are not 'closed systems' and there are many external forces such as
taxation, time lags in production, product substitutes and foreign trade that distort the
simple model. Most subsequent economic theory also considered that demand, not supply, was the primary catalyst for market activity.

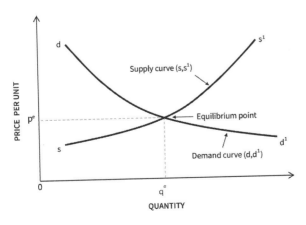

Figure 2.2 Price and quantity functions for commodity 'A'.

Nevertheless, Smith's contribution is important. It was the first comprehensive treatise on economics and his theory was widely accepted and remained dominant for over 100 years, probably until the Great Depression of the 1930s. During this time up to the 1930s, the rise of liberalism and capitalism occurred and the Industrial Revolution and urban-based production changed the economy dramatically. National governments, indisposed to major involvement in the economic activities of their countries (apart from raising income through taxation and debt through bonds) willingly accepted Classical Theory. Governments also saw the theory as a sound, almost moral base for the later rise of colonial imperialism – the aggregation of wealth by the ruling elites – and, in the US, the rise of corporate capitalism.

Smith failed to recognise the emergence of the Industrial Revolution, despite living through its early stages. He was, nevertheless, well recognised for his academic achievements and was rewarded with the post of Commissioner for Customs in Scotland – a somewhat ironic appointment given his defence of free markets and abhorrence of mercantile middle men and externalities.

During the following decades, a series of economists, political scientists and social commentators (David Ricardo, Thomas Malthus, John Stewart Mill and others) further developed elements of Classical Theory and, in many cases, its related political liberalism (Psalidopoulos 2000).

With regard to land (property) economics, the analysis of David Ricardo (1772–1823) is particularly noteworthy. Much of his work relates to long-past political and economic debates such as the repeal of the trade-restricting Corn Laws. However, his theory for establishing property rent and, therefore, value, contained important observations on that sector.

Using rural examples, he defined rent as the difference in production costs between individual tracts or classes of land. That is, he viewed rent as a comparative function of productivity. His comparative theory states that rent does not exist if only one grade of land is available and used for a single purpose (in this case, cultivation). However, when availability of the most desirable land becomes scarce, rent will be charged on higher-grade land because the only alternative is to use land of lesser quality. The level of rent that subsequently became available to the owners of prime land would be equal to the amortised cost of bringing the lesser land to the same level of production and/or overcoming other detriments compared with prime parcels. Clearly, Ricardo's theory takes no account of variables including legislation, time lags and the imperfect knowledge of both buyers and sellers in the market place. However, the work contains important, fundamental concepts about comparative analysis and market operations worthy of contemporary consideration.

Toward the end of the dominance of Classical Theory, theorists including the English economist Alfred Marshall (1842–1924) adopted more advanced and detailed approaches. In his *Principles of Economics* (1890), Marshall extended economic theory beyond the traditional classical approach and to finer analysis of market mechanisms. He emphasised the importance of mathematical rigour, particularly in marginal analysis, and the growing importance of social and political influences on both production and demand (Groenewegen 1998a, 1998b). His variations on the classical themes became known as Neoclassical Economics. One of his principal students was John Maynard Keynes, who used this basis to formulate the next mainstream economic theory, Keynesian Economics.

Marginal analysis is an important concept that deserves elaboration. For most producers and consumers in a particular market, the issue of whether they should participate is one of medium- to long-term interest only. Consider again the wheat farmer who has a farm, machinery, et cetera, and is ready to produce grain. In the short term (i.e. this season), his decision is not really whether or not to be a wheat farmer; rather, incremental, or 'marginal', considerations are more important. The farmer will ask, for example: what variety of wheat should be planted? or, in dollar terms, if a further tonne of fertiliser is applied (i.e. a 'marginal cost') will there be a proportionally larger increase in production (i.e. a marginal revenue)? Additionally, the farmer has major fixed costs (the land, machinery, et cetera) to which variable costs (labour, fuel, seed, fertiliser, water, et cetera) must be added in the correct proportions for optimum production and, therefore, maximum profit. This example illustrates the importance of marginal analysis in real-world, applied economics. These matters are further discussed in Section 2.5.

Likewise, for the consumer of wheat (e.g. in the form of bread), the decision is often not about whether or not to buy bread, but what type, what quantity and at what price to pay – again, marginal or incremental decisions. This consumer opinion, trend or sentiment usually varies with increasing levels of consumption of that commodity (i.e. a 'marginal propensity' to consume).

By the 1930s, the overall classical approach to economics had lost credibility. The Great Depression had shown that a collapse in demand, not supply, was ruinous for an economy, and that no underlying 'merit good' was associated with market equilibrium. In the Great Depression, for example, equilibrium in the labour market occurred where a third of the workforce was without gainful employment. Clearly, left to its own, unfettered fluctuations this economy could not restore sufficient activity to meet economic, community and political expectations. Especially in extreme circumstances, economies clearly needed significant assistance in the form of direct political and government action (i.e. at a macroeconomic level) to return the economy to a more acceptable state. Importantly, but outside the scope of this text, the late classical period also witnessed the rise of several radically different economic and political models, notably Communism (as expounded by Karl Marx [1818–1883] and Friedrich Engels [1820–1895]), and National Socialism, which emerged to challenge the capitalist paradigm for much of the twentieth century.

Keynesian economics

John Maynard Keynes (1883–1946) was a British economist and a student of Marshall with the basic liberal, free-enterprise outlook of his peers. Nevertheless, his principal work, The General Theory of Employment Interest and Money (1936), radically changed the nature of economics. Keynes believed that the great economic depression of his time proved that Classical Theory and reliance on the market mechanism were flawed, especially in large, complex, inherently volatile markets. Instead, he believed that economies required management, particularly by government. Government 'management' could best be applied through fiscal strategies (e.g. government taxation and government spending) and, by implication, through managing aggregates across the economy. Keynes is, therefore, considered the founder of Macroeconomic Theory.

Keynesian theory held that economies could be stimulated to grow or contract by the levels of demand within the economy, and not by simply supplying goods and services as promoted by Classical Theory. Once the level of demand became obvious, producers would then act, in the primacy of self-interest and with the motivation of potential profit, to meet that need. Keynes observed that to demand goods and services, consumers needed three things: the money to pay for them; utility (i.e. a priority use for that good or service); and, very importantly in Keynesian theory, the confidence to actually make the purchase. 'Consumer confidence' was regarded as a critical issue, particularly in the purchase of major capital items such as houses and other durables (Keynes 1936).

The size and importance of the public (government) sector had grown substantially by the 1930s. This trend has continued, in most developed countries, to the point that government and the enterprises it controls often represent 30 per cent or more of national economic activity. Keynes believed that size, leadership in many markets and the underlying regulatory power of governments gave them the singular ability to manage aggregate economic demand in positive and negative ways. Aside from the normal activities of government, this function could be performed as a precise instrument for managing levels of demand within the economy.

This approach is known collectively as 'fiscal policy' – that is, the manner (in both amount and targeting) by which a government collects and spends tax revenue. This policy provides the opportunity for governments to run deficit budgets (i.e. to spend more than the tax revenue collected in a given period) in order to stimulate demand and growth. Borrowed funds (typically secured by selling government bonds) usually make up the shortfall in this case. In contrast, to slow economies that are seen as 'overheating', governments can run surplus budgets – tax more than they spend – reducing the amount of disposable income available to consumers and therefore reducing overall consumer demand.

Keynes also correctly noted that the announcement and later implementation of a government fiscal policy would almost certainly have far greater impact than that predicted for that action. First, the announcement of a government strategy or economic management initiative sends clear signals to members of both the business and household sectors in various markets. Both producers and consumers, particularly in large-scale markets, would typically adjust their behaviour before any government action took place and well before actual impacts were manifest. (This is known in economics as the 'announcement effect'.)

Second, when the government's action actually takes place it will have a greater positive or negative impact (depending on the policy) across the economy than its original monetary value. If, for example, a government wished to stimulate consumption, it might decide to increase the payments of welfare recipients by, say, $10 per week. Each of the recipients at that income level will almost certainly spend their additional $10 on consumption of goods and services. The $10, newly introduced into the marketplace, then circulates in the payments made by the producer to staff, suppliers and so forth. The $10 will be 'recycled' through the economy and will, thereby, often produce a significantly greater economic impact than the value of the original injection. This concept is known as the 'Keynesian Multiplier'.

The concept of multiplier effects gives rise to a range of other questions – how big might such a multiplier be?; how wide would it extend? and how long would it last? Unless these matters were adequately assessed, the ability to precisely manage the economy by fiscal policy would be problematic. Critics of the Keynesian approach would suggest that, particularly in the longer term, such effects will be small, noting that eventually governments will have to recover/pay back deficits expended – so effectively reversing any earlier positive multiplier.

This assessment may be overly harsh and does not take into account the timing of actions that may be important, both in their announcement effect and targeting.

In economic terms, the multiplier is the ratio between the total shift in aggregate demand and the initial shift in aggregate demand from the initial fiscal action as it moves through the economy (O'Sullivan et al. 2008). In practice, the size of the ratio will vary depending on particular circumstances. Certainly, a simple cash injection (such as the transfer payment example above) may well have a final impact significantly greater than the original expenditure. Other examples of government spending, say on a major infrastructure or building project, may well have lower multiplier effects given implementation lags and leakages because of goods and services purchased elsewhere for the project. The effects of government taxes and charges on the project and the propensity of recipients either to save or on-spend newly acquired income all influence the size and impact of the multiplier.

Clearly, while flow-on effects need to be recognised, their scale and impact will vary with each initiative (Bishop 2009). Similar comments apply to negative multipliers – where government fiscal actions are aimed at reducing aggregate demands or providing disincentives to certain activities (e.g. reducing government expenditures or increasing taxes). The impact of any multipliers in these cases would be particularly affected by time lags and the breadth of the relevant fiscal applications.

Importantly, the Keynesian Multiplier presumes that fiscal management will always manifest as positive or negative changes in consumer demand. While this assumption is generally true, there are exceptions. Like all forms of income, increased welfare payments into households can either be spent on immediate consumption or saved. Although more disadvantaged groups, such as welfare recipients, could reasonably be expected to have a high marginal propensity for consumption, this is not always the case. For example, through the 1990s, the Japanese economy was in severe recession. The Japanese government accepted large and, indeed, dangerous levels of public debt through public works and very generous increases in transfer (welfare) payments to stimulate the economy, as Keynes had suggested. Unfortunately, the Japanese consumer base was aged, had little propensity to consume and were very conservative. Consequently, the recipients simply saved most of the increased payments. Therefore, these hugely expensive Keynesian measures failed to effectively stimulate demand and, in fact, worsened the overall economic situation by adding considerably to public debt and putting upward pressure on interest rates.

Keynes, of course, was not without his detractors and critics. The Austrian Friedrich von Hayek (1899–1992) is probably the most prominent. Hayek saw considerable danger in the Keynesian approach because it presumed a level of current and accurate knowledge, which neither the government nor anyone else could possibly possess. Further, he believed that the significant market interference suggested by Keynes would set up aberrations and distortions with many unintended consequences (Hayek 1944; Shearmur 1996).

Hayek considered that only the price mechanism could be used to share and synchronise personal behaviour and markets, and anything that interfered with this mechanism was inherently dangerous. While not suggesting that a total laissez-faire approach was appropriate in complex contemporary economics, Hayek argued that, in a capitalist economy within a liberal, (relatively) free society, underlying trust in the operations of the market mechanism was required; and that simply believing 'governments know best and will get it right for all of us' was remarkably naïve (Wapshott 2011).

Generally, however, the Keynesian approach offered governments an attractive opportunity (apparently) to fine-tune their economies with precise and targeted actions that could steer the economy through the natural instability of free-market activities. Though Roosevelt could not really be described as a Keynesian, his 'New Deal' policies of the 1930s exemplified the approach of stimulating consumer confidence and spending by extensive government work programmes. However, the underuse of resources that occurred during this period was not really resolved until most Western economies were stimulated by wartime production at the end of the decade. (Even though Roosevelt's New Deal was of great value politically, economists still argue as to whether it relieved or, in fact, prolonged the Great Depression in the US.)

The idea of a greater economic role and increased intervention by governments became increasingly attractive, not only for stimulating demand to recover from the Great Depression, but also to facilitate post-war rebuilding (e.g. the Marshall plan for Europe) and rapid growth thereafter. Under the Bretton Woods Agreement of 1944 between the leading economic powers of the day and led by the US, a new economic world order was established which firmly embedded greater government economic involvement through such new organisations as the International Monetary Fund (IMF) and the World Bank, among others.

For the next 30 to 40 years and, to a greater or lesser extent until the present, all Western governments accepted and applied Keynesian philosophies. From the 1950s to the 1970s (a period known as 'the long boom'), the theory appeared to work reasonably well (Skidelsky 2009). After this time, however, the success of pure Keynesian principles started to unravel.

First, governments in many countries could usually stimulate the economy during difficult times by tax relief, increased government spending on public works and/or increased welfare payments, for example. However, if the economy was overheating, the political imperative of re-election often made increasing taxes or reducing spending near impossible in practice. What government, for example, would reduce welfare payments just before an election in the name of 'good economic management'? Furthermore, the economy and its various components, even at a macro level, had become increasingly complex. In reality, the approach of targeting and fine tuning was often illusionary as various fiscal measures conflicted with, or negated, other policy initiatives. A range of unintended consequences also emerged, as Hayek predicted, because of the economic aberrations and exceptions that government intervention created.

Furthermore, in many contemporary economies, the Keynesian proposition that supply will automatically follow demand has proven flawed in practice. For example, opportunities in the form of demand for certain minerals or for some explicit type of production may emerge quickly, particularly from global markets. Governments may act to stimulate production through subsidies, infrastructure provision and a number of other government spending measures. Nevertheless, a range of other issues, including lack of available skills, time delays and bottlenecks in providing infrastructure and/or availability of associated private sector investment, could delay or limit the impact of these initiatives.

Perhaps the greatest challenge to the application of pure Keynesian economics came in the 1970s with the emergence of what became known as 'stagflation'. As outlined above, Keynesian economics is largely based on the principle of managing a continuum of demand levels across the economy. If overall demand is too low, resources (e.g. labour, input materials, machinery and capital) will be underutilised, and the economy will stagnate. At the other end of the continuum, if demand and consumption in the economy are running too strongly, there will be upward pressure on prices, (i.e. inflation) and a consequent loss of market confidence, adding further instability. Keynesians held that fiscal management could manipulate demand between these two extremes. In this scenario, stagnation and inflation cannot occur simultaneously as they are at opposite ends of the demand continuum (Vines et al. 2013).

In the early 1970s, however, major Western economies (particularly the US) began to suffer from simultaneous low economic growth (economic stagnation) and high inflation. Several factors contributed to this situation including the impact of the first oil crisis and rising energy prices (i.e. 'imported inflation'); the manner in which the US Government decided to finance the Vietnam War; and the loss of gold parity in setting many currency and exchange rates. Importantly, the situation proved that the fiscal management of demand alone could not ensure economic stability or guarantee sustainable growth.

Additionally, the presumption that governments somehow always knew 'which economic levers to pull and how hard', and that they would always act in an economically rational, politically balanced way with long- term objectives in mind, proved to be a fairly heroic idea, both in Australia and internationally. Hayek's earlier predictions were proving correct.

Of course, the fundamental elements of Keynesian economics are still relevant and important. The government does represent the largest and most powerful sector in all contemporary economies. The manner in which it regulates acts as a market leader and both taxes and spends (producing either budget surpluses or deficits) have a profound effect on the economic well-being of those countries. In the contemporary environment, however, the precise and correct application of fiscal theory is extraordinarily difficult and cannot be relied upon as the only mechanism for providing economic stability, growth, wealth creation and a reasonably equitable distribution of prosperity (Stanford 2008).

Many observers of a Keynesian approach today might suggest that effective direct intervention in the contemporary market should focus on managing market externalities – those external factors that influence a market or pricing mechanism. Externalities may flow from the market, but neither the consumer nor producer pays or is directly held responsible for them. Some externalities have negative effects overall (e.g. pollution, congestion, waste). Others, such as the cascading effects of research and technology and well-targeted infrastructure investment, are broadly viewed as positive. In this approach, governments could ensure that the full cost of negative externalities is captured and paid for, and that the production of positive externalities is incentivised and facilitated. This, however, is far from the original concept of a Keynesian approach.

Interestingly, when faced with serious economic situations, Western governments remain remarkably willing to use fiscal measures, as Keynes suggested, to stimulate demand and effectively 'buy their way out of trouble'. The reaction of OECD governments to the onset of a rapid and major downturn associated with the Global Financial Crisis from 2007 to 2008 represents a case in point (Skidelsky 2009; Roubini 2010). A similar government intervention ('Keynesian') approach has quickly emerged in response to the COVID 19 pandemic.

Milton Friedman and the rise of the monetarists

The third important philosophical idea discussed here is known as 'Monetarist Theory', associated with, among others, the work of American economist, Milton Friedman (1912–2006). Friedman's influence on economic thought and leadership was, together with Kenneth Galbraith (1908–2006), probably the most important in the second-half of the twentieth century. Though not dismissing the obvious truths underlying the Keynesian approach, Friedman considered it naïve and, for the most part, impossible to effectively apply. Likewise, he saw many of the links between various components within the economy (such as employment levels and inflation under Phillips curve analysis) as simplistic and unstable. Rather, in the 1970s to 1980s, he developed what became known as 'Monetary Theory'.

A Nobel Prize winner, his work over 30 years at the University of Chicago's School of Economics effectively changed economic policy globally. His influence was reinforced through his various appointments as a key advisor to various US Government administrations.

Friedman was basically an economic conservative who, similar to Hayek, retained a fundamental belief in the operation of free markets for the efficient and effective distribution of goods and services. His confidence in markets, however, did not align with the philosophical trust placed in them by Smith and the classical economists. Friedman contended that, as far as possible, markets should be allowed to operate freely, albeit with some pre- established 'rules of the game', to avoid extreme outcomes. He objected to the Keynesian notion that economic management was a precise art with which the economy could be continually fine-tuned by a benevolent and all-knowing government using fiscal means.

Friedman and others suggested that a more generalised and, therefore, less precise method of economic management was required. This approach would set general economic parameters and, as far as possible within these boundaries, let market mechanisms work. He correctly observed that the single common factor in every capitalist market was the use of money, by which all aspects of production, goods and services could be measured, valued and exchanged. He argued that, in the economy of any country, the government's best role was to control the volume and, therefore, the price and velocity of monetary circulation. Thereby, all markets could be influenced. Obviously, this approach became known as 'Monetary Policy', and a country's central bank became the principal instrument for its implementation.

Australia provides a good example of this policy in action. A central bank, the Reserve Bank of Australia (RBA), was established in 1959 with the objective of providing a stable and reliable currency to ensure a sound economic environment for Australian firms and households, as well as encouraging high levels of employment. (The overt priority of the bank's charter, the promotion of high levels of employment, is an important observation and has had a greater influence on bank policy and decisions than is sometimes recognised.)

The RBA is a government organisation with a charter to independently set the interest rate at which it makes funds available to its wholesale customers, the retail banks. Those banks add their own margins to this rate, which sets the retail rates both for loan funds to individuals, households and businesses and to accept deposits from savers.

Major acquisitions such as houses and cars, which are often debt-financed, are obviously sensitive to changes in RBA interest rate policy and, subsequently, to retail interest rates. Emphasis in the media on the regular RBA reviews of interest rates testifies to this fact. Announcements and changes in interest rates by the RBA also have wider impacts. For example, a decision by the RBA to raise interest rates sends a clear signal to all market participants that monetary policy is being used to try to slow the economy somewhat in the short to medium term. This decision would have an almost immediate dampening effect on overall levels of consumer confidence and sentiment, although the actual effect of the policy change takes time to permeate the economy. The reverse would be the case if the RBA wished to stimulate the economy by relaxing wholesale rates.

Monetary policy is an indiscriminate 'blunt instrument' and slow to have full impact. Nevertheless, the simple logic upon which it is based has advantages over the complex distortions of a Keynesian approach.

Changes in monetary policy and interest rates move in a particular direction and have several effects. For borrowers, according to most media reports, there is a simple attitude to changes in interest rates whereby: 'tightening monetary policy means higher interest rates, which is bad; easing of monetary policy is good'. The real situation is more complex.

If a firm or household has savings, rising interest rates are in fact beneficial. All other things being equal, a rise in interest rates also encourages inflow of foreign capital into the country. As politically sensitive as short-term movements in domestic interest rates may be, the RBA's agenda is largely set by global events. Many of the forces that affect interest rates relate to the balance of trade, fiscal settings and, particularly for Australia as a significant primary producer of resources, world commodity prices. The RBA has little or no control over these forces. As well as its commitment to high levels of employment, the RBA's objective is to provide stability and confidence in the currency, particularly in the medium- to long-term. Private sector investors, producers and consumers need this level of confidence to make the decisions that fuel future growth, employment and prosperity.

In contrast to the Keynesians, who identify demand and demand-management as the critical factor for economic stability, monetarists emphasise confidence in the consistent value of money as the primary short- and long-term stabilising influence. The control of inflation is essential for monetary policy to be truly effective. Both consumers and producers need confidence that the falling purchasing power of the currency will not erode their financial decisions, particularly in the long term. Monetary policy requires low levels of inflation as a key prerequisite for sustainable economic growth; high levels of inflation, however, encourage short-term decisions and speculation.

Similar to the central banks of a number of other OECD countries, the RBA provides a cornerstone of Australia's economic system. Its power, influence and independence are well recognised and respected by the private and public sectors alike. Nevertheless, the Australian economy is quite small – only about 1.7 per cent of world economic product (Austrade 2019) and particularly exposed to the prices of globally traded commodities. Consequently, the RBA's ability to maintain a stable economic and monetary environment is limited by global trends and significant changes in the economic activity of much larger trading partners such as the US and China. Additionally, Australia's large banks now source increasing proportions of their funds direct from off-shore markets, instead of the RBA. As a result, the RBA's ability to directly influence retail interest rates is diminishing. As exemplified by the behaviour of Australia's dominant four retail banks over recent years, it is by no means certain that interest rate changes set at whole-sale level by the RBA, either as increases or decreases, will be passed on at the retail level.

As noted above, the impact of changes in monetary policy take time to permeate the economy. Governments, pressured by political realities and accustomed to the 'quick-fix' solutions offered by Keynesian measures, are rarely willing to allow the time necessary to achieve full impact.

All capitalist governments use a mixture of Keynesian and monetarist management strategies, which can cause further complexity as various strategies integrate. Nevertheless, no governments operate only with Keynesian measures; most rely principally on a monetarist approach for general economic direction and stability. Though controversial, the economic approaches that became known as 'Reaganomics' and 'Thatchernomics' in the late 1970s embraced Friedman's monetarist ideas and his encouragement of smaller government. Although it is erroneous to infer simple causal relations in such complex matters, many link recent rapid growth, development and wealth creation, as well as the emergence of China as an economic power, to the freeing of markets – the fundamental tenet of Friedman's philosophy.

Furthermore, it is interesting in this context to note that many of the ongoing problems of the European Union (EU) emerged from the centralisation of financial management under a single currency (the Euro) and, with that, the detachment of major components of monetary policy from the economic and political control of individual member countries. For each individual country (not just the aggregate groups), the appropriate and specific application of monetary policy is critical to economic propensity and political stability.

2.4 Further concepts in economic theory and management

In the preceding sections, a number of concepts, such as markets, and demand and supply, have been discussed in a general sense. To provide a better understanding of these ideas, some of their important characteristics are outlined below. They have obvious relevance to the operations of property markets and dealings. They are further discussed in more detail in Section 2.5.

Markets

A market is the point of transaction between sellers and buyers, (in economics, normally called 'producers' and 'consumers'). The behaviour and reasoning of those involved in a market are at the heart of microeconomics: who buys and who sells?; what are the incentives to buy or produce more, or to buy and sell less? Markets have existed for thousands of years. They were fundamental to early barter economies in which, for example, farmers may have produced more than their own requirements (i.e. a surplus) and taken the surplus to a market (place) to exchange it for other goods and services they required or, later in history, to exchange it for a medium of exchange, money.

In contemporary environments, literally millions of 'markets' (i.e. transactions) occur at any point in time and, if left to operate freely, they will be inherently unstable as buyers and sellers come and go and available options and incentives change. The internet provides a radically different version of a market – 'a marketplace without the place'.

However, the economic fundamentals and the reasons why producers and consumers participate in this giant market remain unchanged. A key determinant of market activity is the reaction of both producers and consumers to price. Depending on the goods or services, many factors (identified variously as 'incentives', 'disincentives' or 'variables') influence decisions on production and consumption. Price is, however, usually the most important determinant and often used to explain economic theories and models.

For example, consider a potential customer entering a shop (a 'marketplace') to buy clothing. The person will be motivated (incentivised) to buy depending on: their need for this article (its utility or usefulness); its quality; whether the person likes the style (taste and fashion); the manufacturer and the place of manufacture (brand recognition); the 'pitch' by the sales assistant; and probably a range of other general and personal considerations. More importantly, however, the customer immediately looks for the price tag to assess whether they consider the article good value for money in comparison with other things they might like to buy; whether the price is within their budget (or access to borrowed funds through use of a credit card); and whether the purchase of that item fits with other demands on their finances. This last consideration is called the 'opportunity cost' – that is, what other things the customer has to forgo to secure the item under consideration.

Thus, for producers or consumers of a particular good or service, a complex interaction of variables contributes to the final decision of whether to transact. However, price is frequently the 'deal maker' or 'deal breaker' and indeed encapsulates or reflects many other considerations. For this reason, the price (p) to quantity (q) relationship is usually at the core of basic market theory. (In fact, any variables relevant to the market could be used: price to quality, e.g. or quantity demanded as a function of the taste and preference of a demographic group.)

By looking at the relationship between 'market price' and 'quantity supplied or demanded', certain market principles can be illustrated. For example, the general reaction of consumers and producers to different price points (incentives) can be tracked. Primary data can be collected in market surveys to accurately establish the price-to-quantity relationships in particular markets, which is exactly what big producers and retailers do. All things being equal, if the market price of a commodity increases, producers will have an incentive (i.e. more profit) to produce more of the commodity (i.e. a direct relationship). A supply curve (s, s^1) can then be constructed, which will typically have a bottom–left to top–right shape (as presented in Figure 2.2). Note that, as well as the generic shape, the function will also normally taper towards the bottom–right, generally reflecting, among other things, time and production and other time lags before the producer is able to respond to the incentive of growing demand.

The reverse applies to consumers. As market prices rise, demand falls as consumers receive less for their money (i.e. an inverse relationship). The shape of this demand curve (d, d^1) is typically top–left to bottom–right (Figure 2.2). (This time the function will often taper down towards the right, reflecting changing consumer preferences after early demand is met.) In a particular market, the supply and demand curves intersect at the point of 'market equilibrium'. Market equilibrium is the unique point that lies on both functions – that is, the point at which consumers and producers agree on both the market price and the quantity to be supplied and consumed. This point is also called a 'market at rest'. While this theory may apply in simple short-term analyses with two variables, real markets tend to oscillate around this point and are unstable because of other dynamic market forces.

As noted previously, there are other influences at work apart from this two-function model, price-quantity. These 'externalities' include changes in taxation, law, substitutes and so on. Changes in externalities will potentially shift the entire demand, dd^1, or supply ss^1, curves to the left or the right depending on the nature and impact of those forces in this simple analysis.

The sensitivity of (or responsiveness to) a change in one of the variables to a change in the other variable, is called 'elasticity'. In this model, price elasticity measures how much quantity the supply of a good, or demand for it, changes if its price changes. If the percentage change in quantity is more than the percentage change in price, the product is said to be 'relatively price elastic'; if it is less, the product is 'relatively price inelastic' (Bishop 2009).

For example, the consumption of essential goods (e.g. bread, fuels and energy, water) is fairly insensitive (inelastic) to changes in market prices. Geometrically, graphs of the demand function (the relationship between price and quantity) for these commodities are steeper than those for non-essential discretionary (luxury) goods such as holidays or new cars. Supply functions may also differ in elasticity in various markets due to their own set of considerations that often change over time (Jackson & McIver 2007).

All of these matters appear quite simple and, in effect, common sense – and indeed they are. Nevertheless, they are at the core of economic behaviour and need to be recognised as such, regardless of the complexity of the dealings to which they are applied. They have important implications for real property markets.

Types of markets

Markets were defined above as the intersection point at which a transaction occurs between buyers and sellers of a particular good or service. Some markets are very obvious; produce and fish markets are overt examples, especially in many countries where traditional haggling over price and quality illustrates the volatile and dynamic nature of transactions. Most markets today, however, are more organised and nuanced and obviously are not simply related to the buying and selling of retail goods. The sales of services, houses, cars, intellectual property, and sales through the internet, are all market transactions, regardless of the medium by which they occur.

Despite this apparent diversity, markets can be classified into four broad categories, based largely on the relative power of the producers and consumers within them. They are: (i) perfect competition; (ii) monopoly; (iii) imperfect competition and (iv) oligopoly. The specific characteristics of a huge range of markets generally lie within these broad categories. In particular, two of the categories, perfect competition and monopoly, are extremely rare in the contemporary environment. Nevertheless, the concepts remain important.

Perfect competition

A perfectly competitive market is, in effect, the market type described by Adam Smith in Classical Theory. This market has many buyers and sellers, all of whom have full knowledge of the market and are free to instantaneously come and go as they wish. The commodity traded is homogeneous; that is, all units of that good or service are identical to all other units offered for sale, and consumers do not differentiate between product units. Buyers and sellers do not influence the market, apart from their individual actions. Therefore, all participants are 'price takers' – if they want to buy or sell, they must meet or accept the going market price. In this model, no external forces act on the market beyond the thought processes and decisions of individual buyers and sellers.

In reality, externalities do, of course, exist: taxation; regulations; product differentiation; trade practices; supply lags; issues of location; seasonal variations; product substitutes and patents, to name a few. However, they are ignored in this model of a 'pure' market. For individual producers, the marginal cost (i.e. the cost of producing the unit) will exactly equal the marginal price received.

Obviously, this model represents a highly theoretical market and, even if such laissez-faire markets existed in Adam Smith's time, they certainly do not exist in isolation now. Closest to this perfect model are probably the markets for gold and primary commodities such as fruit, vegetables and fresh fish when sold in 'product market' environments.

In practice, this 'perfect' market price mechanism fails in the contemporary environment because of three principal factors:

* scarcity power (resulting, e.g. from production lags, patents and other impediments to immediate and uninhibited actions)
* incomplete market knowledge of decision-makers (producers and consumers)
* the side effects of market transactions on parties other than producers and consumers.

Nevertheless, this model of perfect competition (and the price mechanism) should not be dismissed as simply a theoretical exercise. In a capitalist society, the model still represents the most efficient method to decide what, when and how much of a commodity to produce, and the selling price of that commodity. Although this process takes no account of other political, wider economic or social objectives that a community or nation may wish to pursue, it does provide an initial economic benchmark against which variations imposed on the market can be measured.

For example, any externality moves a market away from and distorts the 'efficiency base case' of this perfect market mechanism and sets up a range of intended and unintended consequences. While these consequences may be advantageous and supported by the community, any economic and non-economic externalities should be introduced with full knowledge of their effects on the perfect competition model and likely repercussions.

Monopoly

In a monopoly market, there is only one supplier of particular goods or services. Therefore, the monopolist sets the price (is the price-maker) and also determines the quantity supplied. Consumers in this market are obviously in a weak and potentially vulnerable position. In times past, monopolies were not uncommon in Australia. Scattered, low-density patterns of development and small domestic markets favoured the rise of single suppliers through what is known in economics as 'cumulative causation'.

In part, this describes how, in a particular region, investment in a major project or piece of infrastructure might effectively preclude the establishment of later competitive investments because of the limited market size. These situations are sometimes called 'natural monopolies'.

Examples of natural monopolies of various types include steel making; a range of government-sponsored 'single desk' commodity marketing organisations; utilities such as electricity, gas and water; and ports, railways, airports and telecommunications.

Nearly all of these Australian monopolies were dismantled in the face of strong internal and external competition policies that emerged in recent decades. In addition, the monopolist's position is not as desirable as it may first seem. Any aggressive behaviour, particularly in the supply of an essential commodity, is obvious, and almost invariably meets with strong public opposition and increased regulation and legislation by government. Furthermore, substitutes are available for most goods and services, and rapidly advancing contemporary technologies may make the monopolist's product obsolete should they rely on their market monopoly or patents for long-term success.

However, monopoly markets can still emerge in the contemporary environment. Sometimes, for example, because production by other suppliers lags behind the market leader, one producer, temporarily at least, effectively becomes the only market supplier. Such windfalls are usually short-lived as other suppliers soon take advantage of the opportunity. Legislation to ensure competition and outlaw unconscionable conduct by an existing supplier may also facilitate market entry by new suppliers.

Contemporary monopolies also emerge through the exclusive holding of key intellectual property and patents; unique pharmaceuticals or systems developed by software companies such as Microsoft provide examples of this. Microsoft has recently lost a succession of cases regarding what was perceived as anti-trust and anticompetitive behaviour. These decisions have opened downstream markets, blunting the power of this contemporary (former) monopoly. Reliance on patents to protect a monopolistic or dominant market position is also continually threatened by increasingly global markets where attitudes to intellectual property protection and defence of patents vary considerably.

Imperfect competition

Imperfect competition, or as it is sometimes (confusingly) known, 'monopolistic competition', represents a large proportion of the markets in which individuals and small-to-medium enterprises (SMEs) operate in modern capitalist societies, including Australia. Practically all trade and professional services, small-scale retail, most rural undertakings and many forms of manufacturing by SMEs fall into this category, together with many property and construction markets.

These markets are characterised by the involvement of many buyers and sellers, none of whom have power over the markets in which they operate. Therefore, they are all price-takers. External influences and product substitutes abound, and while all production units within a particular market are of an identifiable type, they are heterogeneous; that is, units are discernibly different from one another in quality, size, style and many other characteristics. In these conditions, buyers and sellers have imperfect (incomplete) knowledge of their markets. In addition, although both buyers and sellers are free to enter and leave the market, barriers to entry may exist and decisions or production may lag behind demand, depending on the product. Market positions may be assisted by securing a range of intellectual property rights – but, in increasingly globalised markets, such protection can be both expensive and difficult to defend. Legislative requirements (such as approvals and licencing) may also prove disincentives to rapid market entry or exit.

Producers often try to highlight product differences through advertising and site (i.e. location) goodwill and by personal goodwill. Because they lack significant market influence, producers in certain markets are often forced to compete principally on price to improve market share.

The market for existing residential housing in a particular town or region provides a typical example of an imperfect competitive market. Here, all the 'units of production' are correctly defined as houses, but each one has its own characteristics and prospective purchasers will differentiate between each residence offered for sale. Vendors and their agents will emphasise positive features of their particular property through advertising and other promotion to encourage the buyer to select their property over others. While both buyers and sellers may research a market, no individual has complete market knowledge. Many externalities, particularly interest rates, bank lending policy, taxation, fashion, personal taste and demographic changes, will exist, influencing the market. When demand is strong, production lags are also common, given the time taken to develop new residential land and to increase housing stocks. Housing affordability often creates an entry barrier and, although housing is a basic and essential commodity, these market characteristics often result in significant structural instability and, at times, volatility.

Oligopoly

Oligopoly is, in effect, market competition among a few. Over recent years, these markets, in which there may be only two (a 'duopoly') or, at the most, five major producers, have dramatically increased in size and importance in the economies of many OECD countries including Australia. Oligopolies, besides dominating their particular markets, are typically located within key sectors of the economy. Often they supply final consumer markets and, therefore, are names with which consumers are familiar: in the motor vehicle market (General Motors, Ford, Toyota); in the banking sector (NAB, ANZ, CBA and Westpac); in large-scale retail (Coles-Myer, Woolworths); and in markets for white goods, airline travel, large-scale construction, building materials, petroleum, media, brewing, telecommunications, et cetera.

Oligopoly markets are normally quite stable, giving both consumers and governments (for better or for worse) a sense of comfort and certainty. The small numbers of suppliers engage in activities that maintain their market positions and make the entry of other potential producers very difficult. For example, existing suppliers typically have economies of scale, (often) patents, high brand recognition and established distribution and service networks; they also invest heavily in fixed capital and research and development. Through competition legislation and the work of the Australian Consumer and Competition Commission (ACCC), the more extreme and anticompetitive activities previously available in oligopoly markets (e.g. price fixing, forming cartels and engaging in restrictive trade practices and unconscionable conduct) are now illegal. Nevertheless, in a relatively open market, tactics such as predatory pricing can potentially limit market entry by other producers and, thereby, tend to maintain the status quo.

In these markets, consumers clearly differentiate between products, brand names and advertising, based on product positioning rather than price, which are often important to producers. Competition is rarely on price, except between front-line retailers of products, and this end of the market tend towards imperfect competition in any case.

For existing oligopolies, price competition tends to be self-defeating. If one oligopolist significantly raises prices, the others may not follow as they are already making sound profits and the company that raises prices greatly will almost certainly lose market share. However, if one oligopolist lowers prices, all others will tend to follow to maintain their market share and, therefore, all will lose profitability. Prices, therefore, tend to be fairly stable, but firmly in the control (at least informally) of the oligopolist group.

Although cartels and price fixing are illegal, price leadership in oligopoly markets is common. Price leadership typically exists in markets where one oligopolist is larger than its competitors and will advertise a recommended retail price for a product. The few remaining producers tend to accept this lead, set their selling prices around this level and market stability, thereby, results. This practice is, on the face of it, within competition legislation; however, the group of oligopolistic suppliers clearly controls overall supply and price of the product. Some oligopolies are frequently in positions to secure extraordinary profits, but consumers often accept this situation for several reasons. First, oligopoly markets are typically stable and goods are almost always available. Second, economies of scale keep prices at acceptable levels. Third, because of the scale of these individual firms, significant investment is continually made in research and development to improve existing and to introduce new models and products. Meanwhile, corporate advertising, funded by huge budgets, reinforces the benefits of maintaining loyalty to known brand names.

2.5 The relationship between real property and the economy

Real property represents a critical component of practically all economic activity, together with capital, labour and management skills. As well as this general economic function, real property represents a basic form of holding and aggregating wealth. Property development, financing, management and operations are significant generators of employment across the economy. Additionally, property provides the collateral upon which most financial sector dealings are based.

Traditionally in economic theory, economists use the term 'land' (as in 'land economics') to describe the whole property sector. As explained in Chapter 1, the term includes buildings and other related, fixed investments.

Real property requires recognition and investigation in the context of resource economics including the management of natural resources such as vegetation, minerals and water. Land (i.e. property) assets will present a range of potential land uses and use intensities, with numerous external factors impacting on final allocation. While all these resources and their associated markets have different producers, regulations, production costs and pricing structures, they share many fundamental characteristics and have comparable economic underpinnings.

Real property also has a range of unique characteristics that influence its role and performance in economic activity. Some may seem self-evident, but all need to be recognised if the nature of real property is to be fully appreciated. These are listed as follows:

- In Australia, property assets are typically held in small parcels with tenure rights (often freehold but, in some cases, crown leasehold) reinforced in legislation.

Therefore decisions on use are mostly at the microeconomic level. Despite town planning, building and other regulations that limit unilateral development activities, friction between the rights of individual owners and community demand and expectation frequently occurs.

- Property (and particularly physical land form, vegetation, soil, et cetera) is a finite resource (fixed in quantity and unable to be recreated), which has obvious implications for environmental, economic and community sustainability.
- Property is a physical asset, and therefore its various components are subject to deterioration, decay, erosion and/or obsolescence. Such assets will require often intensive management and maintenance to maintain functionality over time.
- Property is fixed in space, which introduces issues of location and infrastructure. The potential positive and negative effects of the use and development of any property on the surrounding environs must also be considered.
- Property is a heterogeneous commodity; properties differ from each other to a greater or lesser extent, which is recognised by buyers and sellers alike.
- Property assets are commonly divided into a number of sub-sectors – commercial, retail, industrial, residential, rural, et cetera – each with different use capacities and different determinants of supply and demand. Within each sector, criteria such as location, quality, size, condition, and functionality clearly differentiate individual property and property sub-groups.
- As most property secures value through development and use, issues of functionality, income generation and management of the asset-in-use are particularly important for this asset class.
- The basic land component of property always requires investment, in the form of capital and other factors of production, to reach its highest and best use (i.e. highest potential). In this process, the other production factors effectively merge with the land and cannot subsequently be extracted.
- The property (and development) sectors are subject to considerable 'lags' (time delays) in recognition and production in bringing forward property into usable form. These can create (relatively) short-term price spikes and, sometimes, short-term windfall profits for those already holding that asset type.
- The unit price of property is comparatively high, which produces considerable barriers to market entry. This tends to favour longer-term investments and makes buy-and-sell decisions highly significant.
- A large array of legislation and regulation dictates the allowable level of development, influences value and sets overall parameters for the ownership, development and use of real property assets.
- Property is a dynamic asset that physically, legally and economically evolves over time. Adaptability and flexibility therefore need to be incorporated into the strategic and operational management of such assets.

The original acquisition of the land and the construction and fit-out of buildings appear to be the most capital-intensive part of the life cycle of property assets. In reality, however, the operational costs of the assets – for example, services, insurances, taxes, maintenance, re-fits, et cetera, extend over a long period (i.e. the 'asset-in-use') – are by far the higher cost (often over 80 per cent of total asset cost during its lifetime). This again emphasises the importance of quality strategic and operational management for this asset class.

The typical investment profile for property assets involves large upfront capital investment (i.e. the acquisition), which is rewarded by cash flow (income) over a long period. Therefore, the 'time value of money' relation (also called 'time preferencing' or 'discounting'), long-term risk profiles, levels of investor confidence and the cost of money become critical considerations in successful decision-making.

In considering the overall economic parameters of real property, it should also be observed that the individual nature of each property asset does not lend itself particularly well to mass production, economies of scale or the compounding value of consolidated research and development common to other high-value, manufactured goods. These issues can, at times, slow the diffusion of new technologies and strategies. This situation is made worse by the highly linear, segmented production process typically employed and by the often limited interaction between the stakeholders, professionals and tradespeople involved in development of these assets.

As the production factors of land, labour, capital and management combine to produce a certain outcome or activity, the actual land and capital components are typically limiting factors. Again, a rural example might well explain this situation. A farmer, at least in the short term, has a farm of a certain size and a finite amount of money (capital) to apply to it. If funds are insufficient to properly prepare the land and to buy seed and fertiliser, the farm will underperform and not yield the return (i.e. the crop value) of which it is capable and the farm's inherent value would suggest. This is known as 'undercapitalisation' of the resource.

As the farmer applies successive units of capital – for example, in the form of fertiliser and tractor fuel (i.e. the 'marginal cost') – productivity in the form of additional tonnage per hectare, 'marginal revenue', rises. Clearly, this relation does not proceed indefinitely. After some point, the marginal revenue (i.e. the value effect of the additional capital investment) drops below the marginal cost, and the extra investment becomes uneconomic. In this situation, the land resource is 'overcapitalised' and illustrates the economic 'Law of Diminishing Returns', previously described in Section 2.2. In this example, if the farmer continued to apply fertiliser, he would eventually kill the crop and get no return, an extreme outcome of this law.

The rural example illustrates a simple input–output relationship, but many examples of undercapitalised and overcapitalised properties can be identified throughout residential and commercial property sectors. In these cases, however, incremental further investment in the property is likely to be more complex than when the farmer decides to spread more fertiliser on that crop.

Therefore, it can be observed that the concept of adjusting the level and combination of production factors (particularly land and capital) to optimise profits holds true in the property sector. In fact, given the high capital value, long life and risk exposure of these assets, the significance of the initial strategy and investment decisions (e.g. incorporating sustainability into the future) is arguably greater than in most other economic activities.

In overall economic terms, the market for existing property is best described as examples of imperfect competition: the products are heterogeneous; both sellers and buyers are price-takers; and no one has full knowledge of, nor control or influence over, the market in question. However, for new construction, and particularly for major development or supply of inputs such as building and construction materials, oligopoly markets now dominate.

As well as these largely microeconomic attributes, the macroeconomic context for the property sector needs to be appreciated. This wider environment has an extraordinary number of political, physical, social, community and financial components, all of which change over time and continually impact property usage and property decision-making.

At this stage however, several general observations of the overall relationship between property and the wider (macroeconomic) environment should be noted.

First, the property sector is closely linked to and, to a considerable extent, integrated with the financial sector and the development and construction sectors (see Figure 2.3). These links are further discussed in later chapters.

A second general link between real property and the wider economy lies in the existence and relevance of cycles.

In economics, as in all other sciences, much research is based around the observances of events to detect patterns of behaviour that might explain current and past events and subsequently predict and manage future events. Some patterns may involve easily identifiable cycles in the natural sciences (e.g. the climatic seasons or the cycles of animal life). It is reasonable for economists to enquire if such predictability or patterns could be identified in economic events over time, including for sectors of the economy such as real property.

It is generally accepted that such patterns can often be observed, although it is simplistic to consider them mechanical and perfectly cyclic, like a clock face, by which means these patterns are sometimes represented. The Russian economist Nikolai Kondratiev (1892–1938) considered that, rather than cycles, per se, repetition in the economy is related to waves of change. He considered that economic waves were of a very long aperture – perhaps 60 years or more. New technologies – for example, steam, then internal combustion, then electricity – provided the catalyst for these changes (Alexander 2002).

Joseph Schumpeter (1883–1950) also subscribed to a theory of intermittent, disruptive technologies, which he called 'creative destruction'. These innovations punctuated economic activity and, combined with entrepreneurship, regularly re-energised, changed and drove economic progress forward (Schumpeter 1976; McCraw 2009). Numerous researchers have contended that discernible cycles can be identified in many parts of the economy, but they generally agree that, because of the complexity and range of forces at work within the contemporary economy and community as a whole, there is no indication of precise timing or sequencing (Wheaton et al. 1997; Pyhrr & Born 2006).

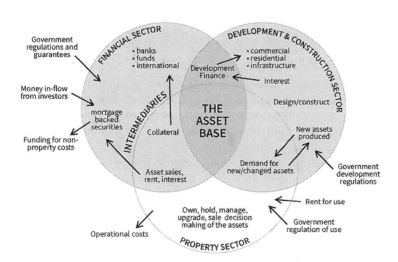

Figure 2.3 **Real property assets within an integrated system.**

Various economic events will almost certainly stimulate specific reactions. It could reasonably be anticipated, for example, that a sustained and significant reduction in interest rates would increase demand for major assets such as property, all things being equal. Unfortunately, in practice, all things may not 'be equal'. A range of other factors including confidence levels and expectations, existing debt levels, production lags, 'crowd behaviour', current levels of investment, government policies or lack of information continually act to distort the predictability implied by cyclic patterns.

These observations are particularly applicable to property markets and any cycles would be likely to be of fairly long aperture, given the large scale of the units of production and the long-term nature of these investments. As a result, property markets/cycles will typically swing over time from unsustainable exuberance to unnecessary pessimism, overshooting from shortage to oversupply and, sometimes, 'boom to bust'. Furthermore, the segmented, heterogeneous and geographically dispersed nature of the property sector may well result in a range of different patterns emerging across different regions. The market for residential properties across various Australian cities provides excellent case studies for this market feature.

This is not to imply that cycles in property markets are so indistinct to be of little or no use. As Reed and Wu (2009) note, an understanding of underlying cycles helps inform markets and the general public of overall market trends; although, to assist in the dissemination of accurate information, recognition of the imprecise nature of these predictors must be included.

In observing certain UK markets, Scott and Judge (2000) suggest that cycles of about eight years could be identified. Clearly, however, major economic or political events emanating from sources external to the normal supply and demand criteria of that particular property market have the power to fundamentally disrupt or elongate underlying cycles.

Because of the complexities of property markets, cyclic trends will be far less rigid than those of the natural sciences. They typically exhibit re-occurring fluctuations, but also suffer from non-periodic irregular events. The attacks on the World Trade Centre, New York in 2001 (9/11), the Global Financial Crisis of 2007 (and beyond), the COVID 19 pandemic and radical changes in the cost of energy all provide contemporary examples of these disruptions.

Furthermore, the overall directions and pre-disposition of the financial sector, the attractiveness or otherwise of other, competing investment options (e.g. the share market, interest rates and bond rates) and the prevailing exchange rates will all affect the level of investment likely to flow into the property sector at any point in time. Drawing on a Darwinian analogy, markets might also be observed to be continually evolving and reacting to changing, external influences while, at the same time, moving more or less in a cyclic fashion in line with underlying economic forces.

Notes

1 Although this analysis may seem theoretical and abstract, the theory can explain everyday behaviour. Consider the following: a person is very hungry and goes out to buy and consume a large meal to satisfy that hunger. It is then unlikely that that individual would immediately order another large meal. Instead, the person may decide to re-prioritise his/her needs and may go to visit friends, go home and read a newspaper, or engage in any number of other activities that previously would have seemed less important when the person was very hungry.

2 In this regard, the observations of an early twentieth-century economist, Joseph Schumpeter (1883–1950), are interesting. Schumpeter described the entire process of economics (not just economic theories) as one of 'creative destruction', in which existing ideas, structures, technology and production were continually being challenged and replaced by discontinuous waves of new knowledge, technology and entrepreneurship. Each new wave will, in turn, be challenged and replaced by successive waves into the future. Though postulated over 80 years ago, this concept provides remarkable insights applicable to contemporary economic, business and social environments.

Chapter 3

The financial sector

3.1 Introduction

Historically, real property has been a highly prized possession and typically attracted significant capital value; it is a 'bricks and mortar', physically enduring asset class. These attributes are recognised and appreciated by the wider community and by lenders who will seek real property, above all, as finance collateral.

The study of real property tends to focus on the important, special characteristics of the assets themselves, but sometimes fails to fully recognise how property integrates with its economic, financial, legal and community environment and the wider expectations of stakeholders.

An analogy may be made of a botanist who closely studies and secures detailed understanding of a particular plant species, but whose work has little or no regard to the soil in which the plant grows, its ecosystem or the overall environment the plant requires in order to flourish. Obviously, while the biologist's work may well have a primary focus on the plant itself, without an understanding of the other critical 'environmental components', that knowledge will be of limited practical use. This is equally true of the property sector. Knowledge, not only of real property per se, but also of wider issues such as physical structure and development, law, town planning and financial dealings are important elements for long-term success.

This chapter considers one of the most important of these 'environmental components', the financial sector – components of which are also colloquially referred to as 'financial services' or 'the banking sector'.

So important is the relation between property and finance that, in a contemporary capitalist economy the boundaries between them are sometimes indistinguishable. The Global Financial Crisis (GFC) of 2007–2008 provided a dramatic example of that.

In the vast majority of property dealings, investments or developments, debt funding will be involved. In return, lenders financing significant loans require that debt be guaranteed through a mortgage against a real property asset. A critical and long-lasting link between the two sectors is thereby established. Again, to use the botanical analogy, in a contemporary economy, money and finance are almost as essential to property as soil and water are to the plant species referred to above.

A general understanding of how money systems and the financial sector operate therefore represents essential knowledge for property professionals in the contemporary environment.

3.2 Defining the financial sector

On the surface, defining the financial sector is fairly easy. It is about money and the business of managing transactions and the circulation of that money.

However, money permeates all markets and, in practice, capital funding and cash flow effectively merge as part of practically all economic activity. In the property sector, for example, the funding that it provides transforms into the new capital investment, funds for operations, maintenance and upgrades and is the medium (i.e. 'cash flow') which effectively holds the entire enterprise together. Particularly in large, capital-intensive sectors such as real property, the two 'production factors' – land and capital – effectively merge into a single entity, namely, that property and the activities that occur there. As discussed in Chapter 4, this can also create tensions between these, increasingly dominant financial drivers and economic performance outcomes and the wider, long-term civic, community and environmental roles that real property and the built form perform.

The OECD (2005) provides a simple definition of the financial sector as: the set of institutions, instruments and the regulatory framework that permits transactions to be made by incurring and setting debts, that is, by extending credit. This system makes possible the ownership of wealth from the growth of physical capital.

By implication, the term 'financial sector' includes a broad range of organisations (both public and private) involved in the circulation and management of money – all types of banks, credit providers, finance companies, stock markets and brokers, investment and superannuation funds, insurance companies and so on.

The definition does not however simply identify the sector in terms of its institutions. It also recognises its key functions within a capitalist economy – wealth creation through the leverage and management of debt. This highlights both the importance of the sector in its own right but also its implications for the successful operations, stability and growth of all other parts of the economy.

The finance sector worldwide is very large and, despite disruptions such as the GFC, has grown exponentially over recent decades, fuelled by globalized trade and overall increases in wealth. In the US, for example, the sector represents 7.4 per cent of GDP, employs over 6.3 million people and earns $B237 in annual income. In the US alone, it controls an extraordinary asset base of almost $T50, when overall financial assets including US pension funds, private equity funds and venture capital investments are aggregated (US Department of Commerce 2019).

Though not on that scale, the Australian financial sector is also substantial. Overall, it represents about 9 per cent of GDP and employs about 450 000 staff (RBA 2019). The four major banks, that control about 77 per cent of all savings deposits into the sector, are each within the top 50 banks globally (Deloitte 2018).

Components of the sector are also impressive. The Australian Stock Market (ASX) has a market capitalisation of $US1.9 trillion and is the eighth largest exchange in Asia and in the top 16 in the world (Statista 2018). The Australian dollar is the fifth most traded currency in the world (IG 2018). Funds under management in Australia total about $T3.6, which include superannuation contributions. This latter group have exhibited a remarkable and compounding annual growth rate – increasing 10.2 times from 1990 to 2016, at that point to represent about 168 per cent of Australia's GDP (ATIC 2017).

Overall, therefore, the sector in Australia could be described as mature, and, by world standards, relatively stable and generally well regulated (IMF 2018). However, because the Australian economy is, in a global sense, small, open and largely dependent on commodity markets, short-term volatility, particularly in currency, is inevitable. Given this inherent volatility also, natural calamities such as prolonged nation-wide droughts or economic aberrations such as rising household debt or asset price fluctuations can produce sudden negative impacts that can take some years to remedy.

3.3 Some history – the development of the financial sector

For much of history, humans have led a subsistence existence – effectively as hunters and gatherers, living 'hand to mouth'. Despite the sometimes valid criticism of contemporary economies and the role of money and financial institutions, there is no doubt that life in the 'pre-market/pre-money' world was extraordinarily dangerous, poor and short.

It was only well after the commencement of farming (about 9500 BC) and the increasing availability of economic/production surpluses that trade commenced (perhaps about 5000 BC). In those early stages of rudimentary markets, money was not necessary as goods were exchanged on a barter system. As markets developed and the volume and variety of trade increased, a commonly agreed standard or unit of transaction became necessary to act as a medium of exchange and as a standard against which all other commodities and goods and services could be measured. This role was taken by currency, with the first evidence of metal coins recorded in the eastern Mediterranean in about 700 BC. The metal value of the coins was about equal to their face value.

As well as being fundamental to the construction of markets, the introduction of currency, and later, banks and the financial sector, also allowed for the critical relation between lenders and borrowers to be properly defined, playing a vital role in aggregating capital. For example, a corporate entity such as a firm, which brings together equity from a number of investors to carry out certain activities, could not really exist without some common unit to measure levels of equity, debt, profit and loss. By the fourteenth century, double-entry bookkeeping was introduced in the trading city-states of northern Italy. This allowed for the better recording of these complex financial dealings and much of the trading success of those small states lay in their ability to best track, administer and control those dealings.

The history of the financial sector is, however, not simply about economics. It is also a story of politics, culture and religion extending back over centuries. The philosophers of the early Christian church argued that money itself provided nothing additional to human well-being. Therefore, it was, in effect, a 'necessary evil' and its use was to be contained solely to a medium of exchange. On that basis, the Church outlawed the use of money for any wider purposes. Known as 'usury', this included borrowing or lending money whether involving interest or not.

Because of the power of the Church and its close links with governments over centuries, this overall philosophy dominated policy throughout that period. Of course, as economies developed, the use of borrowed money became essential for governments and business alike. The impasse was effectively addressed by allowing such activities to be carried out by minority ethnic groups, including Jewish, Chinese and Arabs among others, whom the Church of the time considered were beyond salvation in any case.

Subsequently, and in an odd coincidence, Marx and other communist economic philosophers came to the same negative opinion regarding the financial sector. Marx's theory held that real economic or other advances could be based only on tangible work and, therefore, simply circulating, managing and/or trading money as a commodity had no value-add. Along with entrepreneurship, innovation and most forms of management, monetary activities could, according to Marx, be replaced by central planning and control and by direct labour input.

Both of these views, that of the early Church and Marx were fundamentally flawed, but they were to have significant effects on the development of financial systems over a long period.

While systems of exchange had been evolving over a thousand years, it was only during the sixteenth and seventeenth centuries in Holland, England and Scotland that financial and monetary systems, as recognised today, began to develop. Key events included the establishment for the first trading floors for equities (i.e. forms of stock exchange and futures trading) and the legal protection of private ownership of assets, commercial activity and markets. Additionally, forms of insurance emerged to share the risk of exploration and trading ventures. The profits now available through mercantile activities progressively broke down the Church's previous long prohibition and disdain of usury dealings.

At about this time, bond markets were created. Critically, these merged a range of economic, financial, legal and political agendas. As the role of government grew, the new services they provided had to be funded. Likewise, in the event of a national emergency such as a war, there was a need to secure funding not always immediately available from taxation. The money had to be borrowed normally from the relatively small, wealthy section of the population who historically had little direct political power. However, these groups, through financial systems and the provision of funds, by way of promissory notes and bonds, now ascended to positions of great power and influence. Wars, then and now, are invariably won by political will or military force and by access to money, often borrowed, to develop technologies and pay for the armies. Large sums of money were required for as long as it takes to prevail militarily.

Over several centuries, the bond market grew, fuelled by both government and private-sector demand for capital. The importance, role and size of those bond markets and, indeed, equities markets are often not fully appreciated. Ferguson (2008) notes that, in 2006 figures, global GDP was US$47 trillion. The total value of domestic and international bonds over the same time was US$68 trillion, 45 per cent larger than US GDP.

With increasing size and complexity, the financial sector was no longer simply 'about the money itself' (Ferguson 2008). For example, cash in circulation in an advanced economy such as the US now accounts for only about 11 per cent of 'total money' – in the form of loans, credit, inventory and so forth – existing within the economy at any point in time. Therefore, 'cash money', physical notes and coins, have long since stopped representing 'all the money there is' and certainly, as objects, have quite negligible physical value. Rather, it is the value they represent in trade and conversion to more tangible goods, services, assets and the like, which lies at the heart of the monetary system.

This is increasingly obvious given the use of electronic fund transfers (EFTs) or credit cards. Workers are now typically paid by an EFT; while tax and other deductions are similarly redirected, they pay most bills and buy consumer goods and services using credit/debit cards. While there is no doubt that economic activity and transactions have taken place, it is debatable whether 'money', in the strict sense of the word, as cash,

exists in those transactions. However, in the contemporary environment, these observations, while interesting to contemplate, are not of major economic significance. Like
physical currency itself, it is not that the electronic/digital recording of a transaction has
any intrinsic worth. Rather, the transacted value lies in the fact that all the financial
sector, across the economic and wider community, accept that the dealing represents
a certain monetary equivalent. The stability of and confidence in the entire financial
system lies in that acceptance.

Money (broadly defined) holds a unique place in any economy, particularly under
a capitalist regime. Not only does it have value and can be traded in its own right
(i.e. reflected in interest and exchange rates, monetary dealings and so on), but it also
represents the medium by which all other asset classes are measured and transacted.
Unlike other asset classes (such as real property) it is liquid and instantaneous, flowing
across all sections of the economy, seeking out highest returns and hedging against
risks. While that liquidity is obvious, this characteristic has serious implications when
monetary assets interface with other asset types – such as property. Property assets
are fixed and investment in them is normally longer term. They do not readily adapt
to rapidly changing circumstances nor to the ebbs and flows of money supply and/or
market confidence.

Marxist economics would hold, with some justification, that 'booms and busts' are
endemic to the capitalist system. According to those theories, they result primarily
from the build-up of wealth and free capital that rushes in and out of markets seeking
opportunistic, but increasingly unsustainable returns. In periods of high economic confidence, this inflates an asset price bubble – by now, also supported by excessive debt
funding – and a subsequent collapse as the market eventually realises that inflated asset
values cannot be justified either in terms of earnings or future capital growth (Hollander
2008). Marxist theory would also observe that the role of banks and privately owned investment vehicles, that today include hedge funds and other investment mediums, adds
to this instability. The banks' extremely powerful 'macro' market impacts are driven by
their quest for short-term corporate profits and risk avoidance to themselves. Lenin dis-
paragingly called banks 'capitalism's useful idiots' and given the global economic events
post-August 2008 and in Australia more recently, there would seem to be justification
in that 100-year-old observation.

The financial sector is truly international and has always been on the leading edge
of the globalisation trend (Saul 2009). The nature of finance and money is such that it
can be transferred almost instantaneously, and in digital form, around the world. In
2018, the US represents 23.9 per cent of global economic activity, and has emerged,
particularly post-Second World War, at the epicentre of world economic activity and
the financial systems that effectively control it (World Population Review 2019).

As funds move rapidly around the world, much is made of the debt levels of certain
nations, particularly the US. High levels of debt come with inherent risk. However,
from a wider perspective, it is often simply a sign of money at the global level flowing
to where the highest potential for wealth creation and the lowest levels of systemic risk
exist. Like household debt, whether or not debt is a problem depends on the ability to
repay (which in the case of the US is extremely high) and where capital inflows secured
through that debt funding is expended. It may well be that those borrowed funds are
applied to fund infrastructure or other investment that, over time, will create higher
levels of production and income.

The successful implications of monetary policy across the OECD over recent decades are testimony both to the pervasive nature of the financial sector and, secondly, to the power of central banks across those countries. As current interest rates fall to record low levels however, the future impact of monetary policy to stimulate economic and employment growth becomes problematic.

The dramatic economic events of 2007–2008, and the volatility that followed, exposed fundamental flaws in financial systems that had emerged in a now globalised environment. While the issues were complex and interlinked, the crisis was principally related to:

- inept macroeconomic policy settings, particularly related to interest rates in the US, that encouraged over-investment in housing, underpinned by unserviceable debt
- poor corporate governance, stewardship and statutory control in the development of new forms of financial derivatives established with inappropriate risk ratings and with rising moral hazards in loan/debt management
- the self-serving and sometimes illegal activities of certain banks and institutions in lending and financial dealings particularly in the US and, generally, without appropriate penalty.

Subsequent events, particularly within the European Union (EU), showed that poor governance/debt management practices were not confined to the US nor to housing/ retail markets. Rather, the contagion extended to unserviceable loans advanced to governments and countries, with a range of currency and sovereign risk implications.

All of these issues and their continuing impacts on investments, employment and consumption have seriously eroded confidence in global financial systems (Ferguson 2013). While the GFC may be seen as a product of poor practice and injudicious lending by banks and others, Ferguson sees the malaise as a substantially wider issue. Based on a range of speculation failings and betrayals of trust at local and global levels, the community/society has significantly lost faith in many of its institutions. As well as the financial sector and banks, a range of institutions – many governments, the media, churches, big business and even education systems and universities – have also, to a greater or lesser extent, lost the confidence of their constituencies. Ferguson notes that the restoration of trust and confidence will undoubtedly be a long and slow process.

While these matters demand remedy and change, they need to be considered in context (Shiller 2012). The financial sector was, and continues to be, fundamental to wealth creation, rising affluence and to the effective and efficient use of resources. It is capable of the innovation necessary to meet changing economic and social challenges. Shiller notes the central role assumed by the financial sector had changed. Traditionally, the financial sector had played something of a supporting role to actual physical, economic enterprise (which he called 'manufacture based capitalism'). A fundamental shift leading up to the GFC had been that capital, not the enterprise itself, had become the central element of the dealing ('financial capitalism').

In a wider sense, a balance needed to be re-established between the now dominant and expansive 'transactional (i.e. business/financial) economy' and what is sometimes called the 'civic economy'. The 'civic economy' refers to the way in which the general community (including public services) operates. It is recognised that, unless the transactional economy is healthy and growing, the civic component will suffer from lack of resources and underfunding. However, if there is not a politically structured interface linking the two, the wider objectives of general prosperity, reasonable equity, justice, social cohesion and safety will never be advanced.

The overall point, however, is that financial systems remain fundamental to economic activity and future global prosperity relies heavily on their efficient and effective operations. Recent events should not be the cause for pessimism, but rather for determination to remedy these problems to ensure the sector properly and sustainably manage money and risk transfer. Additionally, it needs to promote innovation and partnerships that better manage sustainable public and private investment into the future (Ferguson 2008).

3.4 Fundamental concepts

Several fundamental concepts, structures and operations underpin the financial sector and its dealings. These are as follows.

Raising capital

At its essence, the financial sector and financial markets are all about raising capital and matching those who want capital (the borrowers) with those who have it (the lenders). Trying to match individual borrowers directly with individual lenders is practically impossible in any sort of comprehensive or sustainable way. Therefore, there is a need for intermediaries of various types to aggregate available monies, match it with demand, manage risks and so on. Of course, within the financial sector, there is not a single type of intermediary and, furthermore, they trade across a number of different markets and market types. This is summarised in Table 3.1.

When any borrower secures funds under an agreement with a lender, the borrower will normally have to provide a formal receipt and a promise to pay it back – that is, identifying how much is owed, when and how it is to be repaid, and agreement as to the additional payment (or interest) rewarding the borrower for entering into the arrangement. These formal promises are known by the generic term 'securities', though, depending on their type, securities adopt a large number of different names including 'bonds' (e.g. 'government bonds'), 'treasury bills', 'debentures', 'certificates of deposit' and so on. All these instruments, however, have the principal purpose of recognising the debt transaction and the conditions that pertain to it.

Table 3.1 Financial sector lending and borrowing (adapted from Valdez & Wood 2003)

Lenders	Intermediaries	Markets	Borrowers
Individuals	Banks	Interbank (clearing)	Individuals
Companies	Insurance companies	Stock exchange	Companies
Funds	Pension funds	Money market	Central government
International/ other countries	Mutual funds	Bond market	Municipalities
		Foreign exchange	Public corporations

In the first transaction, where a security is established, the market is known as the 'primary market'. However, from the lender's point of view, the periodic return of debt represents an asset that, in many cases, can be on-sold, in various forms, to other domestic or international financiers and investors.

This activity is known as a 'secondary market'. Secondary markets also take a huge number of forms as the 'intermediaries' (identified in Table 3.1) attempt to maximise the value of income streams, and manage and/or spread risk, all of which continually change and evolve over time. Sometimes, these secondary (or 'wholesale') dealings may involve aggregating many loans together[1] and, from an aggregation of loans, take tranches (or 'slices') horizontally across that aggregation with various levels of risk (and reward) being identified and allocated to different financiers.

For obvious reasons, these funding arrangements are collectively known as 'structured finance' or 'derivatives trading'.

The borrowers (i.e. those who require the money) are a diverse group including:

- individuals and households with credit card, consumer debt or house mortgages, et cetera
- companies who may want money for expansion of their business, to buy new property, plant and equipment, fund inventory or to provide cash flow and assist with liquidity
- governments, who borrow funds as they try to provide long-term infrastructure and services, but who lack the ability to raise taxes in the short term to meet those, often huge, capital costs. The situation for governments is made worse in times of economic crisis, disaster or war, where additional funds will have to be raised quickly. Additional requirements are the borrowings of local government and public corporations, such as utility providers
- the stimulus packages provided by many OECD governments during the economic crisis of 2007–2008 present examples of these actions, funded in most countries by borrowings.

Borrowers have a number of options to raise capital. The most common, particularly for individuals and small-to-medium enterprises (SMEs), are bank loans. Bank loans are available in a number of forms, including overdrafts and lines of credit, credit cards, personal and business loans and loans for plant, property and stock (inventory). For practically all larger loans, particularly those for the acquisition of major assets, the lender (the bank) will require the added security of a mortgage over real property – sometimes, the property being funded, or a security over other property assets, being held by the borrower. These loans are known as 'mortgage-backed securities'.

The second way that larger companies can raise capital is by way of a bond, which is a debenture or promissory note from the borrower to the lender. Third, a business can decide to seek increased equity into the business – that is, by selling additional shares in that company whereby 'the lender' now becomes a direct investor, securing an equitable interest (part ownership) in the company itself. The lender thereby expects to secure rewards and dividends from that equity, rather than from the interest that would have been secured through either loans or debentures. (The detrimental effect of the additional share issue is that it dilutes the existing equity of the existing owner(s) at least in the share term but in the hope that the overall prosperity and value of the enterprise will increase by a disproportionately larger amount over time.)

Obviously, any such arrangements are not mutually exclusive and it is not uncommon for a particular firm to have a combination of bank loans and debenture/promissory notes, as well as attracting shareholder investment through either a private share offer or public float.

Finally, the overall debt-to-equity ratio and unique risk profile of the particular company (or investment) needs to be taken into account and managed on a case-by-case basis. The risk profile of any enterprise or investment relates to the identification, qualification and quantification of risk through time. Because this involves a large number of components, it is difficult to establish general rules that identify an appropriate or safe ratio. Nevertheless, it is normally the case that the higher the debt-to-equity ratio (i.e. a 'highly geared' project), the greater the risk compared with a lower-geared enterprise – one with a higher proportion of shareholder or owner funds committed. This risk factor is also referred to as the 'leverage ratio' of the firm/investment.

Through these fund-raising and financial developments dealings, company directors must take extreme care to ensure that the company remains solvent, and maintains cash reserves and a secure, reliable cash flow sufficient to meet debts as they come due. The Australian Securities and Investment Commission (ASIC) has regulatory control of these matters and can impose severe penalties on individual directors where a business is found to be trading while insolvent. Property development activities can be very susceptible to cash-flow issues given that significant investment must be made in the initial construction phase before a positive income stream becomes available.

Investment strategies and predictions

Dealing across so many markets within the financial sector provides traders and investors alike with the opportunity to spread risk and seek both long-term and opportunistic/speculative returns. Investors may be individuals, corporations or trusts and funds – effectively individual investors aggregating their resources for mutual benefit. The general strategies and objectives of investors vary – from the longer-term, lower-risk strategies of, say, superannuation funds through to more speculative investors such as 'hedge funds' who rapidly move large volumes of cash across various sectors and markets seeking out high, sometimes opportunistic, returns. The size and near instantaneous movement of hedge funds across global markets, promoted by internet-based trading, has added considerably to inherent volatility.

In all of this, it can be seen that, while obligated to adhere to good governance and corporate law, there are great rewards in being able to reasonably predict market movements in advance of other investors. The time and efforts of economists and analysts across the sector are largely taken up with the study of trends – to identify, particularly in aggregate markets, if some short- or long-term cycle or other pattern can be identified, or indeed, whether markets over time simply follow what is known in economics as a 'random walk' (Malkiel & Ellis 2013).

Moreover, as Hayek (1944) observed, the establishment of complex economic theories and modelling presumes a high level of immediate knowledge and information across the economy, from the most minor transaction through to the study of aggregates. He believed that such predictive modelling must always be used with caution given that, despite improved survey and analysis techniques now available, the accuracy, completeness and currency of information secured will never be sufficient alone to produce definitive conclusions. As noted by Wapshott (2011), this was also the basis of much of Hayek's criticism of Keynes' model of macroeconomics, which also presumed accurate, detailed knowledge of current economic activity to be effective (refer Chapter 2). In a real property environment, this reinforces the importance of direct market evidence to underwrite opinion.

Economics is based largely on mathematics and statistics and, with rapidly increasing computing capabilities and analysis techniques, the financial sector in post-war years, particularly in the US, came to be dominated by quantitative economists (known colloquially as 'the quants').

In a now much larger and more complex financial sector, the ability to use mathematical modelling and statistics to help predict market movements, recognise trends and establish sensitivities was seized upon by banks, portfolio managers and hedge funds, not simply to assist in their buy/sell decisions, but also to help balance and manage their portfolios. The increased computational capacity and skills developed by the US and Russia during the Cold War and space programmes provided an important theoretical and mathematical base for this approach. However, in retrospect, the direct transfer of that capacity into such a different (financial/investment analysis) environment was problematic (Das 2011, 2015).

Overall, quantitative analysis has proved a remarkably important management tool over some decades and continues to fulfil that role. Like any tool, its effectiveness relies on how it is used and managed, the quality and interpretation of the available data and, particularly, recognition of the limitations of its capabilities. Even the most sophisticated multiple analysis, for example, can relate only to a relatively small number of variables. It would be wrong to imply that this analysis can somehow replicate the complexity of the entire financial environment for the asset being considered. Furthermore, the longer the time period over which the projections are made, the more imprecise the predicted outcomes will be.

A series of financial calamities in recent decades, and right up to the global economic crisis of 2007–2008 and beyond, exemplifies how, in practice, the power and value of quantitative analysis was overestimated. Risk profiles can easily under or overestimate non-quantitative variables, particularly government policies, levels of confidence, consumer (and 'herd') behaviour and social trends. The importance of corporate responsibility, governance and adherence to law can be easily overlooked or underrated, given that they have qualitative rather than quantitative characteristics and are, therefore, difficult to accurately assess in mathematical modelling.

It is now recognised that quantitative economics needs to be used in a much more appropriate and measured way. Such analysis provides remarkably powerful methodologies to help predict and manage markets; however, it should never be considered to reflect all market determinants.

Banks and the banking system

It should be clear by now that the financial system and banking (broadly defined) are much more complex than the simple 'deposits, withdrawals and interest' dealings undertaken at day-to-day retail banking levels.

The financial and banking system in Australia is closely regulated with banks requiring a licence to operate. Federal statutory groups, such as the Reserve Bank of

Australia (RBA), the Australian Securities and Investment Commission (ASIC) and the Australian Prudential Regulation Authority (APRA), control the establishment and dealings of financial corporations. At a retail level, there are four major banks that occupy a dominant, virtual oligopolistic, position. There are however a range of other regional and community banks, credit unions and the like that provide comparable services on a much smaller scale.

OECD countries typically have a central bank to provide strategic direction and operational structures and regulations for their banking systems. In Australia, this is the Reserve Bank (RBA), established by the Commonwealth government under the Reserve Bank Act 1959, which took effect from January 1960.[2]

The role of the RBA is relatively independent of government to direct monetary policy – setting wholesale interest rates for the retail banks to align with the objectives set down in the RBA's charter.

While much is made of the RBA's independence, its Board is well aware of global and national economic and political events and, for the most part, is interpreting the same economic trend data as the government/treasury. Consequently, while there may be occasional short-term disagreements, RBA policy in practice is normally similar, or at least complementary, to the treasury and the government of the day.

As well as monetary policy, the RBA's responsibilities include:

- supervision of the banking system overall
- provision and management of bank notes and coinage
- effectively acting as the banker to the retail banks, and, in an important stabilising and confidence-building role, standing as a lender of last resort
- acting as the banker for the government
- facilitating the raising of loan funds for the government
- the controlling of the nation's currency reserves
- liaising and being involved with the coordination of global activities with international bodies such as the International Monetary Fund (IMF), World Bank, Asian Development Bank (ADB) and others.

Given the stated role and objectives of the RBA, it is of no surprise that the control of inflation (i.e. the stability of the buying power of the currency) is seen by the bank as a primary objective – typically with a target range of 2 to 3 per cent each year. Because of the inherent volatility of an open, commodities-based economy such as Australia's, it is important to see past short-term volatility and consider longer-term trends. In this, the RBA has been remarkably successful in meeting those targets of underlying inflation since they were introduced in 1993 with an average inflation rate of 2.5 per cent across that period. It is no coincidence that Australia has not endured a significant economic downturn during that entire period (Koukoulas 2015).

The history of the RBA can best be described in two parts: first, from its inception through to the early 1980s; and a second, more deregulated period, from the 1980s to the present.

The first period, 1959 to (around) 1983, was one of high-level regulation with a highly structured system that effectively precluded outside competition for Australian banks. During that period, the RBA used both regulation (such as Statutory Reserve Deposits) and general agreements with the major banks to ensure sufficient liquidity and to quarantine funds for politically sensitive lending areas such as housing. The retail banks were generally willing to adhere to such tight controls as, in return, the RBA effectively stood as their financial guarantor and, additionally, restricted market entry from potential new competitors.

Through that period, and continuing to the present, major banks also established specialist, free-standing finance companies called non-bank financial intermediaries (NBFIs) to facilitate less regulated financial dealings. Under the Hawke–Keating government in 1983 and 1985, substantial deregulation of the banking sector was undertaken. Initiatives included the floating of the Australian dollar, the granting of new foreign exchange licences and the granting of a number of banking licences for new entrants into the domestic market.

The RBA took a more transparent role as the 'banker to the banks', setting wholesale interest rates, but allowing market forces to largely determine the manner in which retail banks distribute loan funds across various property sectors and, in addition, the retail rate that the individual banks charged their various customers.

It needs to be recognised that the banking system in Australia remains highly regulated, though certainly less than prior to 1983. There are very strong corporate and banking regulations that, for the most part, have struck a reasonable balance over the past 30 years. Good corporate governance and government legislation generally control the parameters of the market, though the power the four major banks enjoy within the Australian economy remains politically contentious.

Arguably, the relationship between the RBA and the four major banks post-GFC is again changing. Australian banks are increasingly trading in and securing funds from international markets. This effectively restricted the level of influence that the RBA could extend over those banks through the application of monetary policy, and the setting of wholesale lending rates within Australia.

The major banks in Australia, as well as providing retail banking and financial services, are also involved in a range of activities, specifically investment, non-retail banking and merchant banking. These activities include corporate finance, insurances, securities trading, investment and loans management, foreign exchange dealings, securitisation and public debt funding and infrastructure loans. As well as the major banks, a range of other international banks, insurance and investment houses, superannuation and insurance funds, hedge funds and others are active in these sectors, at both a domestic and international level.

Because of the large-scale existing customer base and extensive branch networks of the existing four major banks, there appears little likelihood that any other financial corporation could, or would wish to, enter the Australian retail banking sector, at least to any significant extent. However, other internet-based banks are establishing niche markets that may well develop further into the future.

Equity and debt

As noted earlier in the chapter, investment in property can be by way of equity or debt, or some combination of both. An equity position, such as direct ownership or part ownership of an asset implies overall responsibility. The full amount of that investment is potentially at stake and, depending on the circumstances of that investment, even greater liability could be involved. The equity owners will be the last to be paid out, but there is, prima facie, no limit to the potential benefits (i.e. return on investment) that can be derived from the investment.

Depending on the legislative requirements in each particular case, Fund raising through equity may include options such as outright ownership/takeover, partnerships, syndications, joint ventures, the floating of a new company, additional share issues or unit trusts.

There may also be significant potential for leverage through debt funding. For the lender, debt funding involves the provision of some form of funds or loan for an asset or other venture, where the arrangement does not involve taking an ownership or part-ownership role. This normally implies that only the monies loaned are at risk and exposed for a set period of time under known conditions. Benefits (but also risks) are limited to the stated conditions of the agreement. Property loans are often secured by mortgages typically against that asset and registered on the title. The mortgage agreement will normally include provisions in the event that the borrower defaults on repayments. Those provisions may include the rights of the borrower to take an equity position ('mortgagee in possession') to enable the outstanding, recalcitrant debt to be recovered, potentially through the sale of the property. The requirement for real property to be used as collateral in such dealings reflects the financial sector's general view that real property provides a higher level of certainty and retention of value than other asset types, if the need to recover the debt becomes necessary.

As noted above, a range of structures and vehicles, including equities, debt and a mix of others, facilitate the holding of real property and its financing. These include the following:

- Outright ownership (as an individual or in partnership).

  This is by far the most common form of ownership, many of which will be purchased using mortgage-backed loans (securities) – see below.
- Separate investment, construction or development company.

  This structure may be unlisted (i.e. a private company) or listed (i.e. floated on the stock exchange, seeking public subscription).
- Syndication

  These structures typically involve a group (often quite small and comprising high-wealth individuals) that come together (i.e. as a syndicate) to undertake a specific activity – for example, to pool their funds to buy a particular, significant asset. Such agreements will normally have a 'sunset' of perhaps five or ten years. Syndicates often represent a vehicle for passive investment whereby that funding can be made for a specific asset, but the day-to-day management is left to a separate manager or organisation. Often, depending on taxation arrangements, a shelf company actually holds the asset, with the shareholders of that company being the syndicate members.
- Joint ventures (JVs)

  These are normally quite small groups of unrelated parties brought together for a specific task such as the development, construction or operations of a project. Again, the project will typically have a set time frame, but involve active participation by the joint venture partners, with the aim to combine expertise to better manage risk and establish a strategic alliance across their varying skills. In property, JVs might typically involve a financier, a developer/designer and a construction group who, together, can deliver an entire project.

- Securitisation

    This is a variation on debt funding where that debt, once established, is on-sold as a marketable commodity, namely, a long-term 'net' income stream, back into the financial sector. Within the commercial property market, these schemes typically need to be applied to large-scale assets with income secured from high-quality, long-term tenants who can provide certainty of income, including predictable escalations in that income over time.

- Direct mortgage-backed securities

    These are typically the borrowings provided by banks, investment banks and other non-bank debt funders who provide funding into property development or investment, and require collateral through title endorsement (therefore, known as 'mortgage-backed securities'). Often these are direct debt funds in a particular project or asset. Alternatively, debt can be joined together and on-sold with a number of other debts (called 'stapled investments'). These investments can be for a percentage of an acquisition or development.

    In other cases, the funds can be used for an initial, higher-risk investment that provides the financial platform to secure larger levels of mainstream debt funding for the balance of the project/investment. Such first-stage debt is sometimes called 'mezzanine funding'.

- Investment trusts and other investment vehicles

    These represent a related form of property investment and are detailed in Section 3.5.

- Margin loans

    To complete the list, another form of debt funding, known as 'margin loans', is also identified. These are typically used for the purchase of stocks, bonds and other securities, though in principle they could be used as part of an investment strategy into other sectors. A margin loan is a debt-funding source normally secured through a broker, and often for a relatively short term, for the purposes of buying securities. It is itself secured by the client's collateral, normally in the form of a portfolio of securities. The interest rate charged will depend on the level of collateral provided, overall security, and the ability to realise on that. In a stock or bond situation, the borrower is obviously depending on rises in stock value above that margin interest rate to make the transaction profitable. That decision has obvious market risks in what are, in reality, quite speculative ventures.

Inflation

The nature of inflation within a broad economic context was defined and discussed in Chapter 2, but requires further elaboration, given the inherent threat it poses to financial market stability.

Inflation is a monetary phenomenon reflecting a deteriorating buying capacity of a particular currency over time. Once that buying power cannot be relied upon, savings become a poor option, consumption spending and borrowings gain ascendency and a new cycle of inflation begins, which, if unchecked, leads to hyper-inflation. These matters have wider political and social, not just economic, implications. As Lenin observed, there is no surer way to political insurrection and the overturning of societies than to allow inflation to take hold. Far more than war or physical invasion, the loss of structure, governance and certainty around money and currency has been responsible for the collapse of governments and the destruction of sovereign economic power and influence in nations across the world.

Central banks, such as the RBA, are acutely aware of the extraordinary economic and, eventual political and social dangers of the failure to control inflation and the control of inflation tends to be the major focus of their monetary policy. Debt within the property sector tends to be both large scale and long term. Therefore, adjustments to investment rates instigated by the RBA's inflation control targets may, over time, add to the volatility and uncertainty of critical debt funding into the sector.

A strong link exists between inflation and interest and bond rates demanded by lenders. If, for example, inflation over time rises, say, to 8 per cent and the value of fixed interest, on a government bond, is only 5 per cent, then the holder of that bond (i.e. the holder of that investment) has fallen back 3 per cent in real terms, and bonds, therefore, become quite unattractive investments. This observation is particularly important in OECD economies where major pension and other funds will normally hold about 25 per cent of their assets in bonds and can easily move funds away from those critical areas if yields or prospective future yields become unattractive. The link to the flow of major investment into the property sector is obvious.

Major trusts, pension funds, et cetera, have a number of options for significant investment. The attractiveness, or otherwise, of one of those options, such as bonds, will directly impact on the availability of funds into competing investments elsewhere, including property, within a balanced portfolio. Historically at least, long-term investment in real property has been viewed by the investment market as a defence against the threat of inflation and their involvement, either directly or indirectly, in the property sector is, therefore, almost always included as part of a balanced fund. However, almost ironically, if the value of another major asset class such as shares or equities falls considerably, as occurred post-GFC, the overall investment portfolio of the major funds can become unbalanced and may result in some down-selling in (still performing) property assets to rebalance aggregate investments.

3.5 The financial sector and real property

Because of the comparatively large capital value of real property assets, borrowed funds will be required in the majority of cases for their acquisition and development. As a corollary, real property, through mortgage arrangements, provides the most common form of collateral and security that lenders require, not just for property dealings, but in a wide range of other financial arrangements.

Nevertheless, the interface between property and finance also has inherent instability. This is principally because of the liquid and highly reactionary nature of money and finance, including 'near cash' markets such as stocks and bonds, compared with the longer-term, fixed investment represented by real property. The availability of instantaneous information and the global connectedness of financial markets through the internet, short selling and trigger-price selling have magnified these differences in recent years.

Over recent decades, both the financial and property sectors have grown remarkably, both domestically and internationally – not only in scale, but also in complexity. For real property investment, the world has become effectively urbanised with new, concentrated and complex forms of development and property holdings. New forms of infrastructure (e.g. health and civic facilities) are demanded and now provided and funded from more than traditional public sources. This provides additional, asset-based investment options for major funds and trusts. Meanwhile, the financial sector has also been substantially deregulated, and effectively globalised, with an enormous array of financial packages, derivatives, futures and trading arrangements available.

Structural and operational changes resulted from the emergence of an oligopoly of the four banks, together with the introduction of international and local merchant banks, retail loan companies, brokers and investment funds. A further important development in the recent evolution of the Australian financial sector is the significant increase in the scale of the superannuation sub-sector. Superannuation contributions only became compulsory for Australian workers in 1991. It took time for those funds to build, but remarkably, by 2018, contributions had aggregated to $T2.9 or about 210 per cent of Australia's GDP at the same time (ASFA 2019). By 2035, superannuation funds are anticipated to reach $T9.5 – a remarkable figure given aggregated superannuation funds in the country in 2000 stood at only $M500 (Deloitte 2019).

Funds pooled at that scale have no precedent in the Australian financial system. Historically, superannuation funds have had fairly limited exposure to direct property investments, often around 10 per cent, but have considerable appetite for the more liquid units investments provided by quality real estate and other trusts. The direct and indirect capital inflows into the market for major real property holdings and, probably for certain infrastructure projects, will increase considerably into the future and bring major and overall positive impacts on the property and assets sector.

It was inevitable that further integration of the financial and property markets would occur and the complexity and variety of those arrangements would also increase. For the most part, this has been positive. For example, consider a major regional shopping centre, some of which are worth a billion dollars or more. Particularly in a small economy such as Australia, it would simply be impossible to aggregate the funds necessary to develop and own or manage such a huge asset without sophisticated financial vehicles, such as real estate investment trusts (REITs) or directly from superannuation funds. Meanwhile, those funds require a balanced portfolio, which typically will involve a direct or indirect real property component. Ideally, that component should be large-scale, long-term and with predictable cash flows and capable of adjustment and rebalancing over time.

Unfortunately, the closer integration of the financial and property sectors is not always successful, as the 'sub-prime crisis' in the US in 2007–2008 attests. In that case, poor governance, moral hazard[3] and a misreading by major banks of the true risk profile combined to cause near calamity.

Because of the scale and structure of many contemporary investments (particularly in commercial and development property), transactions may well be complex; but, regardless of that complexity, the fundamentals of risk analysis, return/reward and capital value need to be clear and simple to identify and assess. It might be argued that an underlying reason why complex structures are sometimes employed is that the proposal simply does not work easily in its fundamental proposition – a situation which would be grounds for concern.

The vast majority of links between the financial sector and the property sector remain simple – the raising of debt (secured by a mortgage) for an individual or company to secure and/or further develop property and, on an agreed term and interest rate, to return those funds to the lender. It is important to note, however, in contemporary, large-scale investments, the lenders from the financial sector are clearly not interested in the operations, characteristics or idiosyncrasies of real property per se. Rather, they typically seek translation of the income and/ or 'net' profit from that property into a form that is free of the volatility and vagaries of a property investment and as 'clean, certain cash flow'.

If such a cash flow can be established, it is now recognised as a clear and predictable income stream, not as a property asset per se. Such arrangements are the basis of a securitised model with that guaranteed income stream being able to be sold or traded within the financial sector. Under such arrangements, the property investment has, in effect, been fully transformed into a 'financial commodity'.

Several sub-sectors of the property market are worthy of particular mention here, due to both their scale and their links to the financial sector.

Home ownership

Housing represents about one-third of all property and infrastructure investment in Australia and its links with the financial sector are of fundamental importance. Particularly in post-war years, the community embraced the key goal of individual home ownership. A political imperative naturally followed.

In September 2017, the Australian population stood at 24.7 million (about 8.3 million households) with each household accommodating about 2.9 persons, a figure that has continued to fall steadily over some decades (AIHW 2019).

In 2016/2017, approximately 30 per cent of households own their residence outright and a further 37 per cent (approximately) were repaying a mortgage on their homes. A further 27 per cent of households were renting on the private market (up from 20 per cent ten years before) and 3 per cent occupying homes supplied by some government body (down from 6 per cent over the same ten-year period).

One in five households owns a second dwelling for rental or own use (Savings.com 2019). The maintenance of that level of privately supplied investment/rental housing is important for the system to maintain overall stability and reasonable affordability in what is a politically very sensitive area. Ownership and taxation arrangements are such as to provide a major important private investment opportunity while also providing an adequate stock of rental accommodation in most areas.

An interesting and important observation is that the same (physical) property can move quickly in and out of the financial system. At one point, a house can be used as a principal place of residence, not as an income source, and have particular and unique taxation arrangements (reflecting its political sensitivity). However, upon sale or a change in the owner's circumstances, the same property can be rented out on the private market to derive income and, therefore, attract an entirely different taxation regime. That is particularly important given the comparatively small proportion of welfare housing provided by government at all levels.

Housing affordability and housing stress (defined as spending 30 per cent or more of household income on accommodation costs) is a significant problem for those who are yet to start the process of purchasing a home, particularly in desirable capital city areas which are in high demand (AIHW 2019).

The dominance of investment in Australia in domestic housing is an important feature of the financial sector. Increasing house prices, combined with continued high levels of housing demand, have stimulated remarkable long-term growth in the number and size of loans for new-home acquisitions. Additionally, those loans have extended to include renovations, refinancing or accessing equity for other purposes, such as consumption spending. Banks are generally pleased to meet these demands, given the extremely low risk of default in this sector and direct and indirect income opportunities from a range of financial services (Australian Productivity Commission 2018).

There is also inherent conservatism in this system and a general reluctance to change policy directions, led by the large majority of households (approximately 67 per cent) who already own or have considerable equity in their principal place of residence. Political sensitivity surrounding the issue often results in further government stimulation of what is, typically, a volatile and often bullish and emotional market. Because of relatively low levels of domestic savings, much of the loan funding for housing and refinancing must be secured on international money markets, adding, over time, to interest rate volatility.

Housing investment within the financial sector in Australia represents a significant aberration; but, given the nature of the market and the very strong community and political forces at work, little change is likely in the foreseeable future despite obvious supply and affordability issues.

Investment trusts and other investment vehicles

It was noted above in general terms that one of the inherent attributes of a robust financial system is its ability to successfully aggregate funds. This can allow an increasing scale of investment, the spreading of risk and the limiting of liability through mechanisms such as companies, public and privately listed funds and the offers of equity through the stock market.

These characteristics are particularly important for the property and development sectors where many assets are very large scale. The financial sector provides both an interface and a vehicle by which assets of that type can be acquired, developed and managed. A range of these were identified earlier in this section.

Managed trusts provide another vehicle for facilitating equity investment into real property assets or, indeed, other forms of investment. These can be either private trusts, that is, structures established between individuals or corporations without exposure to the open market (called 'unlisted trusts') or, now more commonly, large-scale unit trusts offered on the stock market in a similar way to shares in a company (called 'listed trusts').

In the latter case, the units (i.e. shares) are offered for sale for a particular activity or investment and, with the aggregated investment funds, the trustees and managers of the trust can then undertake often large, strategic investments in accordance with the provisions of the trust. The unit owners can buy and sell their units on the stock market in a way similar to the sale of shares in listed companies. Depending on their charter, property trusts can carry out a range of investments. However, about 70 listed trusts in Australia are exclusively for property investments – Australian Real Estate Investment Trusts (A-REITs).

These trusts are common in Australia and well-regulated through ASIC and the ASX. There is now about 40 years of experience with property trusts in Australia and approximately 12 per cent of all trusts globally reside in this country. Overall, and despite the dramatic impact of the 2007–2008 crisis, A-REITs have a long-term history in Australia of sound yields, capital growth and relatively low volatility. Many of the largest investments in corporate real estate, such as regional shopping centres, major 'CBD' buildings, tourist resorts and the like, are held by these vehicles. Critically too, it has been the internationalisation and globalisation of these markets that has changed the investment landscape, effectively linking over 500 REITs in 20 countries worldwide (Newell & Sieracki 2010).

Like all investment vehicles, there is a range of sub-groups/types often specialising in various parts of the market. Some may be specific to one property or property type (e.g. office buildings, retail and so on). Some may concentrate on the purchase, holding and management of existing income-producing properties; others may be more involved in new development projects; and others may have a mixed portfolio.

There are a number of perceived advantages of A-REITs. These include the ability to attract large amounts of funds and, often, to diversify those funds across a range of investments. They allow for a wide range in scale of investment and make property investments far more liquid and tradable. As noted above, they have, over the longer term, a solid history of performance and transparency together with certain taxation advantages.

However, there are also some perceived disadvantages, including the uneasy relationship between the near instantaneous and volatile sentiment-based movements in equities market, compared with the typical long-term nature of property investment. Post-GFC also, there has been significant takeover and merger activity aimed at reinforcing market confidence and support (Rowland 2010).

Notes

1 Aggregating or joining loans/income streams together is also called 'stapling'.
2 The RBA Act reads:

> It is the duty of the Reserve Bank Board, within the limits of its powers, to ensure that the monetary and banking policy of the Bank is directed to the greatest advantage of the people of Australia and that the powers of the Bank ... are exercised in such a manner as, in the opinion of the Reserve Bank Board, will best contribute to:
>
> a the stability of the currency of Australia
> b the maintenance of full employment in Australia
> c the economic prosperity and welfare of the people of Australia
>
> (Section 10 (2) Reserve Bank Act 1959)

3 Moral hazard refers in this case to the lack of direct responsibility and transparency in investment/financial dealings. Here, the traditional, direct link and contract between the lender and borrower was lost as mortgages were aggregated and on-sold to third parties.

Chapter 4

The use of land resources

History and trends

4.1 Introduction

This text attempts to present an holistic view of the real property sector. At one level, it considers the specific characteristics, opportunities and challenges of individual pieces of property (i.e. 'micro economic' analysis). At a second, more expansive ('macro') level, it considers that individual property and, other properties of its generic type, within its economic, legal, political, social and historical context.

Given the longevity of real property assets, the built environment of today represents the final outcome of generations of physical, economic and community development. An understanding of that background and the philosophies and the forces behind it is integral to an appreciation of property's evolving use. What is considered as the contemporary built environment should be seen as both a product and manifestation of the past and, at the same time, the platform for development into the future.

This analysis is both an expansive and a multidisciplinary exercise and care must be taken to ensure that relevance is maintained and that perception of history does not become a matter of generalisations, clichés or nostalgia. Further, positioning and planning for the future cannot be a matter of wild or unfounded speculation. As the eminent town planner Lewis Mumford (1961) observed (paraphrased): every generation revolts against its fathers, but makes friends with its grandfathers; consequently, the traditionalist (in property and development and indeed all other areas) is typically pessimistic about the future, but overly optimistic about the past. In other words, each generation and its scholars tend to 'construct' a narrative of the past and a prognosis for the future that suits personal and generational perceptions and prejudices, and that tends to support current arguments (Storper 2013).

Often in a retrospective view, researchers seek out precise answers and neat sequences to explain and understand that past. This approach is common and exemplified in the 'procedural theory' for the advancement of knowledge generally. It proposes a sequence of discovery/breakthrough and later its application, say, at the workplace. In reality, development through practical problem solving is equally important and, typically, is anything but sequential or hierarchical (Kealey 2008). In the case of the evolution of the built form, any approach that seeks to establish precise models or exact causal relationships is based on a false premise. The history of land uses and urban development (as with practically any other study area) often exhibits non-sequential, haphazard and opportunistic characteristics, which may or may not align with contemporary, generic perceptions. There are, of course, clearly identifiable underlying trends that have evolved over a long period of time, some overseas and some in Australia. However, the manner in which they have physically manifested themselves varies from country to country and region to region – depending on a whole range of economic, community and political characteristics of the time. The huge diversity of the resultant urban form, which continues to evolve, provides physical testimony to that observation.

Within that context, this chapter considers the interaction between humans, communities and place, how that has evolved in the Australian context and the implications for this interface now and in the future.

This history extends back over 5,000 years or more as the first, organised urban settlements began to emerge; however, its events and developments of the last 200 years have fundamentally produced the built forms and urban settlements that now exist.

Over that latter period, a number of critical forces shaped this, now largely urban environment – they included the widespread distribution and use of electricity; the rise of the internal combustion engine and, with it, the second wave of the Industrial Revolution; breakthroughs in medical science; the general improvement in living conditions, rising consumerism, the remarkable increase in economic development; the advances of knowledge and its communication; and, finally, the growing centralisation of wealth and power. While this presents a rather eclectic group of observations, they have had, individually and collectively, profoundly affected the scale and location of human habitation.

Spatially, those forces have promoted urbanisation, the rise of cities and the concentration of now much larger populations in specific locations. In turn, they present unprecedented pressures on the biosphere and threaten sustainability in the widest sense of that word.

4.2 The importance of place

The physical environment and the built form constructed at a location create 'space' – space that can be physically identified, measured and assessed. That (usable) space also has a functional aspect – an economic input that allows a range of activities to be undertaken within and extending from its boundaries.

However, there is more to this interface than physical characteristics and economic functions. To a greater or lesser extent, every human experiences a personal response or reaction to the particular space, and its configuration that surrounds them. This response may be easy to dismiss as simply an emotional appeal to the senses – the attitude a person may have, for example, to arriving home, or the typical human reaction to viewing a grand landscape; or, in the negative, on visiting a rundown or blighted urban area. However, there is a wider aspect to this interface between humans and their surrounds of which an emotional response is part. Tuan's seminal work (1977) recognised this as the experiential perspective of property, which includes sensations, perceptions and conceptions of what is there and what it could be.

Obviously, there are important functional and economic ramifications of all of these factors. In part, these are the reasons why prospective purchasers will choose to buy or rent one property over another; workers will be more productive in one place over another; a medical patient may have improved rehabilitation outcomes in one physical environment over another; or a customer will walk into one shop or building in preference to another – the examples are almost endless.

This is not to suggest that an individual (or group) perception of place is the only determinant of decisions or economic preference. Clearly there are many such influences, but the experience of all humans would indicate that these emotional and physical responses are of fundamental importance and not to be underrated. While these responses are often pervasive and long term, even the aspects that property marketers refer to as 'first impressions' or 'street appeal' form part of this response.

This type of affinity is long-standing – many traditional land owners around the world, including Australian aboriginals and the islander and sea people in Australia's north and Micronesia, provide good examples, where 'places' take on a level of understanding that is spiritual and religious in a way difficult for others to comprehend.

The concepts of place and function and the interactions with the built form, design and urban space are fundamental to architectural theory and practice in the production of those assets. Owners, designers/architects and developers of property throughout history have been well aware of the power of perception and emotion. Consider, for example, the impression of uniformity, efficiency and scale created by Henry Ford and his designer, Albert Kahn, in industrial buildings, and the work of Nazi architect, Albert Speer, in creating the edifices and physical grandeur of that regime (Tietz 1999).

In contemporary property markets, it may be easy to be dismissive of these apparently non-economic and seemingly intangible attributes. However, that represents a fairly narrow proposition. The fundamental assessment of the value of real property (as discussed in later chapters of this text and laid down in precedent court decisions) establishes the role of 'the hypothetical, prudent purchaser'. To extend that argument logically, that purchaser is an individual who, like all others, will be influenced by both analytical and other, more subjective, perceptions. Some level of subjectivity will form part of any decision-making. Even in large-scale commercial property markets, it should be remembered that decision by corporate boards will invariably include the individual perceptions of board members.

Tuan observes the subtle, but important differences between the co-dependent terms, 'place' and 'space'. Space refers to physical dimensions and the spatial relations between one location and another. 'Place' is perhaps a more abstract term, which relates both to a physical space and to the human relationships and interactions that occur there.

Tuan considers that the human affinity and concepts of 'place' typically has three basic elements.

The first relates to concepts of familiarity – a trait shared with higher-order animals and particularly obvious in tribal and clan behaviour. Where a particular place (and space) provides well for its inhabitants and that community prospers, the area becomes well recognised and defined and the affinity grows. It is remarkable that the contemporary rise of knowledge precincts and clusters closely parallel those observations (Porter 1998).

Second, Tuan notes that this affinity builds over time based on past experience, and the history, traditions and culture of that place, much of which attaches to that physical environment. Iconic examples of this are observable in practically all significant places. Consider Sydney, for example, where the unbuilt environment (Sydney Harbour, the beaches and the major national parks to the north, south and west) and built icons (such as the Sydney Harbour Bridge, the Opera House and the Rocks precinct) are much more important than their functionality alone. Combined, they represent a mosaic of the culture, history and character of the place. All of these (including the built assets) represent for Sydney residents the unique experience of, and their affinity with, the place. That unique combination becomes not simply the basis of the living experience for residents but, at the same time, presents an offer to visitors, which makes it a premier attraction for the entire country.

These ideas align with important works by Mumford (1961) and Jacobs (1961) who both observe that 'place' – in their case, urban living – is more about human interaction and experience at a 'fine grained', human scale, than the physical setting per se. Business and social interactions provide the dynamism and activity that give the buildings themselves evolutionary life, vitality, functionality and value. Consequently, 'affinity' is not simply associated with unique, striking or iconic places, but (more widely) with the locations that provide familiarity, security and functionality – even if that place may not be the most efficient, organised or aesthetically pleasing. The final component in Tuan's analysis relates to how knowledge is instilled and further developed in individuals, community and business groupings of various types at that particular place. A familiarity with place provides a sound and reliable platform to undertake other activities much more efficiently – to already know 'where things are and how they work'. In business, this might relate to development of networks and strategic alliances, supply chains and clustering. At a community level, it might also relate to the comfort of having family and social networks, the support of, and access to, services, the association and affinity with a local street or neighbourhood and overall, a safe, logical and predictable environment.

This can be observed in current urban development trends where innovation and knowledge clusters are underpinning a considerable proportion of contemporary economic growth. This growth occurs where, for example, a central business district (CBD) or other nodes present a vibrant, safe and prosperous concentration of firms and, notwithstanding instantaneous communication available, where personal, face-to-face communications are encouraged and facilitated (Storper 2013).

These trends are reinforced by the fact these business activities are now much less intrusive and therefore can co-locate with a range of retail, civic and residential uses (see Sections 4.4 and 4.5). In addition, for the first time, the functioning 'tools' of business – ICT, the internet and social networks – are common to and used as part of social, community and entertainment activities (Libert 2010). These common platforms and the contemporary business activities are not confined simply to one (e.g. office) location. This means that work, personal and social activities tend to merge into what might be described as 'overall lifestyle', 'networked systems' or a 'work-live-play' mantra now often promoted in precinct and larger property developments.

These firms and clusters of externally connected, trading enterprises take time to mature, but are essential because of their own overt success and the extraordinary multiplier benefits they provide for the remaining (largely service) sectors of that city/region (Moretti 2013; Storper 2013).

Overall, these considerations of place extend more widely than issues of operational efficiency, functionality and the value of assets within it – as important as they are. They also extend to less tangible concepts such as liveability. Though, as individuals, we all would have a general understanding of what constitutes 'liveability', it is a concept that is often subtle, passive and not given the recognition it deserves. This lack of recognition gives rise to a number of important misconceptions.

Perhaps the most common of these is the mistaken observations that link liveability with the density of development at a particular place. There is no real correlation between the two. Some of the most dynamic, valuable and liveable places in the world are also very densely developed; but, of course, the reverse can also be true as many of the world's slum areas attest. By way of other examples, some of the grandest and most spacious places would be considered as very liveable by many, but other vast areas with the lowest density of population are inhospitable, provide no services and would hardly support human life at all.

Florida (2008) makes an important observation regarding the liveability of a particular place. He notes that, across the OECD, the 'rating' of liveability of a particular place is typically determined by a number of interrelated variables: physical environment and climate; work and education opportunities; infrastructure; community services, social networks and cultural activities; law and order, ethics and social inclusion; physical safety and a number of others. Florida's observation is that, for a place to be considered 'liveable', it must have an acceptable 'response to' or 'offer for' each of these criteria, and a failure in one cannot be effectively 'traded-off' against very high performance in another component if overall liveability is sustainable in the medium- to long-term.

It needs to be noted that the concept of liveability is certainly not a matter where 'one size fits all' nor is it static over time. It is easy for the casual observer to equate liveability with a particular rich and progressive centre – Vancouver, Seattle, Boston, London and others – so in effect, creating some model to which all should aspire. The self-perceptions and expectations of communities worldwide vary considerably and should not simply be assessed against those high-profile benchmarks (which invariably have their own particular issues and challenges in any case). More important are the consensus expectations of that particular community, city or region. Those may or may not align with the high, economically driven goals of those quite different leading-edge cities (Montgomery 2014). Many communities in the developing world consider concepts of liveability and achievement in terms of increasing sophistication and economic development together with the preservation of community values (often cultural and ethnic), social networks and support, equity, good governance and the overall concept of 'happiness' (Gittens 2010).

There is more than a little irony in the observation that, compared with a range of OECD countries/cities, many developing world cities and communities actually work well across a range of economic, social and liveability criteria (Montgomery 2014). This is despite their lower living standards and, apparently, chaotic, cluttered and ad hoc development, construction and land-use patterns (Brugmann 2009).

A key finding here is that, while economic competitiveness is an essential criterion, the success or otherwise of a particular 'place' in terms of liveability and sustainability is specific to that location's history, potential and its own expectations. Progressive cities and regions, whatever their level of development, invariably establish consensus on those objectives and parameters as a first priority.

These observations are more than theoretically interesting. The nature of development in Australia's cities, other high growth and 'lifestyle' regions and (particularly) the reinvigoration of otherwise declining rural regions and towns are all best addressed on the basis of liveability and sustainability criteria – of which property and the built environment play an active part.

It should also be noted that urban growth and development are dynamic concepts, the parameters of which change over time (Kotkin 2010; Ehrenhalt 2013; Storper 2013). The move to 'sun-belt' regions in the US, Australia and southern Europe and the remarkable resurgence of a number of 'rust belt' cities including Pittsburgh, Columbus, Charleston and others in the US, and Newcastle in Australia, provide examples of the diverse and often evolutionary nature of urban renewal.

In summary, the broader issues of space and place appear to be under-estimated in considering real property – physical characteristics and function are critical to economic activity; however, to understand property's true 'value-add', the wider context of human conscientiousness, reactions and stimulation must also be accepted and included in decision-making (Saar & Palang 2009).

4.3 An historical perspective

There is an innate tendency for humans to live in groups – originally for security and social interactions and for efficiency in hunting and farming. Later, a range of other, often interrelated reasons to locate in larger groups emerged – including the demand of trade, religious and cultural activities, health and education, government, knowledge development and innovation and economies of scale in production. Significant urban settlements began to emerge perhaps 5,000 years ago in the Middle East but, not too long after and without connection, in locations as scattered as China, India, Egypt and Central and South America (Reader 2004).

Over time, this move to urbanise and the rise to dominance of towns and cities gained momentum. Ferguson (2011) parallels this fundamental shift, particularly over the last two or three centuries to the present day, with the rise of Western civilisation. He identifies the critical components as the emergence of mercantilism, the securing of property rights and the rule of law, innovation and the rise of urban-based industries and breakthroughs in medicine and science. Consequently, world (and particularly urban) populations grew exponentially.

The nature of the long-term impact of successive civilisations varied. For example, the most important legacy of Greek civilisation (3000 BC–100 BC) was its contribution to philosophy and civic institutions; and Arabic influences were related to trade and mathematics. For the Romans (700 BC to about AD 400), it was largely a matter of military control over conquered lands – facilitated by infrastructure and law. Thus, it could be argued that the real, long-term impact of that era was perhaps overrated – another case of present-day perceptions' clouding reality (Kealey 2008).

Over the longer perspective, however, it was trade (and incumbent economic growth) that was the most important catalyst to the development of knowledge, wealth, growth and the rise of significant towns as they would be defined today. Trade, and the transportation and communication systems that underpinned it, rapidly brought forward knowledge and new technologies from Europe, the Middle East and Asia. By the eighteenth century and beyond, the growth of trade and mercantilism generally facilitated the corporate and financial structures of banking, insurance and stock markets, together with the establishment of commercial and property law in places like England, Scotland and Holland. Wealth typically aggregated around those urban ports and trading centres.

That concentration of wealth, trading and transport links combined with breakthrough technologies concerning the steam engine and railways provided the impetus for the various waves of the Industrial Revolution that commenced in the second-half of the eighteenth century.

Industrial progress was fundamentally important in the development of cities and the reconfiguration of economic activity – from an agrarian to a manufacturing base.

Production locations were now predetermined by access to raw material, markets and transportation routes and hubs, particularly around shipping and rail in the first industrial development phase. Based on those criteria, many of these cities were located in 'snow belt' regions with industrial cities of England and later of North America providing the physical manifestation of that.

Never before had production on this scale occurred, especially concentrated in specific locations. This, in turn, saw the rapid aggregation of an urban-based workforce. Regarding urban form and development, the links with economics, philosophy and policy of the time are also important. As discussed in Chapter 2, this was the era of classical economics – of a laissez-faire approach to economic activity and development generally, where government saw little role for itself in areas such as urban development, organisation or the provision of infrastructure. The results were catastrophic. The crowded, unhealthy, squalid conditions of those Dickensian industrial towns throughout the eighteenth century represented the physical outcomes of that approach, and that legacy continues to the present day.

There were of course reactions to these appalling conditions such as the work of John Stuart Mill (1806–1873) and Ebenezer Howard (1850–1928) who recognised the links between the urban form and the health and well-being of the broader community. While these concepts (such as the 'Garden City Movement') were later to significantly influence urban design principles, the effect during that period was very limited.

As industrial cities grew across various locations, typical, if generalised, patterns of concentric growth and development emerged. This was first recognised by Ernest Burgess in studies of land uses in Chicago in the 1920s. Burgess postulated that industrialised urban areas typically built up land use based on circular bands around a central business district (CBD) (see Figure 4.1).

Around that central core there was typically what Burgess called a 'transition zone' – an area of mixed, often competing and haphazard land uses of commercial, industrial and warehousing areas, together with lower-quality residential. Around that, extending out, was the next development circle – of lower socio-economic residential areas, then a better standard, new residential zone, and commuter and peri-urban zones beyond that. Of course, this is a very simplified model and, in practice, the symmetry was often disrupted by topographical features and by major trunk infrastructure routes which are designed to service multiple development

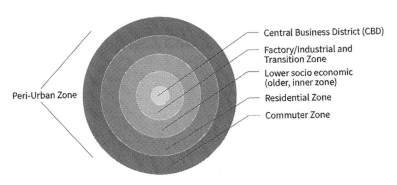

Figure 4.1 Burgess concentric circles of urban development (modified from Burgess Urban Land Use Model Chicago 1925).

sites, and typically converge to the centre. Nevertheless, the basic urban frame, with development radiating out from a central core, can still be identified in most industrialised cities and certainly in the major cities of Australia (with the exception of the planned cities of Adelaide and Canberra).

This basic model evolved over time with the 'zone of transition' often degenerating into slum areas and, sometimes, ghettos, as factories closed and the urban environment deteriorated. Later, and particularly over recent decades, there have been fundamental shifts in the nature and use of many of these urban–concentric zones. In the first instance, CBDs became more concentrated and influenced by a range of commercial, civic, entertainment and multi-residential uses. Limited land supply and increasing land values in the core promoted densification and high-rise development. Meanwhile, factories increased in scale and moved out to peripheral locations, encouraged by both town planning regulations and the availability of improved infrastructure in those new locations.

In turn, these developments caused a major crossover between socio-economic groups. Office and professional workers now tended to gravitate closer to the centre due to the urban renewal and gentrification of those convenient inner–city areas. Others moved to now cheaper, newer dormitory areas closer to industrial workplaces, often in low-density urban settings.

There was also much greater and widespread government involvement and regulation of urban development. More Keynesian, interventionist governments were in control with the obvious need to better direct the huge urban expansions required to accommodate and provide work for the 'baby boom' generation (born 1946–1964), following the Second World War. Nevertheless, the pre-existing problems of the haphazard and congested layout of inner-urban areas remained.

In England, this new approach took the form of 'New Town' developments created under government facilitation from the late 1940s until the 1970s. The trend was also evident in extensive parts of the US where low-rise urban developments were underpinned by the use of privately owned motor vehicles. The application of these concepts and their later ramifications are discussed in Section 4.5.

In a further, more recent evolution, manufacturing and related jobs in outer areas typically have diminished. The next generation of workers/ residents in outer-lying areas often had high levels of education and aspiration and were increasingly attracted to emerging corporate and professional positions that existed within the CBD and the CBD fringe. They were also attracted to economic growth areas, often centred around educational, medical and retail/service nodes. This resulted in considerable increases in commute times, congestion and inconvenience, as workers moved from low–density, dormitory areas to areas of employment. Large-scale, disruptive retrofitting of road and transportation infrastructures was also required.

These recent development trends also had significant impact in urban areas of Australia. The first was the renewed interest, investment and disproportionate rise in property values in inner-suburban areas, typically in a 3 to 4 kilometre radius of the CBD and employment centres.

Second, the differential change in land values both within cities and between them has and will continue to exert significant effects on internal migration flows across Australia into the future.

Further, these developments have created transportation and related issues (e.g. over-reliance on the use of private motor vehicles, rising transportation congestion and cost and adverse environmental impacts) as the 'tipping point' issue for urban development in Australia and, indeed, for most cities globally.

Finally, given the physical constraints and high cost of most established cities, much of the future urban growth will occur in peri-urban areas and particularly in progressive and attractive regional centres with good transport infrastructure and acceptable commute times to a major city (Polèse 2009; Kotkin 2010). In Australia, Geelong, Gosford, Wollongong, the Gold Coast and the Sunshine Coast are examples of this growing trend.

4.4 Urbanisation and the rise of cities

There is nothing new about the tendency for humans to live in close proximity to each other for a range of security, economic, political, social, community and cultural reasons. What is new, however, is the scale and relative importance of this trend, the speed at which it is occurring and its alignment with other contemporary changes in demography, social characteristics and the nature of business, institutions and governance.

This trend represents profound shifts with both positive and negative outcomes.

Leading contemporary researchers, Brugmann (2009), Glaeser (2011) and Florida (2002, 2008) all emphasise human drivers as fundamental to these trends. Brugmann's important work argues that the successful value proposition of major urban/city areas typically has four components:

- Density of living – and the positive impacts on efficiency, economic and knowledge-sharing opportunities and other prospects that such proximity provides.
- Volume – because of the size of population and the number of interactions, there are obvious economies of scale and positive outcome for sustainability.
- Association – that is, the almost organic and sometimes chaotic economic and social networks that are enjoyed by those who physically interact, which represents a unique urban advantage.
- Extension – that is, the development of urban areas and precincts can become recognised nodes and primary connection points in wider global networks by developing and producing weightless commodities potentially traded in almost any market (Leadbeater 2000).

The trend to urbanism manifests itself differently across a range of settings globally. Brugmann identifies four typical urban development categories. First, in the rapidly expanding cities of the developing world, development is largely ad hoc with growth achieved by new building, either legally or illegally, being simply 'tacked on' to existing structures. While the resultant urban environment appears chaotic, this environment works surprisingly well, based largely on social networks and small-scale enterprises.

Brugmann's second category is 'city systems'. These are more advanced, mature communities with enhanced utility, efficiency and productivity. By this stage of development, they have typically imbedded a particular sense of place for property owners and residents alike. Often, they are established around one or a small number of themes, perhaps based on dominant regional production and, thereafter, grow organically with service and related sectors over time.

Brugmann's third is 'city models' that emerged over the past 50 or 60 years, reinforced by town planning, standards and established production methodologies. They are exemplified in the large-scale residential developments that now surround most major urban areas in Australia and elsewhere in the OECD. Although they are, on the face of it, efficient and typically provide good value for money for residents, Brugmann considers these developments tend to serve two masters. One group comprises developers and builders seeking to make a predictable profit and lower risks by reaching economies of scale and providing a homogenous product. The second group, 'the customers', are the government regulatory agencies and utilities companies to whose requirements and standards developers have to conform. Unfortunately, one key stakeholder who plays a diminished role is the final consumer.

Consequently, liveability, the longer-term development of community and wider implications for whole of region development, transport infrastructures and commuting may not be well served by these development models. The quite uniform development patterns still dominating the new suburbs of Australian urban areas largely reflect these forces. Overlying affordability and financial considerations further reinforce uniformity of demand and, therefore, supply of housing product.

The final urban category identified by Brugmann is the 'master-planned city' – which may be an entirely new town, master-planned community or urban village such as those described in Section 4.5. He notes that, typically, these will have some form or theme and attempt to provide fully purposed enclaves with both jobs and liveability (e.g. 'work, live and play' mantras). If properly conceived and sensitively managed, these cities can provide high-quality outcomes. However, there are risks of falling into the errors of the past – overregulation and over prescription of the built form to the detriment of liveability and community development. Whether or not these major developments can be efficiently sequenced to provide the jobs, services and diverse housing in the required volume and similar timeframes is also problematic (Wardner 2013). Furthermore, success in developing diverse, organic communities and social networks can also prove difficult, particularly in the short to medium term.

Many economists and geographers argue rapid global urbanisation is the most fundamental change of the last 50 years and, in reality, encapsulates most of the other geo-political, economic and environmental issues of the era. Paralleling practically all other facets of economic and community life, the very fundamentals of property – its ownership, control, development, management, use and financing – are open to challenge and radical change. Given, particularly contemporary demographic, economic climate, energy and resource usage issues, the future will be quite different – even in the shorter term.

While global forecasts foresee a levelling off in world population by mid-century (Pearce 2010), sound management of urban growth will be fundamental to global and national prosperity into the future. How that evolves will vary widely from country to country and, for each specific region, depends on a number of interrelated variables. These include changing demographic characteristics, settlement history, affluence, technology, immigration, social mobility, government and legal structures and skill levels and education.

For some countries, such as Japan and many within the EU, the dramatic ageing of the population will create quite different issues. However, for others, such as Australia, the US and emerging economies, such as Brazil, fundamental decisions are required over the next few years regarding the form and direction for new urban development. As discussed in Sections 4.8 and 4.9, the gravity of these decisions are their widespread impact on economic, community and liveability outcomes, and that they will be, to all intents and purposes, irreversible.

There have been significant urban turning points before – the post-war baby and (suburban) housing boom from the 1950s; the rise of corporate real estate and CBDs in the 1970s and 1980s and the urban renewal and gentrification trends of the last two decades.

The next wave of urban development will be critical. It needs to accommodate a much more ethnically and age-diverse population. Further, it will need to address the impacts of radically different communications and accommodate a dramatic reduction in the use of traditional fossil fuels for energy and transport. Noted urbanists including Glaeser (2011) predict future urban forms will be based on more concentrated and efficient forms of dense city development; Florida (2002) sees the development of urban clusters around increasingly knowledge-based ('creative') groups as critical; Brugmann's (2009) urbanisation theories promote evolutionary growth patterns that are not based on strict master planning and design guidelines, but rather evolving spatial arrangements to best respond to demands for vitality, economic function and sustainability; and finally, Kotkin (2006, 2010) sees an urban future dominated by new, large, freestanding, cost-effective, low-density developments, such as those emerging in the US 'sun-belt' of Houston, Phoenix and similar areas.

Future urban development will no doubt present a range of variations on all of these concepts, each motivated to create the built form those future communities and regional economies will demand. The actual application (which is the urban plan, not to be confused with urban design) will need to incorporate more localised activities and amenity.

4.5 Urban development – theory and design

In the majority of cases, the urban patterns that emerge across a particular region are the outcome of three, not unrelated, factors. The first is the geographic characteristics of the area – for example, terrain, soils, rivers, drainage, available water, et cetera. Areas that are, for example, mountainous, flood-prone, low lying and make development difficult are, for the most part, avoided. The second element, which typically also has regard to natural features and environment, is the location of infrastructure services – the siting of major transportation routes and trunk electricity, water and sewerage mains and, progressively, the location of major infrastructure including port, dams, airports, hospitals, schools, universities, civic places, et cetera. The third key element is land use, density controls and regulations, together with a titling/ownership registration system applied through government legislation and regulation that, typically, will align with the two other elements identified above.

In a capitalist country such as Australia, the ownership of practically all lands except those reserved for public purposes will progressively transfer to private ownership and, over time, into smaller and smaller individually owned lots. From that point, the ability to re-plan or reconfigure that area is substantially lost. While governments typically reserve the rights to resume property, the large-scale use of that option is not normally practical, given both economic and political parameters. In considering possible re-amalgamation, it might also be correctly observed that market forces will move use towards the highest and best economic return and this may encourage aggregation. Re-amalgamation may occur over time, but only to a modest extent given existing levels of fixed investment and the limited commercial ability available.

The underlying point, however, is that, once made, strategic land-use decisions and allocations are extremely difficult to reverse or even change in the future. Consequently, there is a need, particularly in early planning stages, to ensure that flexibility and adaptability is inbuilt into design, layout and policies, to accommodate, as much as possible, a range of likely future scenarios. Unfortunately, many government planning principles and regulations established variously in areas and regions over past decades have proven almost the antithesis of flexibility.

As noted in Section 4.1, in considering the history and current status of urban development across OECD countries in general and Australia in particular, care must be taken to avoid the idea there are mandatory benchmarks, standards or a typical sequence under which countries, regions and cities develop. Every location is unique, each starting from different physical characteristics, and with its own comparative advantages and available resources. Certainly, there are themes and trends that can be recognised through different eras from the Industrial Revolution through to the rise of manufacturing and from the post-war boom period to the urbanisation and knowledge-based industries of the contemporary era. How that translates to physical development at any point in time will, however, be unique to that particular place.

Regarding urban layout, even from the early period of urban (town and city) development, simple 'grid pattern' road schemes dominated. This still characterises the older areas of many US and other cities, including a number of Australian urban areas (see Figure 4.2). While having a basic logic, these layouts fail to segregate land use or local through traffic and create numerous, at-grade (same-level) intersections. Typically, the resulting urban form is monotonous and featureless and has limited regard to the natural features of the place.

As noted in Section 4.3, it was only through the various stages of the Industrial Revolution from the eighteenth century to the first-half of the twentieth century that industry became urbanised, larger in scale and was location specific; that is, concentrated in areas with access to raw materials, particularly coal and ore, and to ports and rail networks. These were necessary to move large volumes of, first, raw materials and then manufactured goods to markets. A large urban workforce followed with subsequent demands for residential accommodation and other services.

The Australian example was of particular importance given that, during this manufacturing period, substantial industrial areas were developed particularly in western and

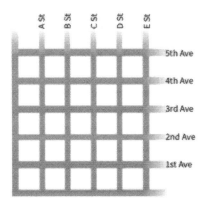

Figure 4.2 Grid pattern road layout (Quora.com).

south-western Sydney, Newcastle and Wollongong in New South Wales; in various lo-cations across Melbourne and Geelong in Victoria; from North Adelaide to Elizabeth in South Australia; and parts of northern and south-western Brisbane. In a number of cases, these towns and suburbs were based on a single economic activity and employer, with many community and social networks focussed around that enterprise (Glover 2015).

Government and town planners had long bemoaned the lack of consistent planning, land use and infrastructure strategies across most industrial cities and the inefficiencies, congestion and poor health outcomes that resulted.

With the exception of Adelaide and Canberra, this was particularly true in Australia where the humble beginnings of the colonies left little time or interest in long-term planning. There was very limited capital investment available and, in any case, in the establishment and development of Australia until the post–Second World War period, a classical economic ('laissez-faire') approach dominated. As previously noted, this approach envisaged only a very limited role for government in urban development. However, concentric models, originally proposed by Burgess (see Section 4.3), and grid pattern road layouts were evident in the organic development of most major cities during that period. With the later widespread use of private motor vehicles, road infra-structure began to dominate the urban frame, with commercial development typically extending in strips along those arterial corridors.

In a number of neighbourhoods and in regional urban areas, development patterns also tended to cluster around commercial and shopping facilities and local infrastructure such as schools and parks. This became known as 'traditional subdivision', with the layouts emerging that paralleled village-type characteristics, but within a suburban setting (see Figure 4.3). It is interesting to note that, while that local neighbourhood design was dis-couraged initially, it later was fundamental to the design of contemporary master-planned communities and new urbanisation principles introduced subsequently (see below in this section). It also became the fundamental design template in the planning of Canberra.

By the 1960s, however, Australian state governments, through their local authorities, were adopting formal town plans based largely on UK development control models. As many cities were now growing rapidly, the overall aim was to encourage efficient organised urban development and to avoid the close location of obviously conflicting land uses. This was a reaction to the failings of earlier industrial cities, which resulted in the close proximity of noxious industrial uses to dormitory (commuter) suburbs.

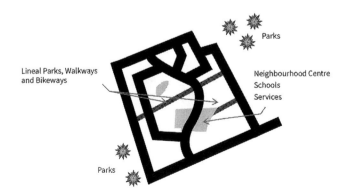

Figure 4.3 Traditional subdivision – schematic layout.

Further, the new system was cognisant of rising growth and levels of investment and affluence and, with that, the growing dominance of the use of private motor vehicles as the preferred form of transport.

This approach (known generically as 'conventional subdivision') segregated the city, town and/or precinct into designated land-use zoning with strict control on land use and, flowing from that, prescribed regulations for built form (i.e. site coverage, building height, plot ratio) for particular developments. Under complementary subdivision ordinances, parameters were also established that prescribed plot size, frontages, infrastructure requirements and so on (typically on a 'user pays' basis). These matters are further discussed in Chapter 12.

Suffice to say that practically all urban development and redevelopment in Australia through the subsequent half century and right up to quite recent times have strictly followed that approach. The road system and the framing infrastructure established a hierarchy of roads from small suburban cul-de-sacs and loops, servicing only the adjoining property, and then scaling up in size and specification through collector roads and distributor roads connecting a number of localities. Overlying that again were sub-arterial and arterial roads and freeways carrying almost exclusively through traffic – either radiating from the original CBDs or developed as ring roads through the outskirts of that urban footprint (see Figure 4.4).

There were significant advantages and efficiencies in controlling development through regulation and it would be hard to imagine any contemporary city operating without development control of one type or another. Further, conventional subdivision and its accompanying land density and controls did achieve at least some of its objectives in quickly rolling out urban development to meet growing demand. However, a rigid, regulation-based approach has of recent years been severely criticised. Each property and project is inherently different, and the attempt to control development under a prescriptive approach will often provide less than optimal outcomes. Further, this approach speaks more too physical development than more holistic considerations of community and economic development, sustainability and resilience in the face of fundamental changes.

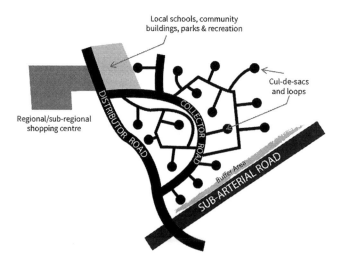

Figure 4.4 Conventional subdivision – schematic layout.

Importantly, the overall road design and typical low density of residential development made the provision of infrastructure and related services very expensive to provide and maintain; and the provision of reliable and cost-effective public transport became more difficult. Social isolation often occurred, given the challenges created in local and cross-suburb transportation. Commute times, almost always in private vehicles, extended considerably and congestion increased as more 'subdivision modules' were added to the urban sprawl being created.

Further, returning to the permanency of initial urban layout decisions, the schemes in question could (understandably) not have predicted the current scale and density of urban developments, or issues of sustainability, energy use and congestion. Nor could they have considered that most commercial and industrial activities are no longer noxious and, in fact, are best located in proximity to, or integrated with, residential environments (Ehrenhalt 2013).

Meanwhile, living conditions, wage disparity and, in some places, racial tensions were concentrating in older, often derelict parts of major cities. Here, the urban form was implicated in the cause of social insurrection in cities including Los Angeles, Chicago and Detroit and elsewhere in the US through the late 1960s and early 1970s; in Brixton in the UK in 1981 and 1985; and later, on a smaller scale, in Redfern, Sydney.

Governments were soon to recognise that ghetto and urban-slum conditions were an important catalyst for social dislocation and unrest. Significant public and, sometimes, privately funded urban-renewal programmes commenced in response. In its early stages, this often took the form of slum clearance, though it became obvious that these schemes often simply replaced one ghetto with another – albeit now high-rise. Worse still, many projects eroded community networks, diversity and identity in those locations.

The work of Jane Jacobs (1961), using the example of Greenwich Village, New York, highlighted the importance of the built form as an integrated and active setting for the community, and that the loss of physical environment, with all its imperfections, would fundamentally change its unique character and interactions. Vibrant communities in whatever precinct or neighbourhood were fundamentally about the people and their ability to interact in a variety of meaningful ways. Consequently, proximity, walkability, diversity of activities and particularly the 'theatre of the street', as Mumford (1961) called it, were the essential elements.

Based on those general principles and driven by government imperatives to address growing urban conflict, a new approach to higher-density urban development and renewal emerged. In the UK it was called 'new urbanism', while similar philosophies in the US were often described as 'urban villages'. Subsequently, a number of inner-suburb redevelopment projects across the OECD, including in Australia, have followed that format.

The key components of 'new urbanism'/ 'urban villages' in some ways reflect a return to traditional urban designs – more finely grained, with an emphasis on detail at a human, liveable scale (Sucher 2013).

Developments based on urban village principles can take a large number of forms; Neal (2003) recognises the key components as:

- development of adequate size and critical mass (i.e. precinct or small suburb minimum) which provided a clean identity and sense of 'place'
- a walkable and pedestrian-friendly environment that accommodates cars but does not overly encourage them

- an innovative and interesting mix of uses and with good opportunities for employment
- a variety of architecture and sustainable built forms – which may be of any scale, but, in an inner-city environment, are likely to be medium density. Even then, however, the diversity in visual amenity is important
- mixed tenure for both housing and commercial uses is encouraged at various price points to promote (though not contrived) a diverse social mix
- provision for basis shopping, health and educational needs within the general boundaries of the precinct
- provision for a degree of self-sufficiency but with porous boundaries within and external to the precinct to allow the free passage and interaction of various individuals and groups
- encouragement to use public transport through appropriate hubs within walking distance
- innovatively designed public, green and meeting spaces for the community and, normally (but not necessarily), with a central point.
- This type of development will invariably look different from earlier forms, but the most fundamental change is the principal driver of human perceptions and needs with the built environment conceptualised around that mantra.

4.6 The Australian context

Australia presents a unique case study of land settlement, development and urbanisation. As recognised in Section 4.1, even though trends and overall waves of development can be identified, they represent only general themes and their scale and physical outcome varies from region to region and, therefore, will evolve in different ways.

Applying these trends to the Australian context, a number of important characteristics are relevant:

- Australia is the only country that is also a continent, sharing no land borders. It is physically remote, has generally poor natural productivity and low rainfall. It therefore enjoys only limited natural competitive advantages, with most of those in mining, appearing in later years.
- It is the only major country that was settled after the commencement of the Industrial Revolution, and the establishment and development patterns of cities and major towns occurred without the need to replace existing urban settlements.
- Contrary to the image presented and often reinforced, Australia is an urban nation – indeed one of the most urbanised on earth. Remarkably, it was urbanised, principally as penal settlements, even before the country was fully explored.
- Based on low-level productivity and rainfall, the population is concentrated in six major cities and along the south-eastern seaboard – a pattern that has become increasingly dominant over time.

Despite inefficient political systems, Australians have a level of common identity, ethos and purpose that has never been challenged or subject to insurrection in over 200 years of white settlement – a remarkable feat in itself. Regarding land use and regional development, however, the constitution creates a range of issues of shared responsibility, financial disparity and lack of uniformity of approach at local, regional, state and national levels. The history of establishment, urban development and urban patterns of Australian cities are reasonably similar – not surprising, given that all were largely developed in the same era. Towns/cities needed to be located on the coast since shipping was the only effective early transportation. All urbanisation started from a very humble base – remote from each other and the rest of the world.

There were, of course, variations on that theme – Sydney had an excellent harbour and 'first point of entry' status; Adelaide was the only free settlement that commenced with a properly laid out town on a level site; Melbourne developed much more rapidly than other cities, following the discovery of gold in 1824. City scale varied considerably, depending largely on the size and quality of their catchments. Water availability, then and now, created an important limitation on development, particularly for Adelaide and Perth.

By the early twentieth century, the construction of railways and the improvement of radiating road systems enhanced the important role of the main towns (capitals), particularly as production outlets for rural and pastoral development, and mines in the hinterland. Most initial development was funded from England, with goods repatriated there – in early years, wool and grains and, with refrigeration from the late 1800s, meat.

Given this scenario, the six cities and their various catchments represented, in effect, city states, often linked more by commerce and migration with cities in the UK than with other Australian capitals. This was reinforced by a British Government that progressively gave self-government to the various states. Based on that history, the cities established then and their subsequent development are reasonably large by world standards but, unlike most other OECD countries, there are few cities of intermediate size.

The rapid growth and urban development that continues in Australia to the present remains concentrated around those cities – either in their densification of their CBDs and inner suburbs, or in the remarkable growth of the urban areas within commuting distance of each capital. The recent growth patterns around Sydney (Wollongong, Penrith and Gosford); Melbourne (Geelong and Ballarat); and Brisbane (the Gold Coast, Sunshine Coast and Ipswich) provide prime examples of those trends.

An important theory that helps explain the continued dominance of the small number of capital cities and their immediate urban areas is called 'cumulative causation'.

In this scenario, a particular place gains some level of early competitive advantage over other surrounding, potentially competing areas. Perhaps that location, when first settled, had access to a good port, had good quality soils or available water. Sometimes that advantage would have been minor, but sufficient to make the place more attractive for incoming settlers and migrants and, later, for business and investment. Sometimes the initial advantage may not be sufficient to sustain a leading position; however, in most cases, the location would maintain that dominant role, which increased over time, particularly when later decisions were made regarding the location of major 'one-off' infrastructures – a hospital, major schools, a university, a power station and so on. Particularly in more recent times, as those centres grew, they also become hubs for talent, innovation and financial investment. This central node attracted people and investment from outside the region and typically drew in free capital, labour and talent from remaining, less progressive parts of the area.

There were, of course, spill-over effects in the development of peri-urban regions and those within commuting distance could benefit from access to jobs in that hub;

however, even as benefits and technologies spilt across the region over time, the central core was more likely to be the recipient ('first adopters') of the next wave of new technologies, innovations and finance and, therefore, again leapt ahead of its surrounds (Horan 2000). As a result, the regions' principal town enjoyed continued dominance and higher growth compared with its catchment and smaller competing centres.

Under the development scenario of cumulative causation, it is almost impossible for lesser centres within the catchment to significantly change their relativity to that capital (or 'central place').

That relatively notwithstanding, urban development and patterns within a town or region are dynamic – where significant changes occur progressively and, sometimes, quite radically over time. In the first instance, the physical nature and configuration of CBDs have changed dramatically over time and are linked to a range of emerging business, economic and demographic forces.

High-rise building development emerged in major cities in the US following the widespread use of electricity; the invention of elevators in Germany in 1880; and the widespread use of air-conditioning from about 1902. The concentrated high-rise pattern in Australian CBDs did not emerge, however, until the 1960s and 1970s. By that period, a more mature commercial and financial sector and, deregulation of the banking sector, provided investment for large-scale commercial projects. In the meantime, the recapitalisation of the manufacturing industry was under way. This encouraged investment in larger-scale, capital intensive factories in new industrial suburbs and precincts outside of the CBD, separating the corporate and management sections of manufacturing enterprises. Other corporations, including banks and larger government activities tended to gravitate to new commercial buildings in the inner city.

Rural production and rural regions have always been important in Australia, perpetuated by notions of a 'bush' heritage and culture. That importance, however, is overrated. Unlike almost any other country, Australia had towns even before it was fully explored and widely settled. With a few exceptions, such as the discovery of gold (and later mineral development), Australia was always politically, economically and socially dominated by its cities, towns and the concentrated urban environment they created. Around the world, the city is the hub of economic activity and that aggregation of power, wealth and urban populations continues to grow disproportionally compared with rural regions (Short 1996; Reader 2004; Montgomery 2007; Glaeser 2011).

As a capitalist, western democracy, Australia has a property sector that bares reasonable comparison to many OECD countries. Although the physical environment, huge size and very low average population density (about 2.5 people per square kilometre) of Australia may appear unique, it is highly urbanised. Nearly two-thirds of the population live in capital cities, and over three-quarters live in cities with populations above 100,000 (Department of Infrastructure and Transport 2015). Remarkably, 80 per cent of economic activity in the country occurs on just 0.2 per cent of its land mass, that being largely around the country's seven major cities (Kelly & Donegan 2015). Practically all of Australia's major cities are located on or near the coast, with 90 per cent of the population residing within 80 kilometres of the ocean, mostly along the south–eastern seaboard.

Real property and built assets contain and surround people, economic activity, communities and society. The development patterns that emerged across towns, regions and states over generations not only tell the history of that place but also provide the platform for current and future use and development.

In considering that in the Australian context, several generic observations are worthy of particular note. First, despite the apparent low density of development and population nationally, the pressures of urbanisation and the competition to develop land in prime locations are comparable to those of North America and Europe. These issues, which have obvious links to issues of optimum population size and density, are matters of ongoing political and community debate (O'Connor & Lines 2008; Smith 2011).

Second, the nature and composition of each of Australia's 60 or more regions are unique and share surprisingly limited similarities to others across the country. Further, regional and growth patterns across Australian regions bear little resemblance to those of regions across most other OECD countries. Consequently, a close, individual assessment of any Australian region's history and geographic, economic and demographic context represents a prerequisite to establishing that region's current opportunities, potential and issues (Eversole 2017).

4.7 Key issues and challenges

Given the experience of practically every other component of contemporary economic and social/community life, fundamental changes to the way property and the wider urban form is held, developed and used appears inevitable and, indeed, is already under way.

Setting detailed strategies and exact plans for the future based on a proceduralist or sequential model is, at best, a fraught exercise and even more problematic in the long term.

The strategic planning approach currently applied to all manner of government, community and corporate activities (including property and development control) had its genesis in military planning and relied on both a relatively stable structure and 'command and control' authority. This overarching approach was dominant since the Second World War and the Marshall Plan for post-war reconstruction, underpinned by Keynesian, interventionist economic strategies adopted by capitalist countries. These institutional, centralist models are now seen as increasingly ineffective in a contemporary environment based on distributed economic units and continually evolving knowledge-intensive businesses and social networks (Morrison & Hutcheson 2019).

Stein (2019) recognises the significant conflicts that have now emerged across the various competing interests which form part of the development and use of land (property) resources. The fundamental changes relate to urbanisation and to the dominant role now secured by financial and 'individual capital gain' models over the entire property sector (refer Chapter 3). Government overview and planning schemes have been swept along with that momentum. Now, according to Stein, the original purpose of planning – that is, to have some sense of how to secure the future – is forced backward in order to capture 'vagabond' capital and to meet the consumer and wealth accumulation demands of current owners viewing real property as simply another tradable commodity.

Underlying all of that are growing issues relating to perceived inequity in land ownership and use and the apparent inability of government to pay for infrastructure and services for a rapidly urbanising population. The works of US economist, Henry George (1879, reprint 2016; O'Donnell 2017) though dating back almost 140 years provides valuable insights here which still enjoys a considerable following among left of centre political groups across the OECD.

George considered that land – its distribution, aggregation over time – was the key determinant of wealth (and, its corollary, endemic poverty). Land was not seen as a 'product' created by mankind as were the other factors of production and continues to exist far beyond normal commercial horizons and defined by rental payments. The laws of economics obviously still apply but, according to this approach, regard needs to be had to land's inherent characteristics that extend beyond the economic and financial and extend to social connectivity, security and identity. He considers that the permission of trading and securing windfall profits from property investments distorts and unbalances markets for essential commodities such as housing.

George recognises that, because land ownership and wealth has long been associated with political power and influence. Consequently, property is often under-represented in the overall 'taxation mix'. Related to that is his valid observation that the current arrangements for government providing major infrastructure from consolidated revenue are both unfair to those in non-benefited areas and also set in place financial anomalies between property owners and tenants. In any case, given of other fiscal pressures and the huge cost of contemporary public infrastructure of all types, there would still be insufficient funds available from such sources.

As part of wide-ranging property taxation reforms, he believed that efficiency and equity would be much better served by the application of infrastructure levies to properties in a benefited area for a period of time until those expenditures are substantially paid down.

The disconnect between planning approaches and current business and community reality represents a significant challenge – with very important implications given the economic drive now evident within urban centres. It would appear obvious that, even on the brave assumption that current urban systems and controls had met objectives set for them, a reassessment of the regulation-based approach is, after almost 60 years, now due – especially given the fundamental changes identified above.

Successful economic development in Australia's regions is based on a combination of trading and service sectors. Trading sectors produce goods and services that are in part sold within the region, but for the most part provide exports from it. The remaining and larger portion are service sectors providing goods and services to that local market. While normally smaller, the trading sectors – now knowledge-based industries, education and health sectors – typically flourish in clusters, often around CBDs and their fringe areas. Service sectors are associated with tertiary education, research institutes, health facilities, ICT hubs and research park precincts. Areas with these types of uses typically generate economic success and leadership that affects the rest of the region.

This utilisation, typically concentrated in a few towns or precincts, while of critical economic importance, bring with them a number of urban challenges. First, a physical precinct needs to be developed that is conducive to those of business types, their interactions locally and linkages to a global marketplace. Commercial and residential prices typically increase in these progressive areas and a new geographic and demographic pattern emerges. Here, those with the education and skill enjoy best access to job opportunities and live within those areas, while others without qualifications receive substantially less income and have limited choices regarding the location and type of housing they can secure (Wasik 2009; Moretti 2013).

These trends establish new dichotomies within regions. They also manifest in the growing socio-economic divide between home ownership and rental accommodation.

The accessibility to residential ownership in progressive major cities such as Sydney and Melbourne and, their most desirable precincts, is decreasing rapidly for potential first-home buyers (Kelly & Donegan 2005). That diminishes the aspiration of home ownership and the previous egalitarian attitude that characterised Australian urban development for over a century.

The growing disparity also leads to the physical separation of these new socio-economic educated strata from other workers, with increased cost, time delays and congestion falling to less privileged groups. The removal of estate duty in Australia some decades ago further reinforces the aggregation of family wealth based on home ownership.

This contemporary example shows how changing economic drivers affect the success or otherwise of new business sectors and opportunities, and how these changes are physically played out within an urban environment that will almost certainly have difficulties adapting to rapid and compounding change.

Despite the belief by a number of economists that the residential market and price levels in these cities are unsustainable over time (Ellis 2009; Quiggin 2010; Keen 2011), there appears limited appetite for change, given the investment in these markets by many existing owners and commitment by the banks. This is reinforced by the general apathy of governments regarding innovation or radical change in urban or housing policies, taxation reform or town planning amendments – in turn, driven by the potential of adverse criticism by the larger proportion of the electorate with existing equity in residential property.

Despite this reluctance to change, however, the situation is increasingly unsustainable given the overheated competition for the most desirable locations, increasing congestion and urban inefficiency, the lack of political will to enforce densification and the cost and difficulty in providing more appropriate public transportation, particularly heavy rail.

Perhaps the most significant trends, however, relate to the economic and environmental sustainability of the urban built form characterised by low-density, car-oriented urban development typically created across much of Australian, US and other OECD cities.

Remarkably, this form of development reflects both demand for much larger households of the past and the homogeneous product typically provided by the development sector under town planning regulations that, for the most part, reward conformity.

Often, while denser and better serviced urban developments are now promoted, issues of congestion, pollution, inefficiency, poor access to services and declining overall urban amenity and liveability are now embedded in many inner-urban areas – problems that will intensify as, invariably, cities continue to grow and energy and transportation costs increase disproportionally.

Returning to Tuan and the importance of place (as discussed in Section 4.2), it is arguable that the uniformity of product and monotony of design and layout across contemporary urban sprawl presents the greater threat to liveability and the holistic urban experience, as described by Mumford (1961) and Jacobs (1961) in Section 3.2.

Travel writer Bill Bryson (2015, p.123) wryly describes the issue as follows:

> [I drive two hours out of Denver,] … the first hour is taken up with just getting out of Denver. It is a permanent astonishment to me how much support an American lifestyle needs – shopping malls, distribution centres, storage depots, gas stations,

zillion-screen multiplex cinemas, gyms, teeth-whitening clinics, business parks, motels, propane storage facilities, compounds holding fleets of U-Haul trailers, FedEx trucks and school buses, car dealerships, food outlets of a million types, and endless miles of suburban houses all straining to get a view of distant mountains.

Travel twenty-five or thirty miles out from London and you get Windsor Great Park or Epping Forest or Box Hill. Travel twenty-five miles out from Denver and you just get more Denver.

4.8 A likely future

While the challenges and issues for the Australian urban form, as outlined in Section 4.7, are significant, they are not insurmountable. Rather than thinking of the future as simply a projection of the past, Australian prospects will be based on knowledge and innovation, the success of the firms involved, together with their associated communities in urban settings.

Each city and precinct will progress on a unique path to those goals in a local and global context.

The contemporary concept of a 'global city' is neither simply defined by scale nor its position within the hierarchy of its own country. That hierarchy (and political geography generally) will, of course, remain important as will physical location (where Australia will almost always be at some disadvantage despite improving ICT connectivity). However, the concept of 'being global and networked' is rapidly changing.

Person-to-person, firm-to-firm interface is critical to the emerging business sector, but primary links between cities form part of the framework. For example, the city-to-city links, say, Sydney to Shanghai, will often be more relevant and practical than, say, Australia to China. Such international links, while politically essential, are normally too coarse-grained to provide the economic, social and cultural interface that underpins knowledge-based and service sectors.

At a second layer, global networks are often concentrated around particular business specialisations. Lacking the profile of the software and hardware development hubs of, say, Silicon Valley, Seattle, London or Bangalore, these networks nevertheless link locations with high levels of specialist expertise in areas including accountancy, advertising, pharmaceutical, banking and finance, insurance, law and management services. The tendency to aggregate is, of course, not new, but the emerging configurations are quite different from global networks of the past, which typically had headquarters in major centres including London, New York or Chicago – with other nodes linked to that centre or cluster. Contemporary networks are typically less structured and have more cities/ locations involved, each with their own approach and point of difference. These linkages are obviously underpinned by near-instantaneous communications, transactional lead firms in the sectors and 'follow the sun' operations. They can draw firms together in strategic alliances that apply local and global expertise to projects, whatever the target location.

Depending on the sector and activities involved, these nodes will often be closely linked with major universities because of access to leading research and a graduate and post-graduate workforce that underpins these activities. This scenario is consistent with the theories of Glaeser, Brugmann and Florida discussed in Section 4.4.

Florida (2008) argues locational preferences are typically about urban form, a range of integrated social, cultural, environment and educational characteristics and direct employment and career opportunities that comprise a complete socio-economic system.

The built form of residential and business real property represents a key component of that holistic offer.

Considerations of short- and longer-term future prospects for economics and real property, while never exact, need not be matters of random conjecture. There are, for example, a number of future projections that, by the weight of existing evidence, will almost certainly prove accurate. These include:

- Change, based on advances in knowledge, communications and networks, will continue, probably at an even faster rate. It will be urban-centred and, in that environment, some regions and countries will be winners and, others, losers. Moreover, even within the same urban area there will be precincts and general locations that will benefit more than others.
- Populations will continue to increase for the foreseeable future and be drawn to increasingly larger cities, though those populations' increases will slowly reduce, particularly in Western countries.
- Across most of the OECD countries in particular, a significant ageing of the population is well under way and this will significantly affect the types and location of the built forms demanded – in residential accommodation and community structures, entertainment and business facilities. Regional differences will continue to emerge in age profiles.
- In OECD countries, traditional heavy large-scale manufacturing will not return – though smaller-scale high-quality manufacture and service industries will develop further with particular emphasis in ICT linkages, education, human resources and social linkages.
- There will be significant changes in energy use and consumption as the cost of traditional energy sources increase disproportionately. This will work to discourage the use of private motor vehicles, though they will continue to be an important mode of transportation requiring accommodation well into the future (Bower 2009; Rubin 2009).
- Water as well as energy will emerge as key criteria for urban development as will global warming and sustainability issues (Weber 2012).

Added to this is a second group of likely scenarios though perhaps a little more problematic in outcomes than the first group. These are:

- The mantra for business, public and social activities will continue to exhibit the dichotomy of 'global-to-local' approach. However, given higher energy and transportation costs, there may be a tendency in some production at least again to consider the localisation of those activities and markets. (The continued dynamic growth of and innovation of ICT, however, makes such observations fairly speculative except in the production of some physical goods.)
- The additional growth in cities and, typically those in Australia, will be outer-residential (peri-urban) areas and regional cities and towns within a one-hour fast commute to major cities (Polèse 2009) – but will include new forms of transport-oriented, concentrated areas.
- There will be a growing socio-economic dichotomy between rich and poor areas, and between urban and rural locations that are well located with high levels of urban amenity and those, typically at more distant locations and less well serviced (Kelly & Donegan 2015).

4.9 Implications for real property and its market

This chapter described a number of interrelated economic, environmental and community/social changes that have emerged over recent decades and continue to change and evolve at a compound rate.

The fundamental challenges for owners of long-term, fixed assets is first to better understand the nature, context and significance of the emerging changes and, critically, to best align these assets with the new demands and the environment, particularly in the medium to longer term.

As discussed in Section 4.7, making predictions in this uncertain future involves significant risk. These should be approached by considering likely scenarios rather than providing prescriptive or detailed plans. As also noted, there are a number of parameters here that appear relatively certain that establish an overall direction.

These parameters include:

- Place still matters. It is innately human to cluster (in tribes, families, groups, teams or whatever) to the aggregate advantage of the group and for individuals within. New technologies and communications can vary the manner in which that occurs, but predictions that emerged in the early stages of internet development, specifically that offices and meeting places would, in time, disappear, have proven demonstrably wrong – and, in fact, the reverse is true. The rise of knowledge-intensive industries will continue to advance the influence of technology, but work will occur in a range of settings and be internet linked. The built environment needs to be flexible and adaptable – to accommodate longer-term strategies and day-to-day operations.
- Changes to the use, functionality and value of fixed assets is typically something that evolves over time – though this is not to suggest that buildings and their uses cannot change over a remarkably short period. They are subject to a range of parameters including the nature of the various motivating drivers, the location and precinct and the adaptability in design of those particular assets. An openness to change and innovation by owners and managers is a key catalyst.
- Even with an existing form/environment, a number of overlays of new urban strategies can be applied – either greenfield or brownfield. Typically, given the impediments of existing urban sprawl, much of that change will revolve around the adaptation to a more sustainable future and be focussed on innovative transportation solutions. Perhaps these transport-oriented developments (TOD's) of a scale and concentration could balance the benefits of local precincts/neighbourhoods, but ensure they are linked within the wider region and to the global marketplace.
- Given the understandably conservative approach taken by the private sector to property investment, an important role for government should be education and exemplification. This includes explaining the nature of new urban forces, the role of the civic economy, precincts, entrepreneurship and innovation and challenging the current inertia created by regulatory planning regimes. This role can be created in part by the promotion of innovation and design excellence and by facilitating the exemplification of these new forms of development.
- Because the fine details of future economic, community and environmental parameters cannot be known, a scenario rather than strategic and operational planning approach is required – one that establishes and adheres to key principles over time; but,

- at the same time, keeps a range of options open that can rapidly seize and maximise benefits from new opportunities. Again, this can only be achieved by a cooperative/ partnership approach to development by public and private sectors, rather than one overly bound by regulation that encourages adversarial confrontation.
- As past history would suggest, radical, largely unforeseen, external events, (for example, the COVID 19 pandemic) should be considered as almost inevitable over time. Nevertheless, in a mature, developed, well organised and comparatively rich country such as Australia, the ability of communities, businesses, systems and built assets to adapt to changing circumstances should never be under-rated. Such adaptations may well involve significant changes in tenure, built form, infrastructure and locational and investment preferences.
- The development sector in Australia, particularly for infrastructure and major projects, is chronically undercapitalised, which causes a range of supply bottlenecks and affordability issues. Innovation in the supply and funding of infrastructure in public/private partnerships and, perhaps, by accessing growing superannuation funds and foreign involvement in partnership arrangements – would help address this problem.
- Finally, and perhaps most importantly, are the arguments raised earlier in this chapter pertaining to the purpose behind the entire property and development sectors – the satisfaction of individual and community wants and needs. It can quite legitimately be argued that the wider issue of personal and community satisfaction ('happiness') of key stakeholders, now and into the future, needs to be accepted as the starting point of the entire debate, rather than simply relying on current economic models that can often view the consumer as the last step in an economic production chain.

Chapter 5

Real property as an asset

5.1 Introduction

In Chapter 2, a number of key characteristics of real property and its markets were identified. Some related to the physical asset and its environment, others to its function, utility and economic and operational performance. It is clear that to fully understand the nature and impact of real property resources, all of these tangible and intangible components and, how they integrate, must be appreciated.

With that as a premise, this chapter considers real property as an economic asset and how it is used and managed to optimum effect.

5.2 Background

Even though property ('the built environment') is so familiar to all, it is difficult to consider it in a dispassionate way – the way in which, for example, a computer scientist may work through algorithms or a pathologist might work through samples looking for themes, issues, problems or links. With property, analysis is often clouded by personal tastes and preferences, emotions, social conditioning and professional background.

An objective analysis of property is difficult and not often addressed in a structured way. It is not that real property is unimportant or unworthy of this attention. In fact, the reverse is true. Buildings contain lives, businesses and economic activity, civilisations and cultures. As noted in Chapter 1, about 44 per cent of all wealth is stored in real property assets (Fiorilla et al. 2012) and global real estate has an estimated aggregated value of $217 trillion (Stein 2019). The development, building and construction industries that create property assets were valued at $10.8 trillion worldwide in 2017 (GlobalData 2018).

Real property is the basis of collateral and underwrites most major financial arrangements. Property assets are traditionally a source of wealth, power, prestige and enjoyment. They are the subject of statutory control, disputes and even wars. In recent times particularly, they have also been viewed as vulnerable political and economic targets in international disputes (Pawley 1998).

The built environment surrounds, looms over and persists beyond the individual and the current generation. Winston Churchill observed that '. . . we shape our buildings, and afterwards our buildings shape us' (Churchill 2013, p. 298). Almost perversely, real property is so intrinsic to human activity and to community and business life, that its wider value is rarely analysed in detail. It is often left to the popular press to provide only the most superficial reporting on what is an obviously complex area.

For capitalist systems to work well, a fine balance needs to be maintained between the property and finance sectors and the remainder of the economy. The Global Financial Crisis (GFC) of 2007–2008 and beyond exemplifies the disastrous outcomes when these interfaces fracture (Roubini 2010).

Real property analysis represents a fusion of applied economics, management theory, tastes and preferences, legal and financial parameters and physical consideration, all interacting to provide, guide and manage those assets. Time dimensions are also critical when evaluating property, and the words 'property', 'real estate' and 'building' should perhaps be seen as verbs rather than nouns – as actions rather than statements – evolving over time.

The study of property development, investment and land economics more generally revolves around the manner in which decisions are made, looking to the future and predicting both income and risk. The typical property investment profile requires large, lump-sum negative cash flows at the start of the venture. The benefits that then flow (e.g. rental income and/or functionality) begin subsequent to the initial investment and emerge in relatively small increments over a very long time (i.e. the economic life of the asset). The early analysis and investment decisions (or 'predictions'), therefore, have long-term impacts on future benefits or returns.

A large proportion of the real property asset base lies in business sectors (i.e. the commercial, industrial, retail, tourism/hospitality, rural and mining, et cetera). In value terms, about one-third of all property investments per annum flow into these 'commercial' fixed assets. Roughly another third is invested in the residential sector, and the final third in all other forms of built assets, principally public infrastructure.

Business-related property assets, in effect, provide a 'platform for businesses', either for an owner occupier or leased to a third party as an input to their enterprise. Over recent decades, digitisation, globalisation and the breakdown of corporate structures, together with changed work environments, have had major impacts on the demand for these assets. Through the early stages of the introduction of ICT and the wide spread use of the internet (i.e. 1990s–2000s), it was envisaged that the demand for office space would reduce and be replaced, to a considerable extent, by remote work locations.

While these options have emerged (often providing different real property opportunities, the demand for traditional office space remain strong with total stocks approximately doubling since 2000 (PCA 2019). The basic human interactions and business dealings are accommodated within office environments have been enhanced but certainly not replaced by technology. Place still matters, but now in different ways.

The greater challenge for property owners and managers relates more to adapting built assets to continually changing demand in both physical nature of accommodation and of the services demanded. For new construction, contemporary design will attempt to 'build in' flexibility and adaptability with the aim of prolonging effective life. The accommodation of such changes in the large proportion of older assets is much more difficult and expensive. Overall sound demand analysis and client profiling combined with innovative design/re-design are essential elements in such activities.

As observed in Chapter 2 there are underlying rules and principles and an economic rationale underpinning the property sector. Based on these principles and having regards for the particular property and project under consideration, overall strategies can be established. Thereafter, suitable management and operational systems can be put in place that align the individual characteristics of that property that can optimise returns and effectively manage risks throughout the life of the initiative.

The key observation here is that a holistic and cascading 'strategy – to – management – to – operations' approach is required to properly secure and advance that asset/property project. It could be argued that, in the past, insufficient time and attention has been given to these most essential structural tasks.

A basic tenet in decision-making in investment properties will be the 'price-to-earnings ratio' (p:e), that is, the relationship between the price paid to acquire or create the asset relative to its ability to earn future income or to create cash flow. To allow for genuine comparisons, these projections must be converted to current equivalents (i.e. 'present value'). These matters, which will be elaborated in Chapter 10, are complex and care needs to be taken neither to over simplify nor to generalise property investment considerations.

5.3 What is different about real property assets?

Real property is acquired and held for a wide range of physical, economic, political and social reasons, including:

- The basic human requirement of shelter.
- The generation of income. Typically, such income can be derived in two ways. First, the property asset can be used in conjunction with other economic factors – capital, labour and entrepreneurial skill – to produce some goods and services of value. A crop produced from a farm or manufactured goods from an industrial property are examples of this. Alternatively, a property asset can be developed or acquired and subsequent income derived from tenancy.
- The accumulation of capital and increase in value through development and/or redevelopment of a property asset for on sale.
- The provision of community services or requirements and the advancement of political objectives (i.e. macroeconomic and political motivations). Examples include publicly owned community assets and public facilities and infrastructure.
- The holding of real property for, at least in part, social and status reasons. The Australian attitudes to housing and home ownership provide a common example here.

In every case, however, real property holdings are live and dynamic assets. The perception of property as stoic, unchanging 'bricks and mortar' is not borne out, particularly in the contemporary environment. The physical and spatial nature of real property – and the evolving political, social, commercial and legal demands placed upon it – require near continuous change and adjustment to the asset base to meet expectations and, thereby, maximise performance and return.

Most other forms of capital assets (e.g. cash, share portfolios, gold, bonds, futures and debentures) are available in units of comparatively small capital value. Unlike real property, those assets do not physically deteriorate or change, and their physical holding may involve little more than the provision of adequate security. Strategic decisions in such asset classes (i.e. when to buy, when to sell, et cetera) will usually be clear-cut and final. Importantly, owners of non-property assets will often be able to make iterative and short-term decisions to buy and sell parts of their holdings. The typical management of a share portfolio provides an example of that incremental approach.

Additionally, most of the asset types mentioned above are liquid or near liquid and consequently, as long as the seller is willing to accept the market price on the day, transactions can often occur almost instantaneously, given the digital trading environment of shares and similar markets now available.

However, each real property asset is of considerable capital value and usually acquired as part of medium- to long-term strategies. Consequently, 'buy-or-sell decisions' are less frequent and of greater importance than, say, the day-to-day finessing (buying or selling) that occurs with share portfolios.

Each property has its own unique 'life' and 'life cycle' which can, by a simple analysis, be divided into four stages (Figure 5.1):

(1)	(2)	(3)	(4)
PRE-DESIGN	ACQUISITION or DESIGN & CONSTRUCTION	HOLDING OF THE ASSET	SALE, MAJOR REINVESTMENT OR RENOVATION
(i.e. concept and project parameters, the *functional brief*)	(i.e. *creation*)	(i.e. *the asset in use*, management and maintenance)	(i.e. *rationalisation*)

Figure 5.1 Development/building life cycle.

Stages (2) and (4) identified in Figure 5.1 tend to attract the bulk of attention, professional input and concentrated effort. On the face of it, these are the areas in which major decisions are made and where significant capital funds are secured, committed or expended within a short time. These stages are often accompanied by the intense activity of owners, analysts, financiers, architects, engineers, builders and other interested parties.

These points of concentrated activity are clearly important, but if viewed in isolation may give a skewed view of the true life cycle and the overall economic performance of property assets. Stage (1) is also critical. Many of the incipient faults in an asset, for example, originate in the pre-design phase when fundamental decisions on location, concept, cost parameters and function are made. The initial acquisition or development phase, Stage (2), typically represents only 20 per cent of the total occupancy and operational cost of the asset during its life, if recurrent costs such as the costs of operation, energy, taxes, maintenance and refurbishment are taken into account. Consequently, good quality, consistent management through this long 'asset-in-use' phase, is essential to overall asset success.

In practice, the stages identified in the simple analysis above are clearly interlinked. For example, design and construction, quality and functionality affect the operational costs and overall capital value. Likewise, astute management while the building is being held maximises long-term net income and, thereby, capital value. It also extends the economic life of the asset and postpones expensive rationalisation or refurbishment phases.

As noted in Section 5.2, the usage of residential, commercial, retail, industrial and other property types are subject to significant and compounding changes in demand resulting from changes in demography (notably age and family size), production and business structures, technology and regulation. These changes will continually challenge the functionality and, therefore, the value of the asset. The greatest danger to large, fixed assets is obsolescence – that is, a reduction in functionality and demand for physical, technical, legal or other reasons. Obsolescence is to property as inflation is to currency: it limits and potentially destroys the fundamentals that secure value. Management, in this context, requires an incremental approach of continual readjustment and realignment in response to changing demands and market conditions, in order to help fixed assets evolve over time. The very name 'fixed assets' should not be seen as some impediment or excuse to avoid this fundamentally important evolutionary process.

Brand's (1994) seminal work in this area recognises the importance of this evolutionary approach to buildings and their management. He considers that buildings should not be seen as a single entity that is impacted by change and responds uniformly. Rather, he notes, that buildings are made up of a number of obviously related, but quite discrete components, each with their own time frames and ability to change. Table 5.1 shows a typical break up for an office building using this approach.

Brand's key observations relate to a time continuum from almost daily events to perpetuity. The very long-term components (e.g. the site and its location, the basic footprint, et cetera) are, to all intent and purposes, unchanging. It is extremely difficult, at times impossible, for subsequent management or further capital investment to resolve incipient problems in those components of the built form.

At the other extreme are the individualistic, immediate and diverse needs of the inhabitants and visitors to that building at any point in time. An owner or manager at any point in time will never fully accommodate this diversity, but good market intelligence and facilities management will effectively lower the overall risk profile and add to responsiveness.

Brand believes that the longer-term sustainability, adaptability and, therefore, success of real property assets relies most heavily on the asset's particular ability to adapt/evolve the 'mid-group' of building components identified in Table 5.1, notably building services and fit out. These will include, for example, the quality, adaptability, capacity and positioning of air conditioning, fire, water and electrical services, service walls and ducting, common areas (lifts and stairways), together with the quality and functionality of surfaces, finishes and equipment within the building. Additionally, the building's and its management's ability to accommodate the changing fit out and partitioning demands of tenants is also a determining factor (i.e. built in flexibility).

Therefore, the meeting of the strategic objectives of a real property asset is as dependent upon its astute management 'in use' (i.e. while it is being held by that owner) as it

Table 5.1 Building components and time frames (adapted from Brand 1994)

Components	Time frame/ability to respond to change
People (occupants/visitors, et cetera)	Very frequent – as different individuals visit/ occupy the building
Furniture/equipment/loose components	Frequent – monthly/annually
Fit out	Changes with churn of tenants as well as business changes – 3–10 years
Building services/finishes	Typically 12–20 years
Extend fabric	Long term – typically 30 years +
Building/footings/structure/frame/ footprint Land/site/location	Very long term/difficult to change Unchanging

is on judicious acquisition and disposal of the asset itself. While a property asset is held or owned, the owner and manager can take actions that improve the asset's position, and thus its capital value – in comparison with both other property assets and returns being secured through that period from other asset classes. That ability for an owner of an individual property to take specific actions to enhance the capital value of that asset represents a key comparative advantage of real property.

Such actions may include targeted capital improvements, cost effective maintenance and improving the quality of leases and tenant management. Active asset management represents an important advantage in the holding of real property assets over other asset classes. For example, if an investor had decided, instead of buying the property asset, to buy gold, fine art, shares or most other types of asset, few (if any) actions are available to the owner to improve the acquired asset's unit value while in his/her possession; the reverse is typically true with real property.

As noted above however, such activities by the owner or manager need to be tailored to the individual property and reflect the unique physical and commercial characteristics of that asset. It is not simply a matter of expending additional funds.

The 'asset management' of any property investment should be viewed as three distinct, yet closely integrated, components (Figure 5.2).

The 'strategic' level involves the overall plans and directions for the asset, namely: What is the purpose in owning the asset? What level of performance is required from the asset now and in the future? In what direction is it planned to take the asset? How does it fit in with the rest of the owner's portfolios? What is the debt–equity position? What is the overall risk profile? Typically, all need to be set for each asset and represent the key decisions made by the owner or portfolio manager. It is remarkable that in practice, in many portfolios, these fundamental questions are not considered in a regular and holistic way.

'Property management', if correctly carried out, translates the owner's expectations and strategic decisions into the organisation, control and administration of the asset, its tenancies, occupation and the provision of building services. In accordance with a formal

Figure 5.2 Component levels of property asset management.

agreement with the owner, the property manager's role is to represent the interests of that owner in dealings with and operations of that property. The property manager represents the owner and, above all, protects the owner's interests. These interests, particularly in the medium and long term, will be best served if all legitimate stakeholders in the building (including tenants, occupants, service providers, visitors and the wider community) find that the building reasonably satisfies their specific needs. Overall, therefore, the property manager holds a pivotal role in 'making the building work' for all stakeholders.

Finally, and at the most practical level, are building operations and facilities management. These are management and operational directives that manifest in physical activities. These activities typically include maintenance, resolving most tenants' issues related to the physical asset, security, cleaning and other building services, together with management of the facilities that seeks to optimise the functionality and efficiency of the building.

Importantly, while each of the three levels of asset management involve different activities and are often carried out by different groups, a unity of purpose and a holistic philosophy and approach is required. This integrated approach is known as 'Strategic Asset Management' and is further examined in Chapter 13.

5.4 Industry terminology

As the property sector has become more sophisticated, the associated professions (engineers, financiers, accountants, architects, solicitors, valuers, agents, project managers and quantity surveyors) have progressively widened their services to meet those growing demand and opportunity. A range of new names and descriptors for the services they offer have followed. Generally, the names are based on the activities outlined above. Industry and colloquial terminology abounds and, in any case, the segregation of tasks is the antithesis of the contemporary, integrated approach.

Nevertheless, in an attempt to clarify some of the common industry terms and the names of services provided, the following general definitions are suggested (Note that in practice, many of these activities may be combined and undertaken by the one individual or group.):

Portfolio management

Portfolio management is effectively the role of the owner. The portfolio may consist solely of a number of real property assets or, more commonly, the property assets will form part of a wider ('balanced') portfolio that also contains assets such as shares, cash and other exchange dealings and intellectual property. Decisions here tend to be strategic, that is, decisions relating to buying or selling, overall performance, portfolio balance, overall risk management, cash flows monitoring and long-term positioning.

Asset management

This generic term (also referred to as 'Strategic Asset Management') encompasses the contemporary, holistic approach to the management of real property assets and will include management and support systems to deliver assets that are fit-for-purpose, remain functional, are adaptable and meet the (reasonable) objectives and demands of all stakeholders. The integration of short-term activities, medium-term planning (e.g. annual asset plans) with longer-term strategic objectives is fundamental to this approach.

Property management

Property management is the process of translating the objectives set by the asset owner into the day-to-day operations and functioning of real property. Typically, and depending on the formal agreement between the owner and manager, it can include responsibilities such as leasing and tenancy management, budgeting, cost management, operational management, contracting out of services such as maintenance, cleaning and security and reporting under the legal requirements of agency to the owner as principal.

Facilities management

Facilities management refers to strategies and technical and operational matters and activities that secure optimum performance from a building asset. The function can include the management of building services, layout, workplace planning, optimisation of expenditures, environmental health, life cycle management and the overall creation of effective and efficient workplaces and assets.

Maintenance management

In most cases, this is a sub-category and operational extension of property management. It refers to the organisation and management of the delivery of services to the building that include maintenance, operations, cleaning, security, building audits, work contracts, scheduling and reporting.

5.5 The characteristics of real property assets

As discussed in Chapter 2, property markets represent an example of imperfect competition. That is, there are many buyers and sellers in the marketplace, but none have market dominance. Consequently, while exceptional circumstances occasionally occur, no individual or firm has full knowledge or control and everyone who enters the market to trade is, in fact, a price taker.

The commodity is heterogeneous, with each commodity unit (i.e. property asset) having its own unique characteristics that are recognised by consumers and who differentiate between assets by the prices they are willing to pay.

Chapter 2 and earlier sections of this chapter also identified a number of characteristics of real property that define it in economic terms. When considering real property's role as a functional asset, the following observations might be made and provide an important basis for ownership and management decisions:

- Real property assets are fixed and physical in nature. Traditionally, this characteristic is considered a strength; financiers still typically value property mortgages above almost all other loan collateral. However, fixed assets need careful management to ensure their continued functionality in the face of rapid changes in demand.
- Property exists and becomes available in large increments with relatively high capital values of the basic units. Therefore key decisions are normally made with a long-term view, and significant barriers to market entry exist. Of recent times,

certain markets have dealt with such characteristics by establishing unit trusts and syndicates, in part to reduce entry thresholds.

- Capital investment in property (e.g. additional expenditures) merges with the asset itself. The additional investment becomes indivisible from the original asset, unlike in other classes of competing investments such as share portfolios.
- Real property assets tend to be inputs in other forms of final production. That is, in economic terms, they represent the land component of production. Consequently, with variations required for owner-occupied housing and for certain publicly owned assets, functionality and the ability to earn net income are important measures of performance.
- The typical investment profile of an income-producing real property includes an upfront, large-scale, lump-sum acquisition cost. The benefits and returns typically extend well into the future, which is very significant in terms of risk exposure and acceptable rates of return.
- Considerable and variable legal and statutory controls pertain to holding, managing and dealing with real property. Examples include restrictions on tenure, dealings, ownership, uses, environmental and economic impact assessments, and taxation provisions. Time parameters are very important because of the long-term nature of property assets. These assets are dynamic, and success requires that they evolve in response to changing demands, preferably in an incremental rather than a radical manner.
- Significant properties typically have multiple stakeholders that may include owners, financiers, managers, tenants, occupants and the community as a whole. All these stakeholders have different demands and required outcomes. Often the property is their only commonality. For long-term success, the asset must meet the legitimate expectations of all of these stakeholders.
- Property assets are usually managed under fairly conservative paradigms of behaviour, decision-making processes and detailed investigations, particularly by financers. This conservatism is reinforced by the typical large-scale and long- term nature of these investments and the level of debt funding that is often involved.
- Cash flows for investment and development properties are complex and require close management using quality systems.
- With astute management, an owner can improve the value and risk profile of a particular property asset in comparison with similar assets and the wider investment market.
- Structured strategic and operational management of 'the asset-in-use' is critical for success.
- In dealings between owners and tenants it can often be observed that the relationship varies over time – particularly in non-residential subsectors. Through most market phases, a prospective tenant will typically have a range of options and will negotiate accordingly. However, once the commitment to lease is made and the tenant starts operations, the power balance often changes. If the business is successful, the 'site goodwill' that the building/location offers will grow and there will be significant cost and business disruption if that business subsequently relocates. Consequently, tenants can have tangible and intangible 'sunk' costs that, all other things being equal, will predispose them to remain long term.

5.6 Income-producing properties

Property assets can be categorised, considered and analysed in a variety of ways: frequently by location (e.g. all properties within a region or town) or by use (residential, commercial, rural, industrial, et cetera). They can also be classified according to whether they are held to generate income, through business activity directly by the owner, rental secured or by development activity.

This latter classification is useful if one is attempting to discover the primary drivers behind the holding of that property, to assess its value and, thereafter, to make reasoned predictions about its future prospects.

With the exception of owner-occupied housing, government assets and assets held by not-for-profit groups, land is generally held to derive income of one sort or another: office buildings, tenanted residential properties, industrial properties, rural lands and development sites are all examples.

In the case of properties held to derive rental income, several underlying investment principles can be identified. In this normally rational, analytical market, buyers and sellers typically focus on the same financial 'bottom line'. Therefore, values secured for particular properties show fairly consistent market responses. For example, a strong relationship between the price that a purchaser would reasonably be willing to pay for the asset and the asset's ability to secure income now and into the future emerges in the analysis of comparable, completed sales. This relationship is called the 'price-to-earnings', or 'p:e', ratio.

The owner ultimately judges the performance and success of a particular asset by its secured capital value (i.e. the market value achievable if the asset was offered for sale). The basic criteria of income-producing assets are, therefore, 'marketability' and 'comparative performance', the latter being the asset's performance compared with other investment opportunities. Despite the complexity of markets and income-producing investments, be they in real property or elsewhere, capital value has only two fundamental determinants. The first is the ability to earn net income after accounting for outgoings. The second is the reward for the risk taken in securing that income. This relationship can be illustrated diagrammatically and described by an equation called the 'capitalisation formula' (Figure 5.3). The triangular shape exemplifies the correlation and interdependence of each of the three elements.

The capitalisation formula should not be seen as an abstract concept. Rather, it simply applies general theory to the decision-making processes typically employed when making an investment of any sort. If, for example, an individual had $10,000 to invest for a certain period, the investor might consider depositing the money with one of the large banks. The risk associated with this option is minimal but, as a consequence, the return on the investment in the form of interest (i.e. the net income) will almost certainly be small in comparison with that from other investment options. While investment in, say, shares for the same period may offer potentially higher returns, obviously the risks are higher and the outcomes are less certain. Depending on the investment choices made, even the original sum invested could also be at risk. This simple example illustrates the interdependence of income, risk and capital value, and the trade-offs between them.

The equation shows the direct relation between net income secured and final capital value. Obviously, all other inputs being equal, the higher the income generated, the higher the capital value that can be asked for the asset if offered for sale.

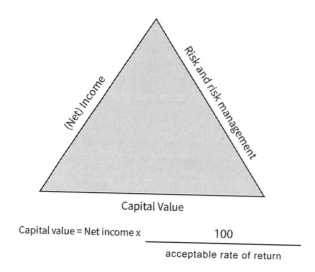

Capital value = Net income x $\dfrac{100}{\text{acceptable rate of return}}$

Figure 5.3 The capital value relationship.

The acceptable rate of return is ascertained by comparison with other investments in the market that have similar risk profiles and the opportunity to attract investment. Note, in the equation above, the capital value is inversely related to the acceptable rate of return (i.e. the lower acceptable rate of return, the higher the resultant capital value of the asset). This relation is not unexpected. If therefore, on analysis, a prudent investor requires a return of, say, 6.5 per cent from a property investment, then the capital value of that property would be greater, again all other things being equal, than that of another property with higher risk for which prudent investors would require a return of, say, 7 per cent.

As simple as this equation may appear, it represents the core proposition for investment in income-producing property and, indeed, most other investment vehicles. Even the most complex of investment and development models and structures are essentially based on the concept of the capital value 'benchmark' and the associated equation. The only key component not directly considered in this construct of capital value is that of time and time-preferencing. These temporal factors recognise that events in the investment and development process occur at different times, which obviously must be accounted for in projections of cash flow and in considering risk profiles. While this complicating factor could be incorporated into selecting the acceptable rate of return – these cash flow projections, in practice, are normally supported by more complex analysis.

Financial models such as discounted cash flows (DCFs) provide much more accurate understanding and predictions. These matters are elaborated on in Chapters 10 and 12.

Overall, the capital value equation does represent the critical underlying relations in commercial investment. Any investment, including investment in real property, is as simple – and as complex – as that. Assessing all the factors discussed above before making an investment decision is critical for success and the proper management of risk in the short, medium and long term.

Finally, it may appear that this analysis, based as it is on income-producing potential, does not adequately reflect the other key objective of property investment – the potential increase in the capital value over time. Certainly, if a property is not at its highest and best use or is held as a short-term, speculative venture, then its income earning potential may be of passing ('holding') interest. In such cases, the consideration of current rental incomes may be of marginal relevance. For all other investment properties, however, market expectations and/ or opinions regarding potential for capital growth are reflected in the capitalisation rate to be used.

It needs to be remembered that the capitalisation rate applied is established through a comparison with analysed, comparable sales. In arriving at the capital value they are willing to pay for acquisitions, the purchasers of those comparable properties have reflected on a range of factors including the potential for capital growth of the asset over time (because of expected rental growth, the improving character of the locality, et cetera). Those with better potential for capital growth will, all other things being equal, attract a lower capitalisation rate and therefore higher capital value. As noted above, more sophisticated analysis and modelling will be undertaken to refine all of this in the case of large, complex properties, but the key elements of income potential and reward for risks that have to be managed to achieve required outcomes remain fundamental.

5.7 Holding and dealing with real property

As previously noted, across the various disciplines involved with real property, confusion can arise because of often quite different terminology in common use. In general use, the term 'real property' is taken to mean land and things permanently attached to the land: buildings and other structures, ground improvements and vegetation. An economist might describe these same assets as 'land' or 'land resources' to differentiate them from other production factors such as labour, capital and entrepreneurial skill. In a legal context, terms such as 'real property' or 'real estate' are used to identify 'land and all things that attach to it' and, simultaneously, to differentiate it from what is described as 'personal property'.

Consequently, a building or other structure, vegetation or a piece of fixed plant (e.g. air conditioning equipment installed and integrated into a building) is normally defined and considered as real property or real estate. Other (usually movable) private property such as cars, loose furniture, computers, household or office equipment are variously described as 'loose plant and equipment', 'chattels', 'personal effects', 'tools of trade' and a number of other collective terms. These items are still subject to ownership and contract and are protected by law, but they are not 'real estate', nor 'real property'.

With those definitions in mind, the following observations may be made regarding the ownership, management and use of real estate assets.

Chapter 7 discusses the Australian tenure system, noting that land is usually held under freehold title or Crown leasehold tenure by any 'legal entity'. A legal entity is defined as an individual or individuals, a company or some form of trust arrangement (i.e. an entity holding real property on behalf of others). All these arrangements are long-standing and well-recognised. For example, almost any individual can hold title alone or as 'joint tenants' (e.g. with a marital partner) or 'tenants-in-common' (e.g. with a business partner/partners). Likewise, a registered company (private or public) may own real property, for example, a commercial or industrial building, either for its own use or investment. Similarly, a trust might be established and hold real property for a range of reasons: under the terms of a will or other document or order (a 'testimonial trust'); as an arrangement to manage the affairs of some incapacitated person, minor or other person unable to do so for themselves (a 'beneficial trust'); as a trust to hold and administer the superannuation funds of contributors; or, as in many trust arrangements, to aggregate funds and purchase assets for the benefit of contributors ('mutual funds'). A wide variety of trusts can be established, the specific nature, purpose, duration, dividend distribution, et cetera, of which are set out in the Trust Deed (i.e. the original legal agreement that set up the trust in the first place).

Given their diversity, complexity and the fact that they invariably involve the holding of the assets of others, both companies (particularly public companies) and trusts are closely regulated by the Australian Government through organisations including the Australian Securities and Investment Commission (ASIC) and the Australian Prudential Regulation Authority (APRA), and by a raft of company and business legislation and regulations.

Particularly strict provisions apply to matters of full disclosure, the content and nature of prospectuses, the role of the asset manager and the fee structures when the legal entity – an investment trust, investment or development company or investment bank – calls on the public to make equity investment. Although the Australian regulatory framework is considered as one of the best in the world, no framework can fully protect against illegal or criminal activities; nor can they 'legislate' against poor management decisions. As with any other investment vehicle, companies or trusts simply provide a legal structure by which investments can be made and managed. While offering certain advantages, they certainly do not guarantee success.

These arrangements are, of course, not confined to real estate holdings. Many trusts and other structures have a range of investments across a balanced portfolio of shares, government bonds, cash and 'near-cash' in bank deposits, as well as property investments. Even within these sub-groups, other choices are available to balance the portfolio, optimise returns, manage risk and meet the specific requirements of the trust deed or company charter. Some trusts, however, are principally established to invest in real property. Sometimes, trusts invest in a specific property or properties or, alternatively, pool funds and spread investments across a range of properties. They may invest directly with other funds or provide debt funding to others investing in or developing real property – so-called 'mortgage-backed securities'. Other variations such as investment funds, syndications, securitisations and real estate investment trusts (REITs) are described in Chapter 3.

The tenure systems and ownership options described above allow both companies and trusts to hold and deal in property. This opens a range of opportunities to invest, hold and deal in property well beyond the common ownership of, for example, a simple residential property by a single person or couple. Because public companies, investment banks and certain trusts can secure funding from various sources, including the wider public and 'offshore', they can aggregate huge amounts of investment capital, far beyond even high-wealth individuals. Unsurprisingly, therefore, most major commercial property assets (e.g. shopping centres, office buildings, major development projects) in countries like Australia are owned by, or through, these entities or vehicles.

Reasons for the emergence of these financial vehicles and some of their potential advantages for investors include:

- Companies, trusts, mortgage-backed securities (through investment banks) and similar entities can be viewed as bringing together equity and financial markets and the property sector. Thus even small investors can buy small amounts of equity in the form of shares in the company or units in a trust. These investors can then access large-scale assets and portfolios and other very sophisticated funds and property management entities to which they could not otherwise aspire.
- These entities provide investors levels of liquidity (i.e. by the sale of shares in a company or units in a trust) that are not normally associated with individual property investments.
- Depending on the structure of the entities, such arrangements may have considerable taxation advantages.
- The entities provide new ways for investors to access equity or debt funds in the property and development sectors. These opportunities are open to small-scale investors as well as major superannuation funds and corporate investors.

Like any structure or strategy, these investment vehicles provide certain advantages but may also involve certain issues and potential risks. For example, property typically represents a long-term investment that, by nature, requires long- term commitment and confidence. Exposure to the volatility of the share market, either through shares in the holding company or through a unit trust structure, is often not easily accommodated – as was starkly demonstrated by the financial events of 2007–2008. Specific fiduciary and corporate responsibilities are also required of certain structures (e.g. maintaining a balanced portfolio, maintaining certain levels of liquidity, overall risk management and responsibilities to report to the market / unit holders).

Over recent decades the Australian property market has matured, become larger and more sophisticated and produced many large-scale, complex property projects and assets. To fund and control these projects and assets, a wide variety of ownership and financial structures has emerged. Some have been particularly developed to accommodate property investments; most, however, are structures that can fund a varied portfolio of shares, bonds, property and other assets.

An investor can secure an interest in an asset, including real estate, in two fundamental forms: an equity position, in which the investment and involvement result in ownership of part or all of the asset; or a debt position, where the investment takes the form of a loan or some other financial underpinning to the legal entity that has ownership or equity. Both positions have benefits as well as potential drawbacks. For an equity holder, risk can potentially extend past the limit of the original investment but, at the same time, all excess profits and benefits can accrue to the equity owners.

Many structures can be established, and it is beyond the scope of this text to describe all that involve both the property and equities markets. Some general structures that are commonly encountered, including syndication, investment and development companies, investment banks and non-bank financial intermediaries (NBFIs), securitisation arrangements and listed and unlisted unit trusts, are discussed further in Chapter 9.

5.8 A changing approach to property assets and their management

Like most parts of the Australian economy and community, property and development sectors have undergone quite radical changes over recent decades. Prior to that, property and development was often viewed as driven only by 'bottom line' profit, short-term gain and opportunism, with little apparent interest in non-financial outcomes such as environmental impact or social and community expectations. While these may well be unfair generalisations, it is true that the sector often suffered from a perceptions of a self-serving attitude and narrow focus.

From there, arguably two, interrelated paths have emerged over much of the recent years.

One is very positive. The sector claims to be a corporate and community leader in issues of sustainability, environmental health, workplace development, ICT facilitation and also provision for changing location preferences. Growing sophistication, education and research within the sector, and the increased number of large, publicly listed corporations and funds involved, have all required high corporate and ethical standards. (These matters are further examined in Chapter 15.)

Overall, the demand for fixed assets such as real property is radically changing. Long-term financial success now demands a rethink of the 'property offer' to the tenants and users of these assets. Historically, holding non-current assets such as real property was virtually seen as a 'merit good' in its own right. However, in the new global and highly competitive markets, property, like every other component of business, must prove its worth and deliver specific outcomes.

Fundamental to this rethink has been a concept known as 'stakeholder analysis'. This concept recognises that any built asset has a number of stakeholders, all of whom have legitimate requirements and expectations, and that their loyalty cannot be guaranteed, particularly in the longer term, if those expectations are not met.

The primary stakeholder is of course the owner(s) and their representatives and financiers, but others include builders, tenants and managers of the facility, service providers, occupants and the wider community. Stakeholder analysis correctly observes that, although an understandable priority is given to securing rental income for the owner, real property is a long-term asset and, if a property does not adequately meet the demands and expectations of any stakeholders, they will withdraw their support. Consequently, the 'bottom line' requirements of the owner will suffer, perhaps irreversibly. The aspects of stakeholder analysis are outlined in Figure 5.4. In other words, and as reflected in earlier observations in this chapter, the approach to the ownership and management of real property assets has now become highly sophisticated and more holistic.

In summary of matters presented in this chapter, contemporary management of property assets rests on a number of key principles:

• The entire management environment (economic, business, social and community) is significantly different from that even a decade ago and realignment is essential. Fortunately, contrary to the traditional perceptions of 'unchanging bricks and mortar', buildings can, and indeed do, change and evolve. Many activities by owners and managers relate to the identification of changes that are required, and the need to continually work to re-align their buildings to meet those demands. Thereby they improve both the competitiveness and the marketability of their assets.

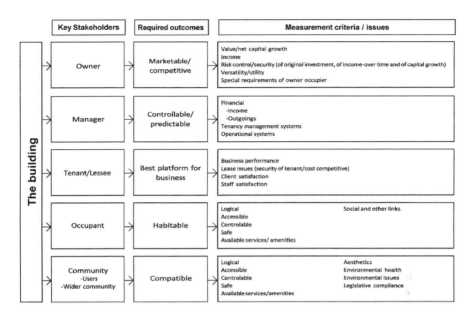

Key Stakeholders	Required outcomes	Measurement criteria / issues
Owner	Marketable/ competitive	Value/net capital growth Income Risk control/security (of original investment, of income-over time and of capital growth) Versatility/utility Special requirements of owner occupier

Manager	Controllable/ predictable	Financial
		  -Income
		  -Outgoings
		Tenancy management systems
		Operational systems

Tenant/Lessee	Best platform for business	Business performance Lease issues (security of tenant/cost competitive) Client satisfaction Staff satisfaction
Occupant	Habitable	Logical Social and other links Accessible Controlable Safe Available services/ amenities
Community -Users -Wider community	Compatible	Logical Aesthetics Accessible Environmental health Controlable Environmental issues Safe Legislative compliance Available services/amenities

(The building)

Figure 5.4 Stakeholder analysis – competing interests in a property asset (commercial property example).

- Given that real property is a long-term asset, investment decisions and management need to take a holistic approach and consider the short-, medium- and long-term implications of even minor changes.
- Real property provides a service (i.e. accommodation) and its value is typically derived from use and function. Consequently, physical, functional, technical and/or legal obsolescence represents a primary and continual threat.
- Property is a dynamic asset and must be considered as 'live' and evolving over time. Therefore, adaptability and reasonable flexibility in design and management are critical in order to meet compounding demand and environmental changes.
- As a general principle, the function of an asset (i.e. what it does and what it aims to achieve) is more important than the form it takes.
- Despite the inherent complexity of the property market and property assets, emphasis must be placed on simplicity and functionality of the asset, always focussing on the basic market and investment parameters rather than sophistication for its own sake.
- Importantly, the asset must not be seen as an end in itself; increased emphasis should be given to 'people and place' to reflect changes in demographics, social structures and the business environment. Thus property should create a 'living and integrated environment'. Rather than providing merely a passive backdrop for residential purposes or a business platform for companies, property must be an active 'participant'. It should value-add, promoting desirable lifestyles or, for businesses, creating a workplace that encourages productivity, reflects the ethos and image of the company, and links both of these to the physical environment and the contemporary corporate and community objectives such as sustainability (Week 2002).

- Considering the above, commercial property management involves much more than 'selling place, space and time'.
- The complexity of many of these assets demands sophisticated, continual and incremental 'fine tuning' of management systems.
- While overall portfolio or corporate guidelines are often important, each property is unique and therefore property-by-property strategies and plans are required.
- A stakeholder analysis approach, as outlined above, is an essential part of a contemporary approach to the ownership and management of real property.

Chapter 6

Legal and government parameters

6.1 Introduction

Early chapters in this text have emphasised that a sound knowledge of real property also requires a detailed understanding of the physical, economic/financial, legislative and community environments that surround and influence those assets.

For any parcel of land, one of the key components is the framework provided by legal and governmental systems in place in that particular locality. This chapter provides an overview of those components, their role and operations and how they affect the holding, use and dealing with property assets. Paralleling those earlier chapters, substantial references are included to facilitate further, more detailed study in specific areas.

6.2 The nature and origins of law

Laws are the rules and operational arrangements by which a community, society or other body organises itself. They establish required standards of behaviour and dealings by and between members of such groups.

There would be general agreement in an advanced, democratic, liberal society, such as Australia, that these laws would:

- protect individual rights
- balance those rights with the individual's obligations back to the community as a whole and to other individuals within it
- provide a framework for how government (and other 'community structures') should operate.

This understanding is usually referred to as 'the Rule of Law'.

Laws should provide security and protection for individual citizens and their property (i.e. possessions). This, in turn, creates a level of certainty in commercial dealings and investment and, across the community, promotes stability, equity and well-being. Good laws also provide a basis for community organisation, the efficient provision of services and (depending on the circumstances) protection against extreme physical and economic events. How successful laws are in achieving these, sometimes conflicting, outcomes is a matter of most political debates (and, indeed, election outcomes).

Laws and their attendant support mechanisms and organisations are continually evolving, influenced by emerging issues, changing community standards and the policy directions by government.

The legal system in Australia is historically and philosophically based on that of England, which evolved over many centuries (Hinchy 2008). Following the establishment of the Australian Federation in 1901, and increasingly in recent decades, Australia has taken a more independent stand from the influence of Britain and British law. These trends reflect the evolutionary nature of the Australian legal system and, in many important areas, it is now quite divergent from systems and structures in Britain and other countries, such as the US and Canada, that also have their genesis in British law. At the same time, there is close alignment with legal systems in New Zealand (Meek 2008; Parkinson 2010).

All of this should be seen as normal and beneficial as Australia sets an individual course in its own environment.

There are three primary sources of what would normally be defined as 'law', outlined as follows (Meek 2008):

Common law and equity

The common law is that body of law which developed in Great Britain over many centuries and brought to Australia by the first and subsequent settlers. Originally it was founded on 'community standards'; that is to say, that there was a general understanding across societies that certain things were right and acceptable and other things were not (Meek 2008). For example, even if there were not written laws (e.g. the Criminal Code), a consensus view existed across society that unprovoked assault or stealing another person's goods was unacceptable. Many components of common law related to the rights and obligations of individuals and between society and those individuals. As time passed, the laws were committed to formal written form. For example, the Magna Carta, issued in 1215 as a written proclamation, was the first great charter of English Law. It established the doctrines of the right to a fair trial and that there be no arbitrary taxation by the King and nobles.

In due course, those rules were developed into formal law (statute) by parliament – the Civil Law (related, among other things, to property rights and commercial dealings) and the Criminal Law. Subsequently, these were interpreted by the courts as to the meaning of particular laws and words used (i.e. 'statutory interpretation') and on the basis of cases that had been previously been considered and determinations made (i.e. 'case law' and 'precedent').

Over time, a further branch of the law, known as 'equity' also emerged, also dating back from medieval times. This philosophy revolves (as the name would suggest) around concepts of fairness and even-handedness.

Equity is different from the other components of the law identified above. As courts applied both common and statute law to specific cases and disputes, they found it necessary to also apply a basic 'fairness test' to ensure strict application of laws to a case, provided equitable outcomes. For example, where there is conflict, the maxim 'equity shall prevail' applies – that is, equity principles are applied as opposed to operations of other branches of the law.

Underlying equity principles provide something of a check or 'fail-safe' to courts. This does not imply that the specific laws are in some way weakened. Rather, it guards against perverse or unintended outcomes from the application of those laws. They therefore provide something of a check or 'fail-safe' and may also provide grounds for appeal for parties aggrieved by an administrative or judicial decision.

All of this also enshrines the requirement of basic fairness in dealings by government, between citizens and where matters of trust and conscience arise. Thus, in the contemporary environment, many pieces of legislation particularly reinforce concepts of equity with specific reference to required process and procedure, provisions for objection, mediation and appeal, establishment of the role of an ombudsman and the outlawing of 'unconscionable conduct'.

Statute law

The second source of law is the most obvious: statute law (also known as legislation). These are the laws, rules and regulations established (depending on the area of responsibility) by the Commonwealth government or a state government (called 'an acts of parliament'). The power to make these laws originates from the respective state and federal constitutions (see below). In closely defined circumstances, these rights to establish and enforce rules can be transferred or delegated to other bodies. The role of local authorities to control certain activities in their areas, as provided for under state legislation, provides a common example of this.

Statute law covers a huge variety of laws, rules and regulations ranging from the criminal code through to commercial, property and administrative law and the rules governing civil actions (called 'torts', i.e. interactions and conflicts between individual citizens).

Because of the complexity of contemporary societies and economies such as Australia, the need for laws, rules and regulations to reasonably manage and control activities increases. There is, of course, the continuing debate as to whether the community and businesses are in fact 'over-governed' (i.e. that there are too many rules); however, that aside, there is no doubt that the volume and complexity of law is increasing at a significant rate.

The Courts

The third source of 'law' comes from what is known as statutory interpretation and precedent; that is, a build-up of knowledge, history and experience as courts interpret statutes and common law in adjudicating specific cases brought before them.

Statute law normally can provide only general rules and/or direction and parameters for action. The application of statute is left principally to the government of the day as administrator, and to responsible courts or tribunals (and subsequent appeal mechanisms). Those bodies interpret and apply statues to the particular case in the event of dispute or lack of clarity. In making these determinations, the presiding judge or member will normally provide a written decision to explain his or her reasoning in coming to that particular decision and interpretation. That decision thereby adds to the 'body of law' to assist in future cases (i.e. providing 'precedent').

Obviously, all other things being equal, the more senior the court, the more weight or credibility will be given to that judgement in later matters. The time sequence of decisions and any intervening legislative and/or statutory changes are also important considerations in the application of precedent.

6.3 The development of law and government in Australia

Exploration and visits to the Australian mainland and offshore islands had occurred for several centuries before the political and legal 'claiming' of the Australian continent occurred in August 1770 on Possession Island in Torres Strait. There, James Cook claimed the area mapped for King George III of England and called it 'New South Wales'.

Though politically motivated to warn off the interests of the French and, to a lesser extent, the Dutch, the legal basis for the claim was that of *terra nullius*. That is, the land that Cook had identified and partly explored was (on the basis of Cook's reports and the conventions of the day) demonstrably not controlled or settled by anyone else and was therefore, to all intent and purposes, unused and vacant. It was, therefore, available for claim. Of course, this concept, which applied to the Australian continent in August 1770, was successfully challenged 200 years later in the Mabo and Wik cases, though only to the extent of securing Native Title rights over, and interests in, un-alienated land – a separate issue from the topic discussed here (Bradbook et al. 2011).

From 1788, six separate settlements and colonies were progressively established on the Australian continent, effectively fully owned and administered outposts of Britain. This was done for very pragmatic reasons: to locate convicts, obtain a politically strategic position for Britain in the Asia-Pacific region and later, to secure commodities including wool, wheat and gold. All of this needs to be considered in the context of the mercantile and colonial period and the political alliances and hostilities among European powers through the late eighteenth and early nineteenth centuries.

With British settlement came British law, administration and social order. Progressively, through the middle and latter half of the nineteenth century, the British government granted self-government and their own constitutions to each of the states. Importantly, the constitution for each state transferred practically all law-making powers to that state (i.e. 'laws for the well-being and good government of the citizens'). Over time, even repugnancy provisions to British law and the rights of appeal to British courts were effectively abandoned (Meek 2008).

Australia's successful emergence, first as independent states and then as a Commonwealth, without the need for revolution or war remains as a very significant achievement. In a practical sense, by the mid-eighteenth century, the need and ability to 'dump' convicts had diminished and the cost of running far-flung colonies was becoming prohibitive for the British government. In any case, British pastoral and mining companies by then largely controlled the economic opportunities that had emerged in Australia. Overall, the British government was happy to facilitate a political exit while maintaining economic benefits and six independent, but staunchly loyal governments.

While the benefits or detriment of the British colonisation and its later role in the settlement and development of Australia are the subject of continuing political debate, the legacy of the constitutional, democratic government and an independent, structured and learned legal system was critical in Australia's subsequent success and relative peace and harmony.

The system was, and continues to be, democratic and based on a capitalist economic framework. State parliamentary structures and administrations were modelled on the British Westminster system with some necessary modifications: an appointed Governor to represent the monarch; a two-house parliament with a Legislative Assembly (the Lower House) and Legislative Council (the smaller, Upper House of review), comprising members elected through plebiscite (originally for three-year terms); a system of executive and administration (ministers and departments) to run the operations of government day-to-day; and, finally, a separate and independent legal system (judiciary). The leader of the political party that enjoyed majority support from the members of the Lower House was appointed Premier: effectively the political leader of the government.[1]

Each of these groups – the Governor, parliament, the judiciary and the executive and administration – has roles that, while obviously sharing boundaries, are quite separate, with differing functions, obligations and areas of responsibility. This arrangement, known as the 'Separation of Powers' (i.e. the freedom of each of these components to do its part without undue or improper interference from any other section), is critical for the effectiveness and balance of the whole system.

It is important to note that, because of the vast size of each of the Australian states and their scattered populations, governments tended from the earliest times to be heavily involved in the provision of services and infrastructure. This is quite different from the situation that developed in Europe and particularly in the US over the same period. Though there are regular (and sometimes justified) claims that government in Australia is 'too big', there is also a general acceptance that it is often appropriate for government to take a leading or interventionist role to facilitate major physical or economic initiatives which, in the US, for example, would be the domain of the private sector. The construction and operation of railways and, in recent times, the construction of the National Broadband Network (NBN) in Australia provide good examples, but there are many others.

Through the latter half of the nineteenth century, the newly independent Australian states developed economically (based largely on agricultural production and gold), but with surprisingly limited inter-relations between the states: each seeing itself as much a direct satellite of Britain as an associate with the five other independent states. (The overall pattern of development and the infamous decisions on different railway gauges typify this approach.)

Nevertheless, (and after protracted debate) there was finally sufficient interest by 1901 for the states to join a federation, the Commonwealth of Australia. The Commonwealth was established under a separate Australian Constitution, again with a non-political appointed representative of the monarch (the Governor-General), two houses of parliament (House of Representatives and Senate), executive and administration and a separate judicial system (Hinchy 2008).

Several key points need to be made about the Constitution and, indeed, the whole philosophy that established the Commonwealth.

In the first instance, the powers transferred to the Commonwealth were quite limited and defined (Hinchy 2008). They principally related to the full responsibility for external affairs, defence, communication and currency, and the obligation to ensure free trade between the states. Revenue to fund the Commonwealth's activities was to be secured largely from customs and excise duty, and political and financial balances were put in place to protect the rights of the smaller states. Some powers were to be shared; however, all other powers were residual – that is, retained exclusively by the individual states. This included continued responsibility for the provision of most services, the criminal justice system, land and tenure management and so forth. Most important of all, the significant powers of taxation that the states ceded to the Commonwealth in the Second World War were never given back. This resulted in, for example, the establishment of the Loans Council and other mechanisms where the states had to go 'cap in hand' to the Commonwealth for funds.

Within the Australian Constitution, too, several provisions later proved critical. One was the 'External Affairs Powers' provisions (under Section 51), which gave the Commonwealth an exclusive mandate to enter into international agreements, treaties and the like. Once any such agreement was established, the states were compelled to conform to the provisions of that agreement, even though it would have otherwise been a matter exclusively within their jurisdiction. The decision of the High Court of Australia in the 1987 Gordon–below–Franklin case exemplifies that interface. In that case, the court confirmed that, because the Commonwealth Government had entered into an international agreement (the World Heritage listing of this wilderness area), the Tasmanian government was prohibited from undertaking infrastructure and land use activities that otherwise would have been within their direct power and responsibility.

Another provision (Section 96) allowed the Commonwealth to issue Special Purpose Grants to individual (or a number of) states. Over time, these grants were used by the Commonwealth with conditions requiring certain actions or activities by the states before the grants were given – in this way, the financial power of the Commonwealth often proved dominant over the legislative strength of the states.

Section 96 powers became a significant issue during the Whitlam government (1972–1975) when local governments and regional initiatives (such as the Albury–Wodonga growth centre) were directly funded from Canberra, effectively bypassing the states which, under the residual powers of the constitution, would normally have had full jurisdiction over property and development matters.

A third key area in the interface between the Commonwealth and the states are known as the 'repugnancy provisions' within the Australian Constitution. These provisions established that, where a conflict or difference existed between Commonwealth and state laws, and where both had legitimate interests, Commonwealth law would prevail to the extent of the disagreement (the extent of 'repugnancy'). Based on these parameters and the evolution of government activity, almost 120 years since federation are characterised by the rise in political and economic dominance of the Commonwealth Government over the states, far beyond that which could have been conceived in 1901. For example, the Constitution envisaged that the Commonwealth Government would fund practically all of its activities by an exclusive right to levy customs duty and excise. These receipts now represent less than 9 per cent of total Commonwealth tax revenue.

As well as the observations regarding the Constitution itself, other forces further strengthened the increasingly dominant position of the federal government – at times almost despite the division of powers and responsibilities laid down in the Constitution. These forces included:

- major political/economic events: the First World War, the Great Depression and (particularly) the Second World War where the federal government assumed very wide powers (and largely stayed involved even after the causal event passed)
- the rise in the complexity and involvement of the role of government, particularly in the economic activity of the federal government
- related to the above, most of the new and growing avenues for taxation revenue (e.g. income tax) were accessible by the federal government rather than the states,

while the state tax base (principally property- and transaction-related) reduced in relative importance
- consequently, the Commonwealth increasingly controlled the tax revenue base and, through the distribution of those funds, was able, quite legally, to 'coerce' the states into actions that it (the Commonwealth) required. These actions may well have been outside the Commonwealth's constitutional control; for example, issues including health, education, land use, agriculture, industry and the environment. Furthermore, the use of the external powers to effectively compel the states to undertake or to desist from certain activities that would have been otherwise matters of their state discretion emerged as important issues in these areas
- over more than 80 years, a succession of challenges to the High Court regarding the power and the role of the Commonwealth (principally with the states as appellants) have almost invariably found in favour of the Commonwealth (e.g. the 'Engineers Case' [1920] and Gordon-below-Franklin [1983]) — so reinforcing by precedent a more expansive interpretation of Commonwealth powers
- the (often) fragmented position of the states that almost invariably spend much of their time and energy arguing among themselves or making politic of key meetings, for example, Premiers' Conferences and Council of Australian Governments (COAG) rather than working cooperatively.

6.4 Federalism today

The federal system has evolved to an almost unrecognisable state/condition over the past 119 years.

The system continues to work in a reasonably democratic and equitable way, and the ability to maintain a fairly cohesive society with considerable wealth generation over such an extended period should not be underrated as an achievement in itself. Few other countries in the world could claim this fundamental success. This may be more the result of the adherence to the principles and basic ethos underlying the system and, perhaps obliquely, to a relatively benign and, sometimes apathetic, populous.

Nevertheless, the current federal system is ponderous, wasteful and inefficient.

Despite this and because of the simplistic resistance to constitutional change and the entrenched, powerful vested interests represented by the states, significant change is unlikely even in the longer term. The failure of the 2000 republic proposition, even though the principle appeared to have majority support, exemplifies these difficulties even in relatively simple matters.

Of more practical importance, the fiscal imbalance between the states and the Commonwealth, together with the differences in scale, rates of development and the comparative wealth of the various states, creates long-term instability for the federation. Increasingly, critics of the system would observe that the states themselves are anachronistic political structures with little regard to the economic, developmental, community or demographic characteristics of contemporary Australia. Instead of the states, the natural 'communities of interest' are Australia's 70 regions. From these could emerge a two-tiered system of government for Australia: the Commonwealth and the regions.

While these concepts have advanced greatly over recent decades and, now, for example, have some manifestations in the supply of utilities, et cetera, the strong forces against political change in Australia are unlikely to allow a full shift to a regional government approach, despite obvious benefits.

6.5 Government and judicial operations

To further clarify Australia's contemporary legal and governmental systems, some elaboration on the day-to-day operations of these systems will assist. These arrangements are fairly similar across the federal, state and territory governments and involve the following general processes and protocols.

In each jurisdiction, statute (i.e. law) is established (and from time to time amended) by parliament and becomes law when assented to by the Crown representative (e.g. the State Governor or, in the case of the Commonwealth, the Governor-General). The parliament then, in effect, passes over the task of implementation, application and enforcement of that legislation, together with financial appropriation (a budget) to carry out that task, to the executive and the administration.

The political party with majority support in the Lower House forms the executive government (the Premier or Prime Minister and ministers) to carry out these duties. That political head and the ministers are individually responsible to the parliament for the performance of these obligations. Each minister is responsible for an individual area of activity ('portfolio') and the administration of certain acts passed by parliament. He or she is supported by an administration (a government department) in carrying out those duties. The name 'minister' – as in 'administer' – reflects this role. Interestingly enough, in the Australian Constitution, the position of Prime Minister or his or her role was not even mentioned by those writing the Constitution – reflecting, no doubt, their view of the more important role ascribed to the law-makers themselves, rather than those who simply carry out the administration.

Parliament meets only periodically and, even then, has a heavy workload. Given the complexities of individual cases and the width and depth of government activities, much is left, within 'the spirit and intent' of the legislation, to ministerial discretion. Over time, therefore, there has been a significant power shift away from parliament and towards the executive and the administration. Additionally, executive and administrative powers were enhanced by the support provided to Ministers by a permanent and non-political public service. That group combined professional expertise and professional skills with long-term corporate knowledge of specialist areas and protocol and precedent that is invaluable in the effective operations of such complex tasks and operations. Though these powers have been tempered by contemporary demands for consultation and transparency, and through judicial and other review systems and the role of the press, the rise in executive and bureaucratic power continues in all Western democracies.

The manner in which the operations of government occur day-to-day is a mixture of the requirements of individual legislation (and the delegated powers given to ministers), the administrative and financial regulations of the government and a range of other protocols and conventions. Depending on those requirements, significant decisions are normally brought to the weekly meeting of the Prime Minister (or Premier) and the ministers (the 'cabinet'). Certain decisions will require the consent of the Governor or Governor-General in Council ('the Executive Council') and its publication in the

Government Gazette or similar (a weekly government publication where significant actions, expenditures or appointments, pursuant to various legislations, are made public).

Of course, in the application and administration of legislation, breaches and conflicts will emerge. Sometimes, these are the result of the government or the administration taking action against an individual or other legal entity; or sometimes, the reverse will occur where an individual is appealing against an action of government or some government agency, which that individual considers illegal, improper, unjust or undertaken without due process or power ('*ultra vires*'). Other disputes can also arise between individuals (i.e. civil actions or 'torts').

To manage these issues, a structured judicial system was established at both state and federal levels. Each of these courts or tribunals has its own area of jurisdiction, control and authority. Traditionally, the hierarchy of courts at a state level ranges through Magistrates Courts, District Courts to Supreme Courts and finally the Full Bench of the Supreme Court (with appeals possible to the High Court of Australia).

The Commonwealth system involves the Federal Court and the High Court of Australia and, eventually, the Full Bench of the High Court. A summary structure is shown in Figure 6.1.[2]

Over recent decades, the volume of litigation and its often specialist nature have greatly increased. As noted above, a range of other specific tribunals have been established at both the state and federal level to deal with those matters and to relieve the workload of the general courts. At the Commonwealth level, the Family Court represents an important example, and Administrative Appeals Tribunals, particularly related to tax and other areas of government administration and jurisdiction, are also in place (Parkinson 2010).

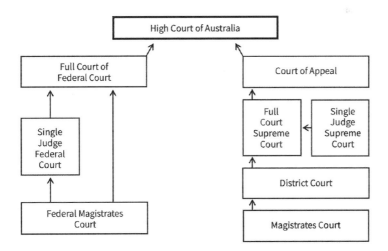

Figure 6.1 Overview of Court structure/hierarchy in Australia. (The diagram is generic and included as an example only, though such a hierarchy is fairly typical of Australian jurisdictions.)

Regarding the states, there are a range of specialist courts and tribunals in areas related to the environment, property (including statutory valuation, compulsory acquisitions, retail shop leases and the like), industrial affairs, mining, health, coroner's court, mining, small claims and many others. In state jurisdiction regarding property, there are typically two main areas involving judicial arbitration. One relates to real property assessments such as statutory valuation, compulsory acquisition and related matters. These are heard by specialist Land and Land Appeal Courts. The second area of involvement, typically named 'Planning and Environment Court' or similar, arbitrate in matters relating to town planning, development and environmental cases. Cases brought to any such court will normally be in the form of appeal against some government decision or proposal.

Such bodies are normally established as courts of 'equity and good conscience'. This means that, while substantiated evidence will always be preferred, hearsay and opinion can also be accepted. In hearings, expert witnesses (such as valuers and other professionals) will often play a critical role in what is known as 'friends of the court' providing independent and reliable investigations and opinions regarding the matter at hand (Parkinson 2010).

Within their areas of responsibility, these courts and particularly appeal courts will frequently represent final arbitration (save for higher appeals on points of law or due process). To relieve the time and cost of litigation, mediation provisions and referral to specialist mediators are becoming more common dispute-resolution mechanisms. Such avenues are collectively referred to as Alternative Dispute Resolution (ADR) and are generally proving to be of significant value in providing expeditious resolution of many cases.

6.6 Government, legal systems and real property

In summary, there is a range of definable items that, in law and in wider usage, are understood as 'property' (Tooher & Dwyer 2008). This chapter and book are principally concerned with land and/or building assets (often called 'real property' to differentiate it from other asset types). It should be noted, however, there are many other categories of property including: plant and equipment; vehicles; production stock and goods; furniture and loose items (often called 'chattels'); intellectual property; patents and so on. The concept of 'property' in relation to these types raises a range of laws, protocols and issues including possession, ownership, use, rights and protection. These vary dependent upon the type of property involved.

The power to create legislation to govern land tenure, to control the use and development of land and to compulsorily acquire it, is almost exclusively held by the states (i.e. 'residual powers' withheld by the states when the Australian Commonwealth and Constitution were established – see Section 6.3). The Australian Constitution did transfer certain, very specific powers regarding land to the Commonwealth. These allowed the Commonwealth to legislate for the tenure and development control systems within Commonwealth-controlled areas: that is, originally the Australian Capital Territory and certain other territories. The Constitution also gave the power to the Commonwealth to resume land (Section 51) provided that it was for a public purpose for which the Commonwealth had jurisdiction and acquired it on 'just terms'. The constitutional obligation to acquire on just terms falls only to the Commonwealth Government in its acquisition activities. No such limiting provisions apply to the states in their ability to take back land that had previously been alienated from the Crown.

The division of powers is, on the face of it, a fairly clear delineation. However, like many other areas of government and of commercial activity, the complexity of the contemporary political, community and economic/ business environment is such that complicating aspects exist.

It remains the case that the state governments maintain the tenure system and have responsibility for planning and development control legislation. However, as discussed earlier in this chapter, there are now a range of overlaying areas of Commonwealth interest which, while not as obvious or direct as state government controls, can be equally important. These include such issues as Native Title, environmental impact (e.g. climate adaption challenges, impact on World Heritage areas such as the Great Barrier Reef), Aboriginal affairs, equal opportunity and a range of other areas where the Commonwealth has either full or shared responsibility with the states.

Also relevant is the evolution of the taxation base as it affects real property and property assets. Traditionally property-based taxation (e.g. rates, land tax, stamp duty, et cetera) was revenue collected and secured by the states and their agencies. While those taxes are still significant, their relative importance within the 'wider mix' of Australian taxation has diminished considerably and, over the past three decades, the emphasis of the property tax base has shifted from taxes on ownership (i.e. state-based, 'ownership taxation') to taxes on income and capital gain derived from property (i.e. federal-based taxation). The introduction of capital gains taxation, clearer negative gearing provisions and depreciation allowances in 1985 and the introduction of a Goods and Services Tax (GST) in 2000 shifted the major impact of taxation in the property sector into the Commonwealth regime. These matters are further discussed in Chapter 14.

As noted earlier, the control and administration of tenure (i.e. how land is held, and how transactions and dealings are recorded) are under the jurisdiction of the respective state governments. These systems and tenure types vary from state to state, and details are further discussed in Chapter 7. Suffice to say that there are a number of general categories of tenure encountered:

- Alienated crown land made freehold (under the Torrens system) and owned by individuals/corporate entities (including community titles)
- Un-alienated crown land, typically land held/used for public purposes: road, reserves and so forth
- Crown Leases of various types and for various purposes to individuals/corporate entities
- Deed of Grant in Trust (DOGIT) lands which are areas set aside for specific, normally public, purposes and often administered 'in trust' by a third party, typically a local authority or similar
- Native Title claims and dealings as an overlay (where applicable) to these tenures.

A summary of Crown land tenures, holdings and lease types is provided in Figure 6.2 and a summary of freehold holdings is provided in Figure 6.3. Note that both are simplified, general outlines only and are based on tenure systems in the state of Queensland. Structures in other states will vary but will typically include similar elements.

These matters are further considered in Chapter 7.

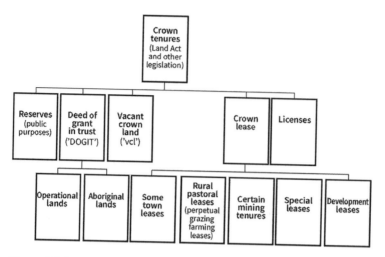

Figure 6.2 Indicative summary of crown tenure, holdings and lease type.

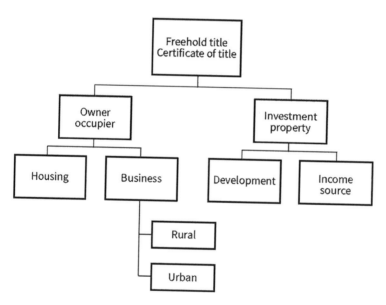

Figure 6.3 Indicative summary of freehold land tenure and uses.

Notes

1 In 1922, Queensland abolished its Upper House (the Legislative Council) and uniquely has a unicameral system of State Government (i.e. one house, the Legislative Assembly).
2 It should be noted that these are only the principal levels in the hierarchy and that there are many other tribunals, arbitrators, reviewers and specialist courts for a wide range of property, development, environment, health, corruption and industrial relations matters and other fields where judicial and quasi-judicial bodies have been set up and operate under relevant legislation.

Chapter 7

Real property

Tenure and dealings

7.1 Introduction

This chapter puts forward more detailed observations regarding the holding of and dealings with real property assets. General framework for this was established in Chapters 5 and 6, and these are advanced to consider more specific legal and operational arrangements with this asset class (both in the form of land and improvements). The complexity of these topics is recognised and, as with other chapters in this book, the objective here is to simply provide a sound basic knowledge of the study area, second, to provide a platform from which further enquiry can be made; chapter references provide details of more specific Australian texts, that provide specialised, supplementary knowledge.

7.2 Land tenure systems in Australia

The word 'tenure' means the way in which an asset (in this case real property) is owned, held and occupied. For any community, society or civilisation to claim order and stability, a clear and universally accepted system of land allocation and ownership must be established and operate effectively. History is replete with examples of wars and insurrections that occurred when these systems were not in place or accepted.

In Australia, tenure systems and their management are almost exclusively state-based. The basic philosophy and approach taken to such matters are comparably across all jurisdictions but there are also important differences, state-to-state. For the purpose of this text, only generic themes and underlying principles are presented. Over many decades, the legislation and administrative arrangements governing tenure matters have evolved differently in each state to meet local requirements. It is essential that enquiries be made, beyond this text, to ensure that the requirements of relevant legislation are understood and complied with as part of any specific actions or property dealings.

In the case of Australia, the land tenure system was established from the earliest times of British occupation and settlement and was based on British legal concepts of common law, statute and equity. In August 1770, Captain Cook, on behalf of the British Crown, took possession of the whole of the Australian continent – 'Terra Australis' – even though it was only partly mapped. He did this on the basis of a legal and political principle known as 'terra nullius', that is, Cook claimed that the land that he had (partly) explored was, to all intent and purposes, vacant and unoccupied and, therefore, available to be claimed.

Under the High Court's Mabo case ruling (1992) and the subsequent Wik case (1996), it was held that Aboriginal Australians were indeed already in possession of that land

and, therefore, had prior claim to it (Bradbrook et al. 2011). These aspects are further discussed in Section 7.6.

On the premise of the validity of the 1770 proclamation by Cook, all of Australia effectively became 'Crown Land' owned by the British government. Progressively over the next century, settlements and, subsequently, self-governing states were established under acts of the British parliament.

The states were given a wide mandate to provide '... well-being and good government for the citizens of that state'. Part of that mandate was reasonably taken to mean the organisation and management of land tenure systems.

In the first instance, but continuing to the present day, the Crown (i.e. the state government) provided leases of land to individuals, firms and organisations for various purposes. These included permission for the lessee to construct a dwelling, to farm, to site a business or other 'special' (i.e. more specific) purposes. In return for leases, the government would charge a rent normally assessed at a percentage of the market value of the land leased, excluding improvements. The period of the lease varied depending on its type, sometimes only year-to-year; however, as the lessees invested more and more capital into the improvement of the properties, more secure tenures were required. They extended from 10 to 30 years (sometimes with renewal provisions) and, eventually, certain leases were established in perpetuity – meaning in effect they were to continue forever, as long as any land rent charged was paid and other lease conditions fulfilled.

Some of these 'development' leases were (and continue to be) established to encourage certain types of land development by the lessee (e.g. to provide residential, industrial or rural improvements, buildings and investment).

Development requirements typically stated the type of improvements required, their value and timing and were included as conditions of the lease. In this way, the government encouraged and directed what it considered to be desirable private investment. These tenure arrangements (or variations) were sometimes used by government to encourage significant private investment around major infrastructure such as ports, airports and large industrial hubs.

These leases (and private leases based on freehold tenure outlined below) establish what is known as an 'equitable interest' in land. This term requires some elaboration as it represents the long-term legal underpinning of all tenure systems. As explained in Chapter 5, common law, statute (i.e. legislation established by Parliament) and, following those, interpretation by the courts established the legal framework for society and businesses to operate. This included provisions for the ownership and dealings with land and property. Underpinning all of this, however, is the legal principle or philosophy called 'equity', which dates back to medieval times.

As the name 'equity' suggests, this principle involves concepts of fairness and even-handedness. It is different from the other components of law identified above, and, where there is conflict, equitable rules override common law principles. As regards property and property interest, 'equity' considerations will be relevant in any dealing or in-court deliberations.

For example, if a person had properly taken possession of a piece of property (either through a lease or freehold), then that person's rights and expectations to use, further lease, or sell, must, in fairness, be taken into account and reasonably accommodated in the operations of common law and statute.

On this basis, underlying equitable principles provide something of a check or 'fail-safe' to courts (and grounds of appeal) – that there must be a basic fairness in dealings by government and between citizens. Thereby, what is known as an 'equitable interest' is established, which (say, in ownership of property) particularly 'flags' to all stakeholders that the interest or involvement described therein represents a serious and overt claim or involvement by that owner or lessee. Equity and fairness reasonably recognise those rights, and expect that such interests will be upheld, by court action if necessary, in dealings and the operation of both common law and statute. Note that in the contemporary environment, much of statute is, in fact, put in place to specifically reinforce concepts of equity. Aspects include detailed provisions for required process and procedure, compensation and disturbance, objection, mediation and appeal provisions, establishment of the role of ombudsman and the outlawing of 'unconscionable conduct' in dealings.

Over time, a range of targeted Crown lease tenures were established including Special Leases, Perpetual Town Leases, Perpetual Lease Selections, Grazing Farms, Grazing Homesteads, Pastoral Leases and Pastoral Development Leases. Progressively, and particularly in recent decades, many of these leases have been converted to freehold tenure (see below).

Of course, not all Crown land was leased to private individuals and companies. Large areas scattered across urban and rural areas were required for public purposes and these remained un-alienated, still vested in the Crown and forming part of the 'National Estate'. When a specific public use is established for a parcel of land, it is identified through an Order in Council and gazettal, as a Reserve, and that classification is attached to the land in both tenure records and cadastral maps. Therefore, say, the public hospital in a town will (normally) be located on a gazetted hospital reserve and shown on maps as 'Hospital Reserve Number #, under the trusteeship of the Minister of Health'. This status will continue in perpetuity or until another specific gazettal amends or cancels it.

There are very many examples of these public lands, from school sites to hospitals to parks, road areas, foreshores, environmentally sensitive areas, national parks and so on. In the case of many reserves, the state government may hand over the day-to-day control and management of the land to a suitable group – perhaps, in the case of local parks or roads, the local authority.

These delegations are organised under trustee arrangements (i.e. in the cases above, the local council is given the land 'in trust') to use and manage, but not ownership per se. At times, the state (or its trustee in the particular case) may not have immediate use for the land and, therefore, may permit a private individual to use it for a specific purpose. In these cases, some form of occupation licence can be arranged. Many examples of the temporary use of such lands exist. For example, a Natural Resources Department might give a licence to a farmer to graze cattle in a state forest or unused road reserves.

On the face of it, this does not create an equitable interest but, rather, a short-term right of use (i.e. the right to do something that would be otherwise illegal).

Sometimes, 'reserve' lands can, for a number of reasons, be issued with a freehold title, but in trust, which limits its use and dealings. In most states, these are called 'Deed of Grant in Trust land' (DOGIT) and, in the past, were used for aboriginal lands and not-for-profit owners including sporting clubs, private schools and hospitals. The deeds are useful because, within the terms of the trust, they can allow normal use and management on a more practical, business basis than quite passive Crown reserves. This type of agreement would normally, however, restrict the use of the land for unrelated purposes and prohibit on-sale.

Finally, regarding Crown land, there are areas owned by the Crown that are neither leased nor specifically reserved for a purpose. These are known as 'Vacant Crown Land' ('vcl'). Tenure and dealing arrangements are maintained as matters of public record by a relevant section of the state government which, depending on the individual state's structure, is sometimes known as the 'Titles Office', Lands Registry', 'Registrar of Titles' or the 'Department of Natural Resources'. These departments are typically also the custodians of the cadastral data base (i.e. all maps and surveys identifying those holdings).

Crown tenures continue to the present day and vast areas of central, northern and western Australia are held under Crown leases – principally pastoral holdings established under the Land Acts of the various states. Over recent decades however, except for reserves, the incidence and importance of Crown leases in urban Australia and more closely settled farming areas have been quite limited.

Australia is the only continent to be urbanised before it was fully explored. As urbanisation, closer development and investment in land occurred, there was a move in the nineteenth century to establish a form of tenure completely alienated from the Crown and, therefore, not subject to rent, specific lease conditions or other interference from government. A system, known as the 'Torrens System', first introduced in Adelaide in 1857, was progressively adopted by all state governments in Australia and, in fact, became a model for many tenure systems globally.

The system creates a land tenure called 'freehold' or 'fee simple' (Tooher & Dwyer 2008). It relies on the physical survey (i.e. identification) of individual parcels of land where, for each parcel, a unique, registered record of ownership is established. In the original 'hard copy' systems, there were two identical certificates of ownership (i.e. the 'Title') – one held by the current owner and the second by the Registrar of Titles (or comparable body within the state government), available for public search.

Under Crown lease and freehold tenure, ownership of the lease or title is available to any 'legal entity'. Legal entity includes an individual (or individuals), company or trust (i.e. a legal arrangement where, for a range of reasons, ownership is held by a legal entity under a Deed 'in Trust' on behalf of others).

Practically anyone with the ability to contract can hold real property in Australia. That implies, however that some individuals or groups (e.g. minors, the incapacitated, bankrupts and certain others) who cannot legally contract may also be prohibited from holding such property. Depending on the circumstances of the particular case and the prevailing legislation, the use of a trust and/or orders of a court may provide for appropriate mechanisms to facilitate dealings.

Furthermore, the ability of foreign nationals and companies to hold property in Australia requires the prior approval of the Commonwealth Government's Foreign Investment Review Board (FIRB).

When fee simple ownership is first established (i.e. first alienated from the Crown), it is known as a Deed of Grant; thereafter, any subsequent dealing, such as further subdivision, creates a Certificate of Title for each property created. Freehold ownership represents unfettered possession in perpetuity to the name of the owner as registered at that point in time on the title. The only specific limitation, as identified in a Deed of Grant/Certificate of Title, is the term 'save for the rights of the Crown'. This means that the state still has the right to legislate in ways that may affect that property's usage and retain the right to take back (resume) the land for some public purpose. The legal principles of equity discussed earlier are obviously relevant here.

The registration system is now computerised ('paperless') in most jurisdictions across Australia. In effect, that unique Certificate of Title represents in a legal sense that unique parcel of land. The name shown upon it (be it an individual, company or other corporate entity) is accepted as the legal owner and that entity may mortgage, sell, gift, bequeath (subject to certain exceptions of succession law not considered here), subdivide, lease or undertake a range of other activities involving that property. It is a remarkably simple system, proven to be sound and dependable – vital characteristics, given the critical role that property and its ownership play in society and the economy.

Freehold ownership related to land is identified by a two-dimensional subdivision. For example, a freehold allotment in a residential street is validated with a Certificate of Title which could be mortgaged, it also could be used for a site for a dwelling (which, in effect, would merge with the land to become a single asset), rented, sold and so on. This type of tenure arrangement represents the vast majority of holdings in Australia, particularly in urban areas, and is applicable to all land uses, subject to town planning controls.

Over time, new applications of freehold title were developed under what is currently known in most jurisdictions as 'Community Title'. These first emerged in a number of states in the 1960s but are now available in quite sophisticated forms across all Australian jurisdictions. This type of freehold tenure can be applied to much more intensive situations including unit and townhouse residential developments or commercial and industrial use, depending on particular zoning provisions. In effect this provided for the freehold titling of buildings (not simply the land that the building occupied). Because that freehold area (called a 'lot') was not necessarily connected to public access (as is a two-dimensional subdivided lot) – and indeed may be high up in an apartment building, not directly touching the ground at all – common areas for access and for other community and servicing purposes were necessary. These were established as part of the building subdivision.

These common areas and common property were managed by a separate legal entity called a 'body corporate', established by state legislation. A body corporate for a particular community title scheme is, effectively, all of the lot owners at any point of time acting as tenants-in-common for those common areas. The workings of these arrangements are reasonably complex and will not be elaborated here. Suffice to say, however, regardless of additional body corporate arrangements, lots under these agreements are identical in law to any other 'conventional' freehold parcel. Where such, new community plans and titles are registered, the original title (over the land) is extinguished.

The Registrar of Titles (under a variety of names, depending on the state government involved) registers all of these dealings as they occur and is an office of public record. Any person can establish, through a search of these records, details on any parcel of freehold land in that state, together with survey details to identify it. Similar registers of Crown leasehold interests are also available.

With this reliable tenure system in place, a whole range of other dealings became possible including and, importantly from a real property point of view, the ability to lease property to another individual for a set period of time. These arrangements are, in fact, contracts and, before considering the property implications of this, it is timely to consider, as a starting point, the meaning of the legal concept of 'contracts' (see Section 7.4).

7.3 Evolving Tenure Models

Over their long application, both the freehold and crown leasehold systems have proven adaptable and have been able to evolve to meet the changing demands of the development sector and wider community for more concentrated and urbanised land usages. Depending on the specific, enabling legislation of the individual state, these have included the introduction of 'volumetric title' – that is the ownership of 'air space' not directly attached to physical land. Some examples of that might include the establishment of property rights to construct, say, over public infrastructure such as rail lines and stations or over public roads in densely developed areas.

Other evolutions have seen further advances in community title structures where, for example, as well as smaller scale group or building title arrangements, an overarching community ownership body can aggregate certain responsibilities of those smaller bodies to administer a larger area such as a master-planned community.

As the cost and demand for housing has increased in many urban areas, tenure arrangements have also been adapted to accommodate changing demographics, housing types and services frequently in a community-based environment. These have often involved the use of manufactured (pre-fabricated) free-standing dwelling units, owned by that resident but on land provided under a licence agreement by the owner of that enclave development. Again, such changes have been facilitated by new legislation across most states.

7.4 Contracts – general parameters

The word 'contract' means an agreement between two or more parties. Some contracts are so straightforward and short-term that they are hardly worth considering. If one goes to the shop to buy a loaf of bread this is, in reality, a contract establishing a transaction: the shopkeeper 'offers' the loaf of bread at a certain price, the customer 'accepts' by paying the money and the transaction takes place and is completed. Of course, many contracts are far more complex and sophisticated, especially in areas such as real property where written documents will usually be required to fully describe the transaction.

Real property contracts typically involve a series of 'covenants' – that is, agreements, commitments and responsibilities by each of the parties – and sometimes joint covenants (i.e. agreements for joint action by both of the parties) (Chambers 2008).

Except where legislation requires it, contracts do not have to be in writing and can be verbal. In practice, however, any substantial agreement needs to be committed to writing to ensure clarity and to protect all parties.

A large body of common and statutory law has been built up around the establishment and operation of contracts and, in the case of property contracts and dealings, a further level of statute is involved. The role of the courts in establishing precedent in the interpretation of contracts over many years is also important in clarifying the operations of these agreements in particular cases.

A detailed study of contract law is clearly outside the purpose of this chapter; however, several key observations should be noted. The abovementioned laws and interpretations establish that, for any contract to exist, six key elements need to exist:

- there must be a clear intent of all parties involved to create a legal relationship
- there must be elements of 'offer' and 'acceptance' – where certain things (either tangible or intangible) are being traded

- there must be consideration passing (i.e. a 'value' for the 'thing' – although this does not have to be 'market' value)
- there is legal capacity of the parties involved to contract and transact (there are limitations on the ability of certain parties to transact, for example, the administrators of a trust, minors – persons under 18 years of age – bankrupts, or people without the necessary mental capacity)
- there must be genuine consent of all parties to the contract
- there must be 'legality of objects' – that is to say, that the activity being established here is legal and not prohibited by other legislation, regulations, et cetera.

As noted above, some contracts can be quite simple (to the point that, in certain circumstances and depending on relevant legislative requirements, a verbal agreement may be shown to be contractual). Other contracts may be particularly complex and run to many hundreds of pages. Regardless of the size and complexity of an individual contract, the basic elements (identified above) are the same.

In business dealings such as those for real property, the contemporary approach to contracts (and to negotiations generally) has shifted considerably from adversarial structures of the past. Breach, remedy, penalties provisions and the like are essential in any contract. However, there is now a wide recognition (reinforced by new contract structures, laws and alternative dispute resolution procedures) that contracts should be seen as documents that provide clarity, assist co-operation and are to the benefit for all parties to that contract rather than overly stressing adversarial or punitive aspects. That the frequent necessity to resort to contract or statutory provisions for remedy may reflect poor attention to detail at the time of contract establishment, or poor negotiation and management skills in subsequent activities.

Clarity and simplicity in drafting are very important for later interpretation of the document. No contract, no matter how detailed, can address all possible contingencies and it is important that basic parameters be clear enough to allow for successful operation during the time the contract has effect. A written contract will normally have a number of set components including:

- the identification of the parties, date and duration
- some statement of intent (called 'recitals' which sets a background of the involvement of each party)
- definitions of words used (to assist in clarity)
- the body of the agreement: listing out the covenants or undertakings from each party and also identifying, where relevant, joint covenants
- identification of the goods, financial considerations and/or property, activities or other 'items of consideration' which are the subject of this contract and details of the consideration passing
- administration of the agreement, provisions for dealing with breach, end of contract and other matters particularly relevant to the case.

Care must be taken in the drafting of any contract to ensure it is able to be interpreted and understood by all parties and that it does not breach Commonwealth or state statute and regulations. Without that level of clarity, the whole exercise can be frustrated and the contract identified as void (not able to be acted upon, as though it never existed);

voidable (while it may be actionable, it can be rendered void under certain circumstances, or by the action of one or more of the parties to the contract); unenforceable (if there is a breach or one party fails to fulfil their part of the agreement, the aggrieved party cannot in effect force the recalcitrant party to action); or the whole agreement collapses because it is found to be in serious contravention of some law or regulation and is, thereby, illegal.

In many commercial dealings, standardised, pro forma documents and templates exist. In property dealings in particular, there are numerous examples of standard sale and leasing contracts, service agreements, agency agreements and the like and, unless there are significant reasons to the contrary, it is sound practice to accept and follow standards known to be successful. The Property Council of Australia (PCA) have numerous property related, standard contracts that include agency, service and maintenance agreements, checklists and standards and procedures for many activities. Likewise, the Law Society and Real Estate Institutes in various states have a number of relevant agency, tenancy, sales and contract agreements available. As no piece of land or circumstance is identical, these documents should be carefully examined to ensure suitability for the parties' requirements in each particular case and to accommodate any specific legislative or regulatory requirements of state and local authority area in which that property is located.

Depending on the provisions of the agreement document, where disagreements cannot be resolved between the parties, the matter will normally be referred to mediation, arbitration or decision by a relevant court. The need for resort to litigation is not common and is expensive and time consuming. Sometimes however it may become necessary. Such occasions may arise, for example, where there is a genuine and serious lack of clarity in the agreement, events have arisen that could not have been reasonably foreseen or where one party is simply recalcitrant and refuses to uphold their obligations. Whatever the cause, it is always essential that such processes strictly follow due process, keep to the facts of the matter and are not allowed to degenerate into personal or emotional attacks. As soon as a dispute emerges, the securing legal advice is recommended.

Common sense and consideration of what is really important to the asset and business success overall should determine how (and if) these matters should proceed. If they proceed, the plaintiff in the case will seek the support of the court, if successful, in the form of a court order forcing a particular action Real property – tenure and dealings (called a 'mandamus writ') or prohibiting, causing or requiring cessation of an action (an 'injunction').

How arbitrators and courts interpret legal documents is a very complex area but, again, several underlying principles of interpretation need to be identified, even in summary.

A fundamental rule that should govern any interpretation of contracts and other documents is what is known as 'the golden rule' or 'black letter interpretation' – that is, 'the words say what the words say' and, given that the other elements of the contract exist (complete with definitions), the words should be accepted as they stand.

Second, an interpretation is that specific agreements and clauses will take precedence over general clauses in the contract. (There are some exceptions, such as legislative intervention.)

Third, it is important to recognise the role of precedent decisions by courts in assisting interpretation. Many issues would have already been considered and determined, and, in litigation particularly, these can assist in the determination of any current disputes or difficulties in interpretation.

7.5 Property leases

Leases (or tenancies) are contracts between two persons or legal entities – a landlord (lessor) and tenant (lessee) – and permit the use of the subject property by the lessee for a set period of time. To be enforceable, any lease will need to contain all the elements of a contract outlined above and also comply with a range of legislative requirements pertaining to property dealings. Again, equity considerations will be important if any matters later require arbitration or litigation.

Leases can be over land, land and building or part of a land or building. While standardised documents will often be used, it is important to remember that, as contracts between two parties, the terms and conditions may vary greatly from case to case, depending on what the parties agree.

Two overall types of tenure arrangements were previously identified in Section 7.2 – Crown land/leases and freehold land (alienated or separated from the Crown and held as freehold by a private entity).

Freehold (fee simple) is held by the registered owner in perpetuity. By the nature of freehold title, the owner can (subject to other relevant statute) lease the property to another party for a set period of time. A lease, therefore, creates a new, equitable interest in land: that of the lessee. It is, however, a terminable interest (i.e. of specified duration) as opposed to the perpetuity interest of the parent freehold title. This implies that, while that lessee's equitable interest is established, it will, at some recognised point in the future, come to an end and revert to full, unfettered ownership by the landlord/owner (called a 'residual interest' or a 'remainder').

The terms 'lease' and 'tenancy' are, in legal terms, synonymous, as both are contracts establishing the lessee's interests in a nominated piece of real property and the rights and obligations of both parties. Colloquially, the term 'lease' is normally used for commercial, retail and industrial properties and the term 'tenancy' for residential. In the latter case, the tenancy period involved may be relatively short and documentation may be a simple pro forma. Nevertheless, the equitable interest in favour of the lessee is established. The agreements (covenants) within a lease or tenancy document are principally as follows (Price & Griggs 2008; Bradbrook et al. 2011):

The lessors, by covenant with the lessee, provides that lessee with the exclusive use of the leased area (legally called 'quiet, peace and enjoyment') or that piece of property for the period of the lease.

In return, the lessee provides covenant (contractual obligations) back to the lessor:

- to pay the rent as determined
- to reasonably care for the property
- to use the property only for specified purposes and within the parameters of subsisting laws and regulations
- not to create a nuisance to any party
- not to sub-lease, assign or modify without prior approval
- to give up the demised premises at the lease expiry.

There may be many other conditions regarding commercial leases, but most are based on the above covenants.

A lease or tenancy can be for a fixed term – that is, a fixed and stated commencement and termination date. For example, a residential tenancy will often be for a period of 6 or 12 months. Because of set-up and business establishment costs, commercial leases will often be for longer periods – frequently three or five years or more. Because these are contracts between two parties, the period of the lease can effectively be whatever length is agreed between them; although, in many jurisdictions, if a lease extends for three years or more, the lease is required by law to be recorded on the title.

Additionally, under the by-laws of some local authorities, land leases of ten years or longer are considered as de facto land subdivisions and may require further approvals. Many leases (particularly non-residential) will, subject to conditions set out in the lease, provide for a 'call option' for a further lease term or terms in favour of the tenant. Consequently, a commercial lease may be described, for example, as a '5 × 3 × 3', meaning that the lease stipulates that, subject to the provisions of that agreement, the initial term is for five years, with the tenant having the option to take up – normally on the same conditions as for the original term other than that relating to rent – in succession, two further lease periods, each of three years.

Other leases/tenancies may be established, not as a fixed term, but on a 'periodic' basis. That means the lease identifies a period – normally coinciding with the rent payment period – for example, fortnightly or monthly. As long as the rent continues to be paid when due and the landlord accepts it, it is accepted that a continuing tenancy is 'implied' and thereby continues without a specified end date. Similarly, if a fixed-term tenancy ends and the lessee continues to pay the previously prescribed rent and the lessor accepts it, then the lease is deemed to be 'held over' – that is to say, it continues, but in much the same way as a periodic tenancy (Duncan 2008). The exact manner in which these arrangements operate will also depend on the particular requirements of relevant state legislation.

Another instrument worth definition is a 'licence'. A licence is a contract between, in this case, a property owner (the 'landlord') and a second party (the 'licensee') that allows that second party to use a specific piece of land or building for a particular purpose. Unlike a lease or tenancy, it does not establish an equitable interest in the land – simply a right to use it for a purpose that would otherwise be considered a trespass – and is not an exclusive arrangement between the landlord and the licensee. Common examples of licences include: a month-to-month car-parking licence in a commercial building; the agreement by one landholder to allow another person to agist livestock; a permit to allow another to grow a crop on the owners' property (i.e. 'share-farm'); or to permit storage of another person's equipment or goods on open hard-stand areas or shed. All of these reflect the concept of a licence being the contractual establishment of a right to use an asset or an identified part of an asset, but not to establish an equitable interest in that asset.

7.6 Compulsory acquisition

In Section 7.2, it was noted that while a great deal of land in Australia was now alienated from the Crown – either by way of lease or freehold ownership to private entities or individuals – there remained the right of the Crown (in its various forms) to take back that land into public ownership and use under certain conditions. This process, is known as 'compulsory acquisition' or, more colloquially, as 'resumption'. This is a complex and highly specific area of the law and valuation practice, enabled by special and highly specific legislation enacted by the Commonwealth governments and by each of the states for their areas of jurisdictions.

Resumption activities by government are serious and sensitive issues both in a legal sense, because of the obvious effects on the rights and property of private citizens, and because of political impacts and implications. Over time, the increasing scale of public works, their effect on now highly urbanised and valuable property and the reaction of well-informed and empowered communities have all heightened political interest and involvement in resumption decision-making. In many public projects, therefore, issues of resumption and their impacts and costs are often as important in determining public works as are construction feasibility and the costs of physical construction.

Being related to land and tenure, legislation pertaining to compulsory acquisition is principally state-based with each state having comparable legislation and regulations covering these activities (typically with titles such as the 'Acquisition of Land Act'). Each state will have a range of government instrumentalities, including the local councils, recognised as 'constructing authorities' and have incorporated into their legislation certain rights to compulsorily acquire property – though for consistency, uniformity and process that legislation will normally refer back to state generic legislation.

The Commonwealth government also requires the ability to acquire privately owned land for a range of activities. These include defence, communications and other matters that, under the Australian Constitution, reasonably fall within the Commonwealth's areas of responsibility. This legislation, titled the *Land Acquisition Act 1989,* follows a similar model to the states, setting out certain purposes for which an acquisition can take place, the processes and requirements of the Commonwealth in its dealings, together with rights of objection and appeal for the disposed owner. A slight variation regarding Commonwealth acquisitions is the provision under Chapter I Part V 51 (xxxi) of the Australian Constitution, which require that Commonwealth acquisition be on 'just terms'.

While, typically, there may be no formal requirement on states to act in a 'just way', the law of equity, legal precedent and the political realities identified above, all mean that compensation must be deemed adequate, either through negotiations or determined by a court of suitable jurisdiction (Bradbrook et al. 2011).

These compulsory acquisitions may be for the whole of a property, for part of it, or perhaps for a right of way (such as an *easement in gross*) which may be required where a public utility such as a pipeline, powerlines or similar facility is secured as a right to cross, but does not actually take the underlying property freehold. In most cases, the public purposes for which land or other property rights can be secured are set down in the provisions of that particular piece of legislation. Typically, the legislation will also describe quite strict procedures governing the Construction Authority's activities. Usually these will include requirements to provide adequate notice of intentions to the owner, time limits, rights of entry, obligation to make good and so on, which must be strictly adhered to in order to avoid claims of *ultra vires* actions against the constructing authority (i.e. that the Authority acted outside their rights and jurisdiction) that would potentially void the entire activity.

The assessment of these matters is often complex and difficult; nevertheless, it is based on well-established valuation principles and court precedent. As well as the value of the land taken, there are a number of heads of compensation claimable in these cases including (depending on the enabling legislation) injurious affection (i.e. diminution in value to the balance or residual land because of actions, activities or works of the constructing authority on the resumed section); severance (i.e. the adverse effects of on the balance of the land of removing part of the original area or, perhaps, dividing it into a number of pieces); disturbance (i.e. the adverse effects on business and other operations on business or other activities on the property, resulting directly or indirectly from the resumption) together with the reasonable costs incurred by the dispossessed owner as a result of the resumption process and related activities (Hyam 2004; API 2007).

Depending upon the provisions of the legislation, claims for compensation may be offset by any increase in value ('betterment') of residual areas held by the claimant resulting from the resumption.

An overall maxim in establishing compensation is a 'before and after' assessment. This implies that, as much as money will allow, the affected owner should be in the same financial position after the resumption activity is completed as before it. This is the difference between the valuation of the property after the resumption and its value immediately prior to those actions commencing. All other things being equal, the difference should represent the quantum of compensation payable to which costs should be added. Because of the time-critical nature of many public works, such as roads, utility pipelines, schools and hospitals, it is normal to instigate resumption processes well in advance of construction activities. In practice, however, the securing of the land will often occur through direct negotiations between the authority and the affected land owner.

The political and operational interests of government, together with the wish to minimise any distress to the owner, are often well served by avoiding protracted legal processes. Consequently, in many public works projects, only a relatively small number of cases proceed through to final compulsory acquisition, and an even smaller number resort to arbitration through the courts to settle compensation claims.

These matters are discussed in more detail in Chapter 11.

7.7 Native title

The preceding sub-sections have highlighted the quality and value of Australia's land tenure systems and their fundamental importance to individual rights, the aggregation of wealth, development and economic and civic security. The fact that these systems have been in place for over 150 years reflects their consistency and effectiveness to the point that they are largely taken for granted.

The validation of the entire tenure system can be traced back to the claim by James Cook in August 1770, taking possession of the continent that he had partially explored on behalf of the British Crown. This 'right to claim' stemmed from an international understanding at the time known as 'terra nullius' (or 'nobody's land'). This held that where an area of land was discovered and that land was uninhabited, then the person/country making the discovery could claim it as their own. (The origins of such an understanding are quite unclear, but the maxim proved convenient to justify imperial expansion by European powers throughout that period.)

Whatever the background, the entire development of tenure systems continued uncontested until 1988. In that year, a challenge by a group of Torres Strait islanders to the concept of terra nullius in Australian history was upheld by the High Court of Australia (Mabo v Queensland (No. 1), HCA 69 (1988) 166 CLR 186). In that judgement, the 'Mabo decision', the High Court recognised that settlement / occupation existed in Australia prior to Cook's claim. In fact, it has been estimated that at that time, there were some 750,000 aboriginal inhabitants and that they had a very longstanding culture and links with that land. In effect, Cook was mistaken, based on the fact that he had seen very few of those inhabitants, had found no evidence of large settlements nor did he have an appreciation of the deep community and cultural systems that existed.

The interface between the different forms of 'land ownership' was encapsulated in Fejo v Northern Territory (1998) 195 CLR 96:

> [Native title]… has its origins in the traditional laws acknowledged and the customs observed by the indigenous people who possess the native title. Native title is neither an institution of the common law nor a form of common law tenure but is recognised by the common law. There is, therefore, an intersection of traditional laws and customs with the common law.

In Mabo however, the Court established that native title was extinguished if the subsequent legal actions of government alienated parcels of land from the crown (e.g. by the freeholding of those lands). Additionally, it was held that, for native title to be confirmed, a continuous association with that land needed to be established.

While the Mabo decision had profound repercussions for property law across the country, it did not go further in giving detailed direction to resolve the complex issues raised. (That of course was only to be expected and not as a criticism of the Court, whose primary role is to interpret law and decide on disputes, not to make law in its own right.)

A second High Court case, Wik Peoples v The State of Queensland (HCA 40 (1996) No. 187 CLR 1 (known as the 'Wik decision'), went some way to clarifying the situation. This ruling held that native title could co-exist with pastoral holding tenures and, where there was some overlap of rights, the native title rights would yield (Brennan 1998).

These early decisions began to address some of the past injustices inflicted on the Australia's indigenous population but, in a practical sense, presented new issues regarding the status of mining rights and the security of leasehold tenures such as pastoral holdings which occupied much of the Australian landmass. The issue of pastoral holdings was particularly important as tenure history searches, dating back to colonial periods, not only established that many pastoral tenures were established to make the land available for pastoral purposes but also envisaged that the indigenous inhabitants would have continuing access and use for their tradition purposes (Reynolds 1996).

It was clear that legislation was required to create a framework around such dealings and to provide a higher level of certainty for both indigenous owners and those with statute-based tenure. Given the national implications of all of this, the Commonwealth Government, with the general support of and complimentary legislation by the States, enacted the Native Title Act (1993) ('NTA'), which was further amended in 2007 and again 2009. This approach sat well with the 1967 amendment to the Australian Constitution which better established a lead role for the Commonwealth Government in Aboriginal affairs more generally.

Key provisions of that legislation included that:

- Various types of native title rights were recognised over land and marine areas. These could be exclusive or non-exclusive rights, depending on the activities involved. (These rights could vary from full use through to access for hunting and fishing, 'walk through' rights and/or use for ceremonial and other activities.)
- Those claiming native title had to seek formal recognition by making application to the Federal Court with the onus of proof upon the applicant.
- A National Native Title Tribunal (NNTT) was established to mediate in such matters and to determine the matter if it had not been settled within six months. Additionally, the NNTT was to maintain a national register of established native titles.
- To enable negotiations to take place and for later successful claims to be managed, a Registered Native Title Body Corporate (RNTBC) was to be established in each case to represent indigenous claimants.
- In negotiations regarding native title claims, mining activities and other uses, an obligation was placed on all parties to act in good faith. In such negotiations, indigenous owners did not have the power to veto otherwise legitimate land use requirements by other parties and once native title rights were lost, in whole or in part, that level of loss could not be reinstated.

While there was recourse to the determination through the NNTT with appeals to higher courts, that route was not generally perused with the first decision on the matter of compensation being determined by the High Court in the Timber Creek case as recently as 2019 (see below).

Rather than seeking a judicial solution, the Native Title Act provided for direct negotiations between the parties involved and, if successful, an Indigenous Land Use Agreement ('ILUA') would be signed and registered with the NNTT. This approach has proven much more popular with some 1300 such agreements in place by early 2019. This represents significant progress in addressing these complex legal, property and cultural issues.

During sensitive, inter-cultural negotiations of this type, fundamental differences in approach often emerge between the respective parties, based on the relative importance placed on economic and cultural priorities and motivations. The nature of decision-making processes and the time frames required are also markedly different between development/investment interests and the consensus approach typically required by traditional owners. In such complex, legal and technical negotiations also, power imbalances can easily emerge between the parties.

With a now improved understanding and greater experience, relevant tribunals are increasingly watchful to ensure that agreements truly provide a resolution that is transparent and equitable to all parties and interests.

An additional, practical issue here is that ILUA agreements are confidential documents with each one being modified to suit those particular circumstances. Consequently they provided little or no guidance as to the matter in which compensation should be assessed.

A period of uncertainty followed wherein a series of court decisions and subsequent appeals struggled with the assessment of compensation in a situation where native title rights had been lost or diminished. On the face of it, the loss of economic rights/opportunities could reasonably be assessed under well-established compulsory acquisition principles (see Sections 11.3 and 11.4 in Chapter 11). Where shared rights (as envisaged by the Wik decision) were involved, a land–value assessment not dissimilar to that applied to engross easement compensation assessment (see Section 11.5 in Chapter 11) appeared to enjoy general acceptance. However, the conundrum here lay in the fact that, almost by definition, there seemed no way that the 'cultural value' lost to the dispossessed nature could truly be expressed in (commercial) monetary terms. Some attempts were made to consider assessment considering anthropological aspects but these proved quite nebulous – leading again to further appeals.

Meanwhile, the lack of certainty and legal delays led to a range of unintended consequences. These included the loss of security for public infrastructure and potential issues confronting development projects. These not only involved lands directly affected by native title dealings but, much more widely, where new public works were to be sited on unalienated crown land.

Almost inexorably, one compensation test case moves forward on appeal for decision by the full bench of the High Court of Australia. The full title of that case was Northern Territory of Australia v Mr A. Griffiths (deceased) and Lorraine Jones on behalf of the Ngaliwurru and Nungali peoples & Anor; Commonwealth of Australia v Mr A. Griffiths (deceased) and Lorraine Jones on behalf of the Ngaliwurru and Nungali peoples & Anor; Mr A. Griffiths (deceased) and Lorraine jones on behalf of the Ngaliwurru and Nungali peoples v Northern Territory of Australia & Anor [2019] HCA 7. The colloquial title became the *'Timber Creek' case*, referring to the remote Northern Territory lands and community involved.

The decision, handed down in March 2019, recognised compensation in three parts: a payment for 'economic loss' (determined in this case at 50 per cent of freehold land value); an interest payment recognising the substantial delays in settlement (though adopting only a simple, not compound, interest calculation); and, of particular significance here, the award of a 'solatium' payment in recognition of the loss or diminution of connection or traditional attachment to the land – that is, compensation for cultural loss. ('Solatium' payments are further defined in Section 11.4 in Chapter 11.) Such cultural loss could have ramifications and impacts wider than those related to the land physically affected – including, for example, the 'walk-through', hunting, fishing and ceremonial rights and practices of the claimants.

Several important aspects of this case require elaboration and are highly relevant future dealings. In the first instance, while the actual compensation payable was (as always in compensation matters) dependent on the circumstances of the case, the quantum of compensation here was very significant: $320,000, or 50 per cent of assessed freehold value, for economic loss, about $910,000 in interest and an amount of $1.3 million for cultural and spiritual loss, which the court considered was at a level that 'the Australian community would regard as appropriate, fair and just' in this case.

Second, it was clear that the actual monetary assessment, particularly of cultural loss, could not be taken as a 'rule of thumb' or to be accepted as the benchmark for other cases. While the decision clarified the concepts and approach, each assessment must be considered on the merits of that particular case.

Third, the Court reaffirms the compensation approach for partial economic loss. This component could be assessed as a percentage of the freehold value of the land affected, that percentage being dependent on the severity of the impact. It is further noted here that the total compensation could be considerably more than land value alone.

All of these concepts, including the reasonable and appropriate inclusion of accrued interest, were consistent with other types of 'compulsory acquisition'. It is the assessment of damage to, and monetary compensation for, cultural rights that is of critical importance here.

It is likely that the Timber Creek decision will figure prominently in future negotiations and litigation in native title matters with higher value of claims being anticipated. In time, it may also act as a catalyst to further legislative amendments. As always in such matters, the underlying and still unresolved question is to what extent this evolution will protect the rights and enhance the overall well-being and prosperity of indigenous Australians in the longer term.

Chapter 8

Property sectors

Urban

8.1 Introduction

Chapters 2 and 3 identified key characteristics of the property market, noting the common, underlying themes but, at the same time, the remarkable diversity of the various functions and uses represented.

From that basis, this chapter considers in more detail a range of largely urban-based property sectors – commercial, retail and residential (as an investment medium) – together with aspects of ownership and management of properties held by government and other organisations for purposes other than the securing of rental income.

In addition, a short overview of a number of more specialist urban land uses is included. The purpose here is to provide an introduction and to set some parameters for further study into those, more esoteric areas.

8.2 Commercial (office) buildings

The rise in scale and importance of commercial buildings and of office space reflects and parallels changes through the development of Australia's economic and business activity within it – from an original agrarian, export-oriented base, through manufacturing and now to one dominated (over 70 per cent) by the service sector (Australian Government, DFAT 2017). That general sector employs four out of every five Australian workers and includes all manner of professional services, banking and finance, government activities, et cetera, practically all typically domicile in office buildings.

From the early 1990s, information and communication technologies (ICT) began to impact significantly on business and community life. Many analysts predicted that these technologies would cause the demise of office buildings and of central business districts (CBDs) that had emerged over the previous half century. It was envisaged that workers would, for the most part, remain at home or elsewhere and simply work remotely as individuals, all connected through telecommunications, particularly the emerging internet.

There is no doubt that the revolution in both accessibility of information and instantaneous global communication has fundamentally changed the nature of business and built assets associated with it. However, rather than diminish, the demand for office space continues to grow. Not only do more people now work in offices in absolute terms, but the proportion of the workforce involved in office-related activities has never been greater. In the two decades from 2000, total office space in Australia has effectively doubled in area (PCA 2019).

Those earlier predictions had not recognised that, particularly in early developmental years, investment in ICT hardware actually caused a drop in productivity. It was only when new technologies can be integrated with human capital, social networks and innovation that productivity greatly increases (Leer 1999). The physical office environment would prove critical to this intersection of the human and technical components of contemporary business. This important interface requires elaboration given its impact on determining physical space requirements.

It is of the nature of humans to meet, act, interact and cooperate in teams to maximise output. The 'tribe' has been replaced by the 'work team', but the principle is similar. These observations align with mainstream sociology theory (Johnson & Johnson 1997; Smith 2008), which notes that from pre-history, clustering into smaller groups, initially family but, often linked to a larger network (such as a clan), is innate to the human condition.

In contemporary societies, families and similar relationships remain very important, though typically they are much smaller in size than in the past. It is no coincidence that other groups, such as sporting teams, work groups or army squads, typically involve between 5 and 12 or so members (Belbin 2010). This group size is sufficient to achieve desired results using the range of skills, cooperation and joint effort from the group. At the same time, it is small enough to allow each member to be familiar with the characteristics of other individuals in the group and define their own place in it. As tasks became more complex and at times esoteric (e.g. within the contemporary 'knowledge economy'), the need for interaction has both increased and become more intricate, supplemented and supported by access to a vast array of information and communication options.

Social environments are dynamic. In recent decades, traditional, social, community, sporting, youth and other structured groups have steadily declined in importance (Putnam 2000). Progressively, these have been superseded by a range of interest groups, often linked to social capital and professional networks using the same internet-based communication for business and social linkages (Week 2002; Florida 2004). Economist Joseph Schumpeter had, decades ago, predicted the role of disruptive technologies (which he called 'creative destruction') in breaking down and re-forming economic, social and political structures (McCraw 2009). The current fusing of new business models and technologies and the rapid rise of social networks and social media provide contemporary examples of Schumpeter's theories as they rapidly supersede and 'leap-frog' past activities (Libert 2010). All of this has implications, both positive and negative, for the importance of place and the nature and role of the built form.

Overall, place still matters, but now in quite different and more complex ways.

The completed and occupied commercial (office) building represents the product and interaction of a number of contributing areas of economic activity. These include:

- the final user/rental market
- the financial/financial asset market
- the market for land and development sites
- the development market – that is, the creation of new assets by the development and construction sector (Ball et al. 1998).

While each of those markets share common elements, their drivers and determinants are different and this diversity contributes to overall market instability. This sets up a level of friction and a less than cohesive supply chain of these assets, issues being made more challenging by the high capital values and long production timeframes involved.

Most businesses benefit from close physical association ('clustering') with complementary firms and with clients. Therefore, in an advanced country such as Australia, office buildings aggregate in the major cities and particularly in CBDs of those urban areas (Ball et al. 1998). Following the typical urban pattern in Australia, office accommodation is concentrated in the major cities with around 70 per cent of total office space located in the CBDs of those major cities (PCA 2015). The remaining space (approximately seven million square metres) is located in fringe-CBD areas, regional nodes and office parks in the suburbs of major cities (PCA 2015).

It would also be observed that new construction has typically occurred in waves, sometimes with timing variations between the major cities and depending on economic conditions and prospects and local supply and demand parameters at the time.

Total stock in major Australian office markets stood at just over 17.9 million square metres as at January 2019 (PCA 2019). There are about 4,300 office buildings with over 500 square metres in Australia, and the approximate distribution (by net lettable area) across the major urban areas is shown in Figure 8.1.

These figures for commercial activity indicate the continual dominance of CBDs, and particularly in Sydney and Melbourne which, in aggregate, represent about half of total stocks. The proportional increase of fringe-CBD and non-CBD locations has been significant in recent years. Large corporate and government tenants typically prefer the identity and business convenience of a CBD location. However, that choice typically attracts higher rental resulting from high construction and operating costs and the use of high-value sites within inner-city locations.

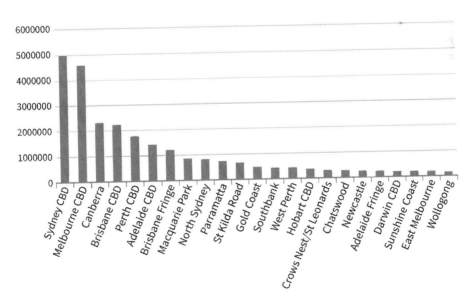

Figure 8.1 Australian office stocks, 2019.
Source: PCA (2019).

There may well be rental-cost advantages in non-CBD (e.g. 'fringe' or more remote) locations, and contemporary, high-speed internet communication can, in part, offset the lack of physical immediacy available within the city core. There is an increasing trend, especially for smaller firms, to recognise the real alternatives and options that non-traditional locations can provide.

Additionally, non-CBD locations often prove more convenient for staff and provide a wider choice in design. Improved accessibility and availability of parking in these locations can also provide a superior work environment for many contemporary firms. This can include a better opportunity to cluster with similar or complementary firms, universities and clients. Consequently, as urban congestion inevitably worsens and new CBD sites and developments become increasingly difficult to secure, the further development and use of non-CBD commercial locations, business and research parks, and even commercial activity in well-located provincial areas is anticipated (Kotkin 2010).

Even these trends may themselves be evolving. Over time, many desirable fringe-CBD locations may become as expensive and congested as the CBD themselves. Furthermore, disproportionate increases in fuel and transportation costs into the future are likely to bring forward other location options for commercial activity and physical clusters, perhaps including more distant nodes within major cities or well-located provincial cities (Kotkin 2006, 2010).

Commercial property assets provide a critical component and physical environment for business activity and represent one of the major sub-sectors of property investment and asset and property management. These assets are diverse in location, scale and quality, ranging from major CBD buildings of 40,000 square metres or more, with a capital value in the hundreds of millions of dollars, through to small, regional or provincial office buildings. Similarly, the categories of investors involved range from some of the biggest superannuation and investment firms in the country (and globally) through to individual investors for the smaller assets. Obviously, this diverse group will have varying requirements and expectations.

The quality of office buildings in Australia is assessed through a rating system defining Premium and A, B, C and D grades. The system, which is widely recognised and accepted across the sector, was established by the Property Council of Australia (PCA) and assesses such factors as size, design, configuration, location, amenity, and services including communication technology, security, lift, air conditioning and mechanical services. The proportion of buildings in each grade in the Australian market is indicated in Figure 8.2. Overlaying these ratings are more specific environmental and energy measures that are important within the market place, and increasingly subject to mandatory disclosure upon sale in various jurisdictions. These matters are further discussed in Chapter 15.

More than 81 per cent of Australian office stock is older than ten years, accounting for around 17.5 million square metres. Much of this aged component is rated in the lower categories of office stocks. These buildings are, therefore, particularly vulnerable to obsolescence, given the dated services and energy-management systems typically available, and rapid changes in tenant demand over recent years. The higher vacancy rates normally experienced by these buildings (despite their comparatively lower rental levels) reflect these, often endemic, problems.

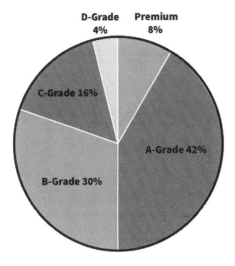

Figure 8.2 Australian office stocks by grade.
Source: Ernst & Young (2015) and PCA (2015).

The income-producing ability of these assets now or into the future represents a fundamental performance criterion regardless of the scale of the investment or type of investor. More specifically, the investment performance criteria of any commercial property asset will include the optimisation of return, in both the form of net income and the sustainable capital growth over time, and the management of risk in securing those returns. Even in the case of an owner-occupier, these criteria still apply – though, in that arrangement, in the form of opportunity cost.

The following observations provide a summary of the commercial property sector and further details of investment and management criteria particularly relevant to these assets.

Some history

As discussed in Chapter 4, Australian cities have reasonably comparable layouts, developed over 150–200 years, each with a fairly concentrated CBD radiating out to surrounding industrial and residential suburbs.

In their present form, however, CBDs are remarkably new constructs with many of the buildings produced during a wave of development through the 1970s and 1980s – a period sometimes referred to as the 'rise of corporate real estate'. For a number of reasons, during that period, banks, insurance and superannuation companies, and major corporations and government departments saw the need to establish a greater presence and scale of operation in CBD locations, with retail and non-commercial users tending to relocate to new nodes in the outer suburbs. Furthermore, in response to growing office demand, a substantial investment market developed to provide rental accommodation for large and small firms gravitating to the CBDs.

Nevertheless, while underpinned by solid corporate and government demand, the market for commercial buildings is inherently cyclic and volatile, particularly in smaller CBDs and regional areas. There are a number of reasons for this including:

- the relatively high cost, complexity and time lags in producing new stock
- the large scale of these individual developments which, can singularly, can add significantly to vacancy levels
- external economic and political events that will influence (either positively or negatively) underlie business confidence and, over time, will affect the demand for new and existing space, locational preferences and the types of buildings, services and amenities demanded.
- remarkable changes in the nature and function of CBDs resulting from increasing costs, development, congestion and competition from a range of civic and new residential uses. As costs for office accommodation increased, other options in fringe-CBD areas, office parks and so forth became increasingly more attractive, particularly with the communication links now available.

Perhaps the most important change within this sector is a fundamental shift in business structures and work practices across Australia and all other OECD countries. Major corporations and administrative bodies remained the dominant users of office accommodation. However, over a remarkably short period of time, another group of user firms and their staff emerged and became increasingly important as office tenants – perhaps representing 20 per cent or more of all users of office accommodation. This group would be commonly defined as 'knowledge workers' (Florida 2002). That term is somewhat misleading as all work and human activity requires a level of knowledge and skill. Here, however, it refers to those contemporary firms and their employees whose business relates to the creation, manipulation, customisation and adaptation of knowledge as a primary product.

Because knowledge workers are scattered across a range of existing and emerging business sectors, statistical details are difficult to secure. Nevertheless, their rapidly increasing scale, role as decision-makers and, in turn, their increasing influence on the development of commercial clusters and offices are all quite significant (Week 2002; Florida 2004).

Clearly, many ICT-oriented firms provide examples of the types of businesses and the workers involved, but there are many others, ranging from professional services through to spin-off research and development companies and other services. Important to the owners and managers of commercial property assets, the nature and demands of these firms and their employees are quite different from those of traditional firms. Typically, these firms are smaller, much more volatile, adaptable and, often, much less location specific.

Their work practices, business activity and value are normally based on intellectual property and human and social capital. Their employees are likely to be professional, individualistic and have strong social networks related to work and their profession. They typically spend much longer periods at the workplace, are well educated and seek a physical environment conducive to their work performance. They are articulate in expressing their requirements and vocal in complaint. Under these scenarios, work, social and other activities tend to merge, to a greater or lesser extent, often involving the same networks and similar internet-based communication. This integration has not previously been encountered and presents a new consideration for those providing contemporary built forms to accommodate new businesses and their employees.

Importantly, once these workers enter an office space, their activities are quite different from the often procedural and administrative activities of their peers elsewhere. In 'knowledge work', the processes are more aligned with manufacturing than traditional office work, as those involved come together in teams and, at other times, work individually. This can occur at home, travelling or with their clients, to 'produce' the required outcome – be that a computer program, a marketing submission, a new product design, professional opinion or a diverse range of other products.

The creation and maintenance of a work environment that accommodates this emerging and important sector represent a significant challenge to owners and managers but, with that, bring potential opportunities, given their high earning capacity and rapid growth in scale and influence.

The trend represents an opportunity for property owners and/or investors to shift from the passive, simple provision of 'place, space and time' in a building under commercial lease. Particularly with these new types of commercial tenants, there is an ability to 'value add' to the business and work environment to attract greater loyalty, improve the quality of tenants in the building, increase rental income and potentially sell additional services to tenants and their employees.

A well-designed office can create an environment that is conducive to both team and private space environments described above; moreover, through style and design, it can help reflect the overall approach and philosophy of the firm, noting particularly that many firms now create intangible, 'weightless' product and services. Consequently, their office provides one of the few physical manifestations and images of that company (Week 2002).

Also related to this trend is the growing interest by firms and employees in ecologically sustainable development (ESD), energy efficiency and other demonstrated sustainable practices. (Refer Chapter 15.)

If commercial offices can provide these environments and management that align with the wider aspirations of these firms and their employees, then a partnership arrangement and level of loyalty can be developed that is more enduring than the simple, legal base provided by a commercial property lease. The sophisticated and integrated use of ICT needs to form a part of that 'offer to tenants' (Hefferan 2006).

Similarly, the area, precinct or cluster in which the building itself is located, together with the external image of the building, provide a physical interface and image for the firm to the outside world. For example, location in a major, 'institutional' building in the financial district obviously carries with it a different image for the firms located within those buildings from, say, a small low-rise, contemporary, open plan office building in a fringe-CBD location, or a business or university research park.

How all of these factors evolve further post the COVID 19 pandemic remain to be seen. On the one hand, through necessary and facilitated by available ICT now being used to potential, options for working remotely have been confirmed and advanced. On the other however, the practicality of long term lease commitments, the need for corporate cohesion and staff demands for association and team interactions will continue to favour office environments of various types. Individually tailored, hybrid solutions can be anticipated.

Commercial leases and changing tenant demands

As with any other income-producing property, income streams for commercial buildings are normally protected by leases between the building owner (landlord, lessor) and those occupying the building (tenants, lessees). Details of the parameters of commercial leases are included in Chapter 6.

While sound leases are critical to the securing of investment finance and the eventual marketability of the asset, the emerging business environment and 'new wave' of tenants and workers provide property owners and managers with new opportunities to attract and hold tenants and maximise the opportunities they present. Obviously, this requires an asset offer that is well located and of a design, style and fit-out conducive to contemporary business. However, an understanding of the particular requirements of those businesses and workers is important in developing a 'business partnership' that extends further than the simple landlord-tenant relationship – sometimes adversarial approach of the past. These requirements include good communication and information flows, response to complaints, individualisation of space, security, sustainability, parking and quality air conditioning.

8.3 Retail

The nature of the sector

Retailing forms part of the service sector and contributes about 7 per cent of Australian GDP, though its full impact cannot be exactly quantified both because of its inclusion in the wider 'services' category and overlap with distribution-related activities (Baker Consulting 2018).

This issue of close definition, however, does not distract from the critical importance of the sector to the overall economy and community and the property sector in particular. For example, retailing employs about 1.3 million people, a little under 11 per cent of the entire workforce (Baker Consulting 2018). Australia wide, there are 1630 shopping centres aggregating about 26.5 million square metres GFA space and accommodating some 65,000 shops (SCCA Urbis 2015). In June 2014, these centres had an estimated annual turnover of $130.7 billion.

Given that the sector is embedded into all Australian communities, its operational and financial success has a profound effect on the circulation of money and on related multipliers (refer to Chapter 2). It also represents a primary client to the development and construction sectors and, on completion, provides an attractive and historically sound investment of a scale attractive to large trusts, superannuation funds and corporate investors (Ball et al. 1998).

The asset base is geographically spread across the country and states, obviously enough, paralleling the location decisions of the wider population. In those urban areas, about 91 per cent of retail trade is undertaken in regional, sub-regional and neighbourhood centres, with CBD retail trade now representing only 5 per cent of activity. Outlet and 'big box' models, imported from the US over the past decade, while increasing in importance, still only represent between 2 and 4 per cent of total turnover – again depending on definitions for classification (SCCA Urbis 2015).

Online retail transactions represent a continuing challenge to 'bricks and mortar' retailing in Australia and, indeed, to regulation and taxation authorities. This still represents a relatively modest proportion (about 7.25 per cent) of total retail sales (Statista 2019). The key observations here is that there has been a wide acceptance of this form of retailing across the market for many consumer goods and with maximum penetration focussed on a range of consumer durables. Additionally, the rate of growth of this sub-sector, at about 14 per cent compounding, is far greater than that quite subdued state of general retailing over recent years (Trading Economics 2019).

Compared with practically any other sector, the performance of large-scale retail property assets has been remarkable over several decades, exhibiting a consistent history of reliable cash flow and compounding capital growth. This growth is based on remarkable long-term, compounding increases in annual consumer spending across Australia – on average at 5.71 per cent over some 36 years to 2019 (Trading Economics 2019). For superannuation funds and corporate investors, major retail centres have proven attractive large-scale investments, either in single ownership, syndication or through Australian Real Estate Investment Trusts (A-REITs).

Additionally, because of the scale and importance of the retail sector, there are very close economic and other linkages with the construction, development, and property investment sectors, each dependent on the other for the effective production and management of the assets, and for physical accommodation.

Some history

Historically, significant retail uses had been located within the CBD core; but, by the mid-1960s and through to the mid-to-late 1990s, a new wave of contemporary, freestanding retail centres were developed, principally in key suburban locations across all Australian cities. These centres were typically large-scale, freestanding and housed one or two major traders with a range of specialty shops. The design enforced vehicle-pedestrian segregation.

Within that model, a general hierarchy of centre size was established and reinforced by town planning of the time, whereby the catchment of smaller centres would be aggregated into the overall catchment of a major regional centre. In general terms, that hierarchy is summarised in Figure 8.1.

Several observations need to be made regarding this Table. First, the hierarchy arrangements are such that the catchment area of one large centre (say, a sub-regional centre) may overlie the catchments of a number of smaller centres (e.g. neighbourhood) from a lower category. Therefore, for example, the approximate catchment of, say, five or six neighbourhood centres may loosely form the catchment of a sub-regional centre and so on. Because of the different target markets and specialisations, there may be limited competition between centres from different levels. A local or neighbourhood centre, for example, may be located in close proximity to, say, a regional centre with practically no trading impact on either of them.

Second, the concept of this hierarchy should be accepted as providing a general classification only. There will be many centres that align with a single category and also range across a number of them.

For example, CBD retailing is not included as a separate classification, given these precincts are unique to each city and are, typically, diffused with a range of other commercial, entertainment, civic and cultural uses. In any case, within the fabric of contemporary Australian cities, the CBD retail component is not dissimilar to the role of a centrally located 'regional' centre. Finally, given the diversity and level of specialisation of parts of the retail sector, the 'other' classification identified in Table 8.1 is increasingly difficult to qualify and quantify with uses ranging from 'big box' and outlets centres through to boutique, tourism and other specialised clusters.

Table 8.1 Typical retail centre hierarchy

	Regional	Sub-regional	Neighbourhoods	Other
Typical components	One or more full-line department store, a full line discount department store, one or more supermarkets and 100 plus speciality shops. (Can also be a centre with two full-line discount department stores – all other characteristics being equal.)	At least one full-line discount department store, a major supermarket and 40 speciality shops.	A supermarket and up to about 35 speciality shops.	Includes emerging, specialist and freestanding centres ('big boxes'/single traders, outlet centres, et cetera). Typically draw trade more widely than an immediate catchment.
Typical size	30,000–50,000 m^2	10,000–30,000 m^2	<10,000 m^2	Varies. Typically freestanding.
Approximate catchment population[a]	300,000	120,000–200,000	25,000	Varies but often wide.
Approximate % of consumer spending	Approximately 27%	Approximately 33%	Approximately 31%	Currently a relatively small proportion of total spend (<5%) but rapidly growing sub-sector.
Number of centres in Australia[b]	78	291	1,120	N/A

Source: Based on SCCA and Barker Consulting reports, 2018, and PCA/Urbis reports, 2015.
a Also impacted by population density.
b Plus 96 CBD centres.

Overall, therefore, the hierarchy concept has probably lost relevance somewhat in recent years with the emergence of a number of specialty or hybrid centres that do not align with conventional groups.

There has been a range of drivers for the development of retailing – a major business and property sector. Some were economic, some community and social based, while others were facilitated by land-use planning in Australian cities. The most important of these drivers are as follows:

- rising affluence (rapidly increasing levels of consumption, confidence, generally high employment levels and the rise of two-income families)
- changing demographic characteristics (the impact of baby boomers reaching full productive capacity, changes in family composition, the rise of women in the

workforce, the influence on consumer patterns of migration and migration settlement, urban shifts within cities and urban growth)

- the changing roles of CBDs (becoming commercial/corporate hubs and therefore
placing cost and construction pressures on retail activities in those locations)
- rising mobility (with increasing affluence, a rapid increase in the number of private
vehicles, which allowed for cross-city transportation)
- the rise of oligopolistic major traders (facilitating economies of scale, range, convenience and price in an environment unavailable to old-style 'main street' strip centres)
- the encouragement of many town plans of the time to develop a retail hierarchy and
regional nodes to counter congestion resulting from radial city development
- the aggregation of major funding sources made available through the introduction
of compulsory superannuation, the progressive deregulation of the financial sector
and the development of financial vehicles (such as Australian Real Estate Investment Trusts, A-REITs) to channel such funds into major assets.

Since the mid-1990s, however, there have been distinct changes in development patterns within the retail sector and, while continuing to grow significantly, the sector is
now more mature, with few greenfield regional developments emerging.
Some of the characteristics of this new phase are as follows:

- a more complex and sophisticated market (more interested in choice, and width and
depth of offer; changing consumer demand seeking point-of-difference in size and
location, including specialist products and a rise in the demand for fresh produce)
- near saturation of traditional retail outlets in many areas with few greenfield sites
and few anchor tenants available
- difficult and protracted process of obtaining development approvals for major new
centres
- the rise and increasing impact of e-commerce
- little potential for new generalist centres, but a move to 'theming' such as home-
maker centres, smaller, targeted inner-city developments, retailing integrated with
tourism, entertainment and residential and 'big sheds', 'edge city' or 'category
killer' centres, including clearance and discount centres
- fewer restrictions on trading hours and the integration with entertainment and
'third place' community centres
- demographic change, rising affluence, different demands and attitudes among
young adult and teenage groups
- threats to the status and position of department stores as anchor tenants
- challenges to the further growth of mature centres, given the ongoing and changing consumer patterns of their established catchments
- an ongoing commitment to refurbishments to maintain and upgrade/expansion
market share.

Retail centres as property assets

Large-scale retail centres are among the most complex of property investments and
assets. They are very large scale, with commensurate, complex cash flows. Large centres are challenged by a range of strategic and operational issues that arise from a

diversity of tenants – from some of the largest corporations in Australia through to sole traders – and by the complexity that arises from a huge number of daily visitations by consumers.

Unlike an office building, there is a strong symbiotic relationship between stakeholders who own, manage or occupy a retail centre. Regardless of the centre's location, its target market, theming or any other characteristic, success for all will rise and fall on one objective – the maintenance and enhancement of retail sales. Consequently, there is a common understanding that, for each stakeholder and tenant to succeed, the centre as a whole must likewise be successful in drawing customers in the first place. Even the best of businesses will struggle in a centre that cannot attract the necessary volume of daily trade. Furthermore, because of the integrated offer presented, the failure of an important trader or group of traders to perform can have catastrophic impacts on others – depending on the nature of that business, its location in the centre, et cetera. All tenants have responsibilities here, but the performance of the major traders in the centre (the 'anchors'), their reputation, offer, target market and so on are critical. The location of the anchor(s) regarding other traders and clusters in the centre as well as pedestrian patterns is also important.

Each centre presents a unique combination of location, identity, theming and targeting, tenancy mix, and width and depth of offer. However, the underlying elements in all centres remain the same. In considering the viability of any existing or new centre, the following are key criteria:

- The economic and demographic profile of the catchment area (aspects include: economic, employment, stability, history, likely future growth or decline, age and socio-economic profile, racial groups, education, types of employment and potential for adaptation for the future).
- The physical location and characteristics of the site and improvements (including zoning, road networks, access and egress, floor plates, layouts and ability to expand).
- Competition (identification and appraisal, scope for expansion/change).
- Theming and layout of the centre itself – design parameters and financial and performance criteria (see below).

The design and layout of the centre need to meet the specific demands of that catchment group and, additionally, have in-built adaptability to evolve with changing requirements. In general terms, the following design considerations are important:

- The provision of retail services as a convenient and enjoyable offer for the customer group. In particular, the anchor tenant(s) need to be of particular attraction to that population, with specialty stores targeted to the same socio-economic groupings. The overall concept must be inviting and attractive, with a layout that is convenient, safe and logical for customers moving through the centre.
- The design needs to satisfy tenant demands as to shop type, design, size, shape, location and the ability to amalgamate, reconfigure and upgrade over time, without major dislocation to the centre or building services, et cetera.
- The centre needs a level of uniformity and theme, with a style, concept and design that flows throughout to help promote its identity, which can prove a particular challenge as the centre progressively expands over time.

- The centre needs to be aesthetically appealing both within and external; to have identity through good exposure, easy access and the ability to accommodate large volumes of consumer parking that is safe, logical, segregates pedestrians from vehicles – with easy access to shops to facilitate transport of goods by customers.
- The centre needs to comply with existing town planning and development and other statutory requirements and, where possible, to have secure expansion potential.
- As well as convenient parking and amenities, larger centres particularly need to provide public/civic spaces and activities relevant and attractive to the catchment demographic. Cleanliness, physical safety and security are important issues in pro-visioning these amenities.
- The centre needs to be affordable to all, not just to consumers; and, for tenants and asset owners, provide maximum space at optimal cost, considering ongoing oper-ational costs. With that in mind, particular attention in design and construction needs to be paid to high-quality surfaces and finishes. This will minimise the costs of ongoing cleaning and reduce the physical risk profile.
- The quality, location and adaptability of air conditioning services and the size, loca-tion and adaptability of service centres are also important considerations.

Assessment and performance

Retail centres are income-producing properties and, consequently, are subject to the same performance and assessment methods as other income-producing properties, as outlined in Chapter 10. Capitalisation of income and the achievement of a particular internal rate of return (i.e. hurdle rate) represent key criteria.

As noted above, practically all cash flows, rentals and income in a shopping cen-tre emanate from retail sales. The ability of individual tenants and the ownership and management of the centre to maximise retail sales is critical to the overall objectives of competitiveness and marketability of the asset, which are assessed through comparison of rates of return (ROI), capital growth and effective risk management.

More specifically, but based on all of the above, some criteria typically used in the assessment of centre performance are:

- capital value per net lettable area
- initial yield, percentage return on purchase price/value derived from current pass-ing net rental income (if known)
- moving annual turnover (MAT), that is, the average of the last three annual perfor-mance figures and a rolling assessment of that, year to year
- rental level achieved over the year, either per square metre, net lettable area or per square metre gross floor area
- dollars dropped per customer visit (i.e. the average expenditure for a person enter-ing the centre 'average spend per customer')
- dollar turnover per square metre, net lettable area or gross floor area
- vacancies, including trends in the number, size and type of vacancies.

While all of these provide key performance indicators, a range of other criteria need to be considered in developing a full financial and operational analysis of these complex and dynamic assets. Additional considerations are:

- customer and tenant surveys and demographic and demand surveys across the primary and secondary catchment areas – in all cases looking longitudinally at studies in the past and identifying even subtle changes in demand and sentiment
- the potential that the centre has to expand and refurbish – an important criterion as it provides the centre with a level of flexibility and adaptability into the future to respond to changing market demand and the actions of other centres competing for market share
- the level of churn (i.e. turnover) of tenants and the tenure and lease conditions of major/anchor tenants.

Tenancy issues

The performance of individual retail tenants and the relationship between them in providing a full offer to the customer base are vital to the success of the entire centre. Each centre will have its own configuration as to the best types of services offered, their location and the grouping of individual tenants within them. These services will evolve over time; nothing in retailing remains the same for long and must continually be upgraded and modified to refresh the offer and maintain its attraction to continually changing consumer preferences.

Obviously no centre can offer all retail services; therefore, decisions need to be made as to the overall theme or recognised specialisation of the centre – be that fashion, fresh food, entertainment, prepared food, white goods and consumer durables and so on. These decisions on types of services are sometimes called 'width' and 'depth' of the offer, that is, the width relates to the range of services be offered, and once that is decided, the amount of comparison shopping and complementary businesses within that general category will be provided (i.e. the depth of offer).

Some tenant businesses are based on high-volume, low-margin goods and services. Others will supply non-perishable, larger goods that require extensive showroom areas and have a lower turnover and, therefore, they are unlikely to pay the high square-metre rental of other traders. The sale of whitegoods is an example.

Professional services that rely more on personal and professional goodwill than locational factors may be unwilling or unable to pay for the higher rents of other more intensive site-specific uses. Further along the continuum, charitable institutions, second-hand shops and the like are typically able to pay only very modest rents. A good indicator of the success or otherwise of a particular centre lies in monitoring any changes in tenants, such as a general trend over time from a high intensity/high rental business to a lower intensity, larger/cheaper space user. Not only does this affect rental income, but it can also profoundly affect tenancy mix and the overall offer provided by the centre. Some traders within a centre, particularly those selling foodstuffs, require large-volume daily trade to ensure the quality of their merchandise is not diminished. Consequently, these types of shops, particularly butchers, bakers and those trading in the food court, are often the first to be affected by a downturn in trade and centre patronage. Significant turnover of tenants or vacancies also represent a serious, early under-performance indicator for most centres.

Retail centre management and related matters

The strategic and operational management of a retail centre requires significant skills beyond those normally required of asset and property managers. The scale is significantly larger than most other property investments and, as noted above, performance for all stakeholders lies in the protection and enhancement of retail sales. Therefore, centre managers need to be familiar with marketing concepts and the retail sector in general and develop a good rapport with tenants to enhance individual tenant and overall performance of the centre. Even in the most successful retail centres, vacancies will occur from time-to-time and clearly need to be addressed as a priority since rental income lost in a particular period can never be recovered. Nevertheless, a careful approach needs to be taken in such cases so that the tenancy mix of the centre, and the immediate precinct within the centre, is not adversely affected, but enhanced. The reactive tendency to simply 'fill a space and secure rent' may not only result in the new tenant being unable to successfully establish the new business but, more widely, may adversely affect the viability of the existing shops that form part of that tenant mix and rely on the other tenants to provide a complementary offer. Other more general property management requirements and activities are included in Chapter 13.

Finally, it is noted that in practically all state jurisdictions in Australia, specific 'shop lease' legislation exists to set out overall parameters for centre dealings and the relationship between owners and tenants in these environments. It is critical that both the asset owner and the manager understand and fully comply with legislation relevant to that centre's location. While there are important differences from jurisdiction to jurisdiction, the thrust of this legislation is typically to protect smaller tenants in centres from a potential power imbalance between the asset owner and the tenant. Smaller tenants often rely on their location and trading conditions being maintained within the centre. A detailed knowledge of the relevant legislation needs to be acquired, but typically the laws specify:

- specific methodologies for assessing 'market rent' – normally identifying that goodwill is owned by the individual tenant rather than the centre owner
- set provisions for allocating outgoings
- provisions of disclosure statements by landlords
- cooling-off periods prior to leases taking effect
- the establishment of mediation procedures and tribunals to assist in resolving disputes
- the emergence of provisions to outlaw 'unconscionable conduct' by any party in undertaking a lease.

8.4 Industrial

Industrial use parameters and zonings

Industrial land and buildings represent another significant component of the property sector, though without the profile of others. They accommodate a range of business types including manufacturing, fabrication and assembly, distribution and transportation and service industries (i.e. largely trade-based enterprises servicing local customers).

For over half a century, local government land-use zoning plans in Australia have recognised the various sub-categories of industrial land. Typically, these classifications have included:

- future industry (i.e. englobo or accommodation land reserved for industrial use into the future as demand requires)
- heavy or general industry
- noxious, offensive or hazardous industry
- light industry
- service industry
- extractive industries (to preserve and protect deposits of mineral and construction materials)
- special zoning or development control areas to accommodate uses such as ports, airports, transport infrastructure and the like.

Each of these categories has its own demand criteria and desirable characteristics. Heavy and general industry, for example, will typically require large sites, regular shapes (as a single lot or for amalgamation), near level topography, sound sub-strata and access to reliable utilities (water, electricity, gas, ICT and waste disposal services). Depending on the commodity produced, proximity to raw materials, location within a supply chain, access to a suitable workforce, and accessibility to ports, railways and/or highway networks may be high locational priorities. The ability to store materials on open hardstand, to accept and manoeuvre trucks on site and the ability to expand operations over time are also often important to site functionality.

Activities classified as 'noxious, offensive or hazardous industry' might include oil refineries, petroleum and gas storage, chemical plants, gravel and sand quarries and concrete batching plants. They will typically attract specific conditions for safe operations including the provision of buffer areas, adequate waste disposal facilities, contamination control and consideration of prevailing wind and noise levels. These uses require close regulation to ensure that environmental and community health and the overall amenity of the location are protected. Issues such as prevailing winds and smoke drift, the bunding of sites to defend against major spillages, the disposal of waste and the sealing of sites to ensure that noxious chemicals do not leach into surrounding lands and waterways all need to be adequately addressed.

While not understating the environmental damage inflicted by certain industrial processes in the past and the potential hazards still presented by many of these land uses, the contemporary situation presents a quite different scenario. In the first instance, most industrial processes in countries like Australia are now much more sophisticated and advanced and with more contained systems than in the past. Second, a whole suite of development, environmental protection and workplace health and safety legislation and regulations across all Australian states now control such developments in a much more detailed and specific ways than through the application of land-use planning controls alone.

Overall, a balance needs to be established. The viability and sustainability of these activities are essential to the physical development and operations of contemporary towns and cities. Their optimum location and efficient access to those urban areas also require protection and management.

Light manufacturing or service industries include small-scale fabrication, motor and other repair workshops, wholesalers, storage facilities and a range of others. These may have a limited negative impact on the surrounding area. Furthermore, given the products and services provided, they need to be easily accessible for the wider community and residents who typically represent their customer base.

It might be argued that current statutory planning and strict use segregation often reflect industrial activities of past eras. The declining state of many traditional industrial areas and the alternative uses for these areas provided by, for example, residential and 'big box' retail projects would attest to these changing demands. 'Industrial' uses are now typically smaller and less intrusive than those of the past, and these businesses may find little benefit in the austere built form provided by traditional industrial suburbs (McDonald & McMillen 2010). Overall therefore, increased integration rather than segregation of land uses is now required but this requires some fundamental shift in planning philosophy.

As for industrial building design and development, parameters such as size, shape, access, egress, load-bearing capacity of floors, maximum and minimum slab-to-roof clearances, crane services, fire rating and certifications for various types of industrial usages need to be considered. It should be noted also that much of the equipment used for goods and storage transportation (forklifts, mobile cranes and other specialised plant) have limited ground clearance and little or no vehicle suspension. Consequently, steep ramps, slopes and imperfections in floor slabs and hardstand areas can seriously limit operations.

Regard should also be had to the demonstrable changes in the ratio of office/showroom space to factory/storage areas now required in many contemporary business/industrial uses. Previously, and still reinforced by many town plans, only relatively small proportions of space were allocated to offices and showrooms. The rational for that was that such property uses would require a larger proposition of floor space used either as manufacturing/assembly areas or for finished stock available for immediate delivery. In many regions, tenant businesses are now more likely to be involved in service industries and distribution with a much lower demand for manufacture or storage on site. With significant advances in communication and the sophistication and speed of distribution networks under 'just in time' models, the holding of large on-site inventory stock is less likely – thus significantly changing the built form demanded.

Wider impacts and networks

At a national, state and regional level, all of these industry sub-groups are very important at an economic and community/services level. Many secure export income and most provide major opportunities for investment and employment. Often, a range of industrial processes and services will be required, aggregated around a common theme (i.e. a 'cluster') or to complete a production process (i.e. a 'supply chain').

For any business, including traditional and contemporary industry, there are advantages in locating in close proximity to similar and/or complementary businesses. There is a range of reasons for this. Clustering often provides market identity and allows access to a greater workforce pool, specialists, equipment or logistical facilities that may aggregate in the same locality. Often firms will be involved in the same supply chain, providing services to one another or to a downstream major final producer. These clusters or precincts are encouraged by the town planning scheme for that locality (O'Sullivan 2009). It should be noted that while a 'cluster' will normally imply some physical proximity, supply chains may take different forms. Often they will be localised and immediate but, when a larger national, or even global scale linkages are involved, the network may be widely dispersed. For example, final producers may draw various components though a number of streams, eventually all arriving at the point of final manufacture/assembly. The automotive components sector provides a long-standing example of that type of arrangement, but more contemporary examples of design-to-manufacture, global enterprises are now common across many sectors.

The full scale and importance of all of these uses and assets may not be immediately obvious, given the diversity and sub-classifications of the economy involved and the interaction between them. For example, manufacturing itself represents only about 6 per cent of GDP directly, but the use of 'industrial buildings' extends through services to mining (10 per cent), construction (8 per cent), retailing (4.1 per cent) and many other components of the services sector (including transportation and logistics) (RBA 2019).Though the majority of these sectors in statistical analysis are not defined as 'industrial', their links, interface, link or interdependence with industrial buildings / assets are obvious (Kelly 2015). The importance of such assets continues to grow rapidly based particularly on the rapid changes in logistics, sales and distribution systems described below.

The increasing scale of development and the tendency of these land uses to cluster have implications for urban design, development and the provision of services and infrastructure. Given the increasing importance of skilled, human resources to these activities, the accommodation of the required workforce together with their community services and commuting arrangements also require closer consideration than was typically the case in the past (Glover 2015).

Real property ownership models are also evolving. Traditionally, properties which accommodated large-scale industrial undertakings were held by owner/occupiers, or by an associated holding company. Such arrangements had (and continues to have) considerable advantages given the long-term and specialist nature of many of such assets. Other smaller scale undertakings were held as either owner/occupiers or as lessees of industrial sheds, hardstand areas and associated offices, storage, et cetera, procured on the private market.

Recent decades have seen an increasing number of large industrial properties (including improvements) held be investment trusts. Typically, individual, trust-owned properties are leased to major manufacturing, distribution, service or other operating company under very long-term, net-net-net leases. There can be substantial advantages for tenant companies under such arrangements through the freeing up of the company's capital while also potentially providing significant taxation advantages. Additionally, new large-scale development concepts have emerged in the form of multi-tenanted industrial, business and technology parks and precincts. Their scale, long-term, secure income streams and multiple tenancies have also attracted large trusts and investment funds seeking to diversify their property portfolio. The new wave of distribution centres described below provides further, major investment options for such groups.

Like other sectors discussed in this chapter, industrial land uses and property assets should not be considered as either homogeneous or static and, in fact, few sectors are more diverse or exhibit more radical and continuing change. Industrial property assets range from huge mines, processing and manufacturing plants, sugar mills and meat works, through to distribution and transportation hubs and small, local, service industry clusters. They are geographically scattered across the country, with some obliged to locate close to raw materials or primary production, port, rail or road network or large quantities of water or energy. Some need to locate close to customers; and others, because of the nature of the product or skills involved, may have a range of location options available to them.

Often the unit value of the raw materials and of final production will be the major determinant of industrial location. Consider, for example, an industry such as coal production in Australia. Here the unit value of the product is relatively low and, at the same time, Australian labour costs are high by world standards. The only way that such a sector can remain viable here is to identify and extract only the highest quality resource and to invest in a capital intensive, highly mechanised bulk coal extraction, transportation and processing systems. Thereafter, production volumes must be maintained at very high levels and transportation costs per tonne kept to an absolute minimum. Under those parameters, very large-scale mines with easy access from the mines to bulk rail and port facilities, such as those near Newcastle, New South Wales, and near Gladstone and near Mackay, Queensland, represent the only viable physical locations.

At the other extreme, 'industrial' uses might include very high value-added activities, say in biotechnology, nanotechnology, instrumentation or one-off electronics circuitry. These are highly skilled areas and may occur almost anywhere – in a university laboratory or workshop, technology park or a range of other possible locations, with the final product easily air-freighted to anywhere in the world within a very short time.

Uses such as these are increasing in Australia and other OECD countries as they move to what are now increasingly known as 'post-industrial', knowledge-based era (or 'the second machine age'), where supply chains are often global rather than local and the use of technology, in the form of Artificial Intelligence and robotics, become fully integrated into economic and business processes (Brynjolfsson & McAfee 2014).

Because of the importance of many of these uses (particularly resources and manufacturing) to employment, exports and regional development, state governments have historically been willing to become directly involved in the sector through the provision of serviced land, transport and other infrastructure and utilities in concentrated industrial areas (often called science and/or technology parks). As industrial sectors have become larger and more private sector developers and investors have been attracted, the involvement of the state governments in such activities has diminished rapidly.

Emerging trends and changing assets

As noted above, the nature of industrial land uses and assets in Australia have changed dramatically over recent decades. While Australia has considerable economic advantages in natural resources, infrastructure and skilled labour, it suffers from high-cost structures, a small domestic market with limited economies of scale, geographic remoteness and volatile exchange rates. The emergence of highly competitive global markets, the rise of new technologies and high-quality, low-cost production in Asia in recent decades have exacerbated this phenomenon particularly for large-scale, export-exposed sectors such as steel and car manufacturers.

In Australia, these factors have resulted in radical industry restructuring and, similar to other higher-cost countries, plant closures have followed in many of these sectors. Remaining activities have typically been aggregated into larger plants, often in centres such as Melbourne and Sydney, where investment in new technologies has improved efficiency but, often, also reduced overall demand for high-cost labour.

All of that accepted, it is wrong to construe that manufacturing and other industrial uses will become near extinct in Australia. This diverse sector is undergoing major and sometimes painful change and there is no doubt that non-competitive enterprises will continue to struggle. Given this scenario, it will be difficult for the owners of these often highly specialised property assets to hold rental or capital value, as these assets typically do not accommodate change well (Glover 2015).

Change, however, is inevitable and must be accommodated by owners, as well as management and the workforce. The changes in industrial processes, near-instantaneous communication and 'just in time' manufacturing means that, for many enterprises, little stock needs to be held-at-hand. As noted above, this significantly alters the proportion of space required for the production, storage and administration/office components of those businesses. Thus, industrial land developers and asset owners need to provide a more flexible and responsive built form – far different from the factory buildings and industrial suburbs of the post-war era (Raymond 2019).

Overall therefore, there is a clear trend in Australian industry (paralleling many other parts of the economy) to move away from the large-scale, heavy manufacture of relatively low-unit-cost goods based on large investment in long-term fixed assets, patents and mass production. Rather, growing sectors are typically of smaller scale, diverse and with specialised targeted production requiring high skill levels and innovation. These uses are typically based on the knowledge, experience and networks of employees and, therefore, not restricted to particular locations as were industries of the past. Under this scenario, high-quality amenities and liveability for employees become a key location determinant for contemporary, 'high end' manufacture and the creative and innovative activities that now form part of those businesses (Florida 2002, 2004).

More positively, a significant number of existing manufacturing enterprises in food processing and other areas have adapted to the new environments and prosper in both local and international markets. Additionally, new forms of manufacture and other business models are rapidly emerging in many regions of Australia and take different forms from industry of the past. Often these new firms are smaller in size and more specialised – focussing on knowledge, knowhow, and intellectual, human and social capital – enabling firms in countries like Australia to have a true, long-term competitive advantage, overcoming inherent challenges of high input costs and remoteness from major markets. An early response to the COVID 19 pandemic has been a call to rejuvenate traditional, on-shore manufacturing. While, with government support, that may possible into some strategic production such as medical supplies, those fundamental challenges for large scale manufacture in Australia remain as major impediments to such expansion.

Perhaps the most important change on industrial land use in the last half century reflects the impacts of both the globalisation of manufacturing and the advent of e-commerce and the overall increase in consumer spending worldwide (Raymond 2019). This trend found a ready home in the US where the practices of 'mail order' had been in place for over 150 years and readily shifted from mail order catalogues to internet shopping. While somewhat slower in take up in Australia (still only about 9 per cent of total consumption spending), its compounding growth of 29 per cent per annum over the 4 years to 2019 attests to its growing impact that it will continue to have into the future (CBRE 2018).

This change is important at a number of levels. In general terms it provides something of a case study of one of the themes of this text – the ability of real property to evolve and adapt in the face of changes in demand. In this case, it represents having significant effects on the nature and location of assets that will house and provide platforms for new forms of 'industry' into the future.

Furthermore, contemporary 'industrial' activities are more easily integrated with other land uses.

This has resulted in new forms of industrial land use – notably large-scale distribution centres (incorporating wholesale/storage, logistics and the arranging of rapid delivery to customer level) These are termed 'omni-channel' systems where supply from a wide range of sources are organised, typically using a 'just-in-time' approach. As well as imported and locally manufactured goods for e-commerce transaction, comparably systems are now established and being expanded for both perishable and non-perishable retail goods. These large-scale storage and distribution hubs seek out large-level, regular-shaped sites, strategically located as regards production sites, airports, rail hubs and ports and, particularly, the national highway network.

Industrial property used for such purposes takes on a much higher functionality and value than was previously considered and seen as 'simply sheds'. Contemporary distribution facilities involve very high levels of sophistication as regards the use of technology at various levels, including Artificial Intelligence (AI), robotics and systems development – all coordinated at that physical site and based mainly at the retail/consumer level, multiple commodities and decentralisation of those supply networks (Raymond 2019).

8.5 Non-income-producing properties and portfolios

The nature of not-for-profit organisations and their assets

The premise of contemporary property theory and studies, and the basis of this book lie in finance and economics, that is, the understanding that property assets are a valuable and scarce resource that will be typically used to secure the optimum benefit from their highest and best use. Consequently, most of the analysis, and many of the decision-making models, centre on the ability to derive income and capital growth as the primary objectives of holding and managing that asset. This is quite a reasonable proposition, particularly for commercial investments. While the residential sector is far more emotive and subjective, basic economic imperatives of capital value, rent and holding costs are very important in the mind of the owners in that sector.

There is, however, one other general category of property owners for whom these normal economic maxims and financial returns on investment are of quite limited strategic importance. This group is generically known as 'not-for-profit' organisations and typically include government, church groups and charitable institutions, universities and schools, hospitals, community organisations – through to public assets such as national parks, public utilities, public infrastructure and so on. While this grouping is both extremely large and diverse, it typically holds property with the intention of using it to provide some service, whether or not that service is profit-making or represents, in an economic sense at least, the best returns for the often huge capital investments involved. Typically, political, civic, charitable and/or religious imperatives dominate.

That is not to say that the property asset and its function are any less important than for commercial owners, but the manner in which not-for-profit organisations consider performance measures and 'successful property outcomes' is quite different.

In times past, the various levels of government were frequently quite inefficient in the ownership and management of property resources. A range of reforms of the public sector in recent decades, combined with a more commercial approach to government business, have largely overcome those past failings with many public-sector bodies now providing close to 'best practice' in strategic, operational and facilities management of their property portfolios.

In many cases, public-sector service providers also operate in a commercialised or corporatised environment where, under Productivity Commission encouragement and National Competition Policy parameters, operations are freestanding and transparent with tax equivalent paid, and separate annual reports and financial statements issued.

For private sector investments, the ultimate deterrent lies in financial failure and bankruptcy. For government/public sector and other major institutions, the concept of bankruptcy does not really exist within the normal commercial sense of the word. Nevertheless, these organisations do face other real, if less quantifiable, risks through political embarrassment, legal action and loss of public trust and confidence which, for them, may be practically as damaging. Second, it should be noted that public institutions have quite different concepts of time imperatives, not simply in the time taken to make decisions, but also in the length of time that assets will typically be held and used. This compares with private sector commercial considerations through capitalisation of net rents, where commercial horizons rarely extending past 12–15 years typically required payback of invested funds. Provisions establishing the effective life of assets for taxation and depreciation purposes also reflect a relatively short period for write-off. For many not-for-profit institutions, such as those identified above, assets are often secured and held without any real consideration of subsequent sale or financial realisation of the asset.

Given these particular strategic parameters, the 'economics and financial' approaches to property assets that underpin much of this book need to be modified to some extent.

Assets and property management approach for not-for-profit organisations

Because these types of properties are so diverse in nature and use, it is difficult to be definitive about ownership strategies and management practices. Nevertheless, some general observations can be made.

In the first instance, effective ownership and management of these assets require a particular understanding of the basic philosophy, purpose, objectives and approach of the organisation owning and operating the assets and that of their stakeholders. This will include knowledge of governance, decision-making processes and responsibilities, legal and tenure issues surrounding property ownership and the taxation status of the organisation.

With that as an overview and strategic direction, the best approach is to manage and operate the portfolio on as close to a commercial basis as possible. Typically, direct rental income will not be available, but other components found in a well-run property portfolio anywhere – property registers, the use of cost centres (so that individual costs and charges can be allocated to the individual asset), together with regular audits, good quality maintenance systems, record keeping and so on – should be established and maintained.

It can be expected that these organisations will frequently make asset decisions that are less than economically optimal and, under these parameters, that may well be appropriate. However, a good quality asset and property management system will at least provide decision-makers with a benchmark regarding the opportunity costs of those final decisions.

In the contemporary environment, it also needs to be noted that government and other not-for-profit organisations are equally liable under various pieces of legislation to ensure the safe operation and use of those assets. Indeed, the level of embarrassment and potential for litigation may well be even higher for institutional owners than a private individual or corporation. This reinforces the importance of good quality management and systems, regardless of ownership.

Many institutional owners are 'asset rich', though the ability to rationalise those assets and maximise capital returns is often relatively limited. However these organisations should keep the ownership and use of all assets under regular review and corporately encourage policies that can liquidate surplus property assets and re-invest in the core, service activities.

Typically too, these types of organisations have limited operational funds and those, understandably, are largely directed to service delivery within the organisation. Consequently, the securing of funds for strategic maintenance expenditures, upgrades and refurbishments, and even to ensure statutory compliance, can be difficult. Again, this emphasises the importance of good audit systems and the rationalisation of assets no longer required. Owners of these portfolios, as with any other, must be aware of the full and wide implications of asset decisions.

In summary, the management of these portfolios and properties needs to:

- emphasise functionality and client service
- ensure full cost recognition and allocation to cost centres (i.e. individual activities and assets)
- recognise the obligation for statutory compliance to buildings, regulations, workplace health and safety legislation and for public liability now demanded of these institutions
- ensure quality, specialised asset management systems, recognising the specific issues and challenges of the particular organisation
- establish a regular asset-review, asset re-alignment and asset re-investment strategies to avoid additional costs of holding under-utilised resources
- use a normal commercial model as the basis for an operational and strategic asset management system and the benchmark for all asset decisions. Whether the final decisions made by directors represent best financial or economic outcomes or not is a matter for them, but at least it can be made with the knowledge of commercial reality.

8.6 Residential

General observations

Development and construction within the residential property sector is, for the most part, simpler than the sophisticated built forms accommodating commercial, retail and many industrial uses. However, when considered holistically, this sector represents one of the most complex areas within the economic, community, political and built environment landscape of Australia.

The complexity here arises from:

- the large scale of the sector, representing approximately one-third of the value of the entire Australian built environment
- the interface of this sector and the assets produced with the economically critical development, building and financial sectors

- the diversity of ownership, in which approximately 67 per cent of all households in Australia own outright, or own with a mortgage, their principal place of residence (AIHW 2019a)
- the electoral interest and sensitivity in securing home ownership and in related issues; this has resulted in public policy, typically by providing government financial support and taxation concessions that favour home ownership, and by establishing a range of long-term, complex economic market (demand and supply) aberrations
- the regional and physical diversity of housing assets and the uniqueness of each individual asset. The sector, on the face of it, meets the fundamental human need for shelter; however, within the overall objective, there is very wide diversity in physical characteristics, location, value and type
- the ability of the same housing asset to not only provide the basic human requirement of shelter and housing, but also transfer seamlessly into a source of income (through rent) and, depending on market conditions, produce a capital gain to the owner. The latter, because of political sensitivities noted above, will potentially attract different tax treatments for different use categories
- the concept of home ownership lies deep in the aspirations and ethos of Australian households. While those perceptions are arguably evolving, they are still fundamental to identity, prestige and socio-economic standing within most of the Australian community.

As at the Australian National Census 2016, there were nearly 8.3 million dwelling houses and other dwellings of various types in Australia. Dynamic forces are continually at work from within and external to the housing sector. These impact and mould both the quantum and nature of demand and, in turn, supply. In such a complex environment, simple causal relationships can rarely be established but, over time, trends become obvious.

The unique nature of the sector is highlighted by its statistical profile. A large proportion (approximately 73 per cent) of all housing stocks are detached dwellings with approximately 26 per cent being home units, apartments, town houses, et cetera (AIHW 2019a). It is important to note, however, that as the demand for and value of land component of residential developments increased, a significant rise in the number of multi-dwelling built form emerged. Tenure allowing for freehold title over multiple dwellings only became generally available in the late 1960s. By 2008, home units and apartments represented about 32 per cent of new residential construction, rising to 47 per cent in 2018 (HIA 2019), the shift most noted in the major cities.

In parallel, significant changes in residential ownership structures are underway. In 2017/2018, 66 per cent of all dwellings in Australia were either owned outright or were owner-occupied properties subject to a mortgage (a figure that had decreased 2 per cent since 2015/2016). Meanwhile, 32 per cent of all households occupied rented accommodation, a figure that had increased by 2 per cent since 2015/2016.

Meanwhile, overall housing stocks have increased substantially with, for example, approximately 56 per cent more dwellings (of various types) being produced in 2018 than 20 years before (HIA 2019). Population during the same period rose by a lesser 37 per cent – the difference largely the result of the falling population density/dwelling. (The average household now contains only 2.55 residents [Profile id 2016].)

As with other property sectors, it needs to be observed that the impacts of change in demand and functionality fall principally on existing stock which, as fixed, capital assets, may not accommodate significant changes well. Further, and as noted elsewhere in this text, unavoidable time lags within the system, mean that adaption to significant change will typically be a medium to longer-term proposition.

Some of the contemporary forces influencing the Australian housing sector include the ageing of the population and their changing locational and house style preferences, the dramatic overall fall in the size and composition of individual households, the drift to the capital cities and, in particular, to their inner suburbs and, finally, favourable taxation treatments of property investments, notably for residential owner-occupiers.

This environment over recent decades has generally favoured the purchase of residential property in Australia over renting, particularly in high growth regions and cities. The continued limitations of available land, rising infrastructure costs and underbuilding in some areas, together with the prolonged period of low interest rates and a range of a range of fiscal incentives have held demand levels high, despite slowing population growth. As a result, housing prices on average across Australia have increased by a remarkable 30 per cent since 2008 (Koukoulas 2015).

A recent influence on the Australian housing market has been the increased activity of domestic and off-shore property investors, particularly in the major cities. By the end of 2014, almost half of residential mortgages were for investment properties, compared with normal rates of about one-third in the past (Koukoulas 2015). Even though that higher proportion has subsequently fallen back somewhat, based on Reserve Bank encouraged changes in lending policies of the retail banks, such aberrations typically over-heats certain local markets, over-inflates prices and puts further pressures on affordability particularly for those attempting to enter the market for the first time.

Significant media attention on housing issues normally concentrates on superficial symptoms rather than root causes and analysis of underlying issues and problems. Furthermore, many of the statistics and analyses are provided on a national or state basis with local influences and characteristics more difficult to observe.

Typical 'headline' issues relate to affordability. This is normally defined as the ratio of average household income in a particular location to the median value of residences sold in that location during a certain period. Housing stress is defined under the '30/40 rule', whereby households in the lowest 40 per cent of gross annual income distribution pay 30 per cent or more of that income on housing costs.

It is also important to note what appears to be some confusion concerning the related topics of 'welfare' (social housing) and 'affordability' (affordable housing). The former relates to the normally government-sponsored support for those in the community who do not have the financial ability to pay market rental for accommodation (because of financial or other circumstances) and who would otherwise be homeless.

'Affordability', however, refers to the ratio of housing costs as a percentage of household income and, with that, includes issues such as value for money, choice, size and housing features demanded, regional variations and so on.

'Price point' as to the various housing options in the area may be more relevant as a guide to affordability (in terms of value for money) than raw statistics measuring overall housing affordability or housing stress.

Overall, housing affordability and stress issues fall heavily and very narrowly on the poor and those attempting to secure home ownership for the first time. For those entering the housing market, government grants and other support measures, together with periods of relatively low interest rates act to reduce the potential for housing stress, even if they are transitory measures.

Variations in regional household income and, indeed, very substantial variations in house prices between towns and regions can easily distort raw affordability statistics. Overall, however it would appear that, despite protestations, issues of prestige, identity and accessibility to government support and taxation allowances ensure that the demand for housing, even among first home buyers, remains generally strong. The very low level of default or even arrears on mortgages on principal place of residence housing in Australia attests to the fact that, despite individual hardship, households can, and indeed do, meet housing repayments and consider it a high spending priority.

While these are important issues, they can easily be taken out of context and proportion. For the majority of households, who either own their place of residence outright or have substantially reduced the size of their long-standing mortgage, accommodation costs in Australia are relatively low. There is, therefore, an understandable conservatism within that large and electorally important group to maintain the housing 'status quo' and not to significantly increase public expenditures or other interventions in these areas.

There is a mismatch here between economic realities and the final housing decisions made by Australian households that appear heavily influenced by social and community expectations. For example, while there were very significant regional variations, the average price of houses across all Australian capital cities from December 1995 to December 2014 rose by almost 300 per cent, or a compounding rate of increase of 7.6 per cent per annum (Kusher 2015). This compares with an average Consumer Price Index (CPI) increase of 2.6 per cent per annum over the same period (ABS 2015). Since that time, it is again difficult to identify any close correlation between CPI and housing price variations, and even the more anticipated links between interest rate levels and housing prices are muted in light of other influences such as other bank lending policies, existing levels of debt, wages growth, community confidence and a range of others.

There is a range of reasons for the remarkable increase in overall housing prices in Australia over recent decades. Some are indeed economic – for example, supply and cost pressures on-land supply and infrastructure, particularly in high-demand regions. Other reasons include related supply lags and intense competition for well-located suburbs, particularly in Sydney and Melbourne, where investors, in what is thought to be a strong and rapidly rising market, effectively crowd out potential owner–occupiers. This reinforces observations made above that a house easily moves between a tradeable commodity, a source of (rental) income and a residence in owner-occupation.

Fundamental demographic changes are also having a significant impact on the type and location of housing stocks demanded. Some relevant statistics highlight these changes: Based on the 2016 census (updated as available), of the 8.86 million dwellings in Australia, almost a quarter had only one occupant, and a further one-third had only two occupants. As noted above, the average Australian household now contains only 2.55 residents, a figure that has been generally falling for more than a century (Profile id 2016).

Even with only the modest growth in total population anticipated over forthcoming decades, the net result of smaller households will be, obviously enough, that the demand for additional housing stocks will continue to rise, particularly in desirable, urban locations and in capital cities. The types of housing demanded will almost certainly change, however, given that some 24 per cent of households consist of a single person (AIHW 2019a). These trends are now beginning to emerge. The size of new dwellings constructed had risen steadily from the 1980s, up over 10 per cent during the 15-year-period to 2012 (ABS 2013a, 2013b). That trend has now been reversed and, since 2012, average new dwelling size in Australia has been falling fairly consistently (CommSec 2018).

The readiness of this market to be influenced and, indeed, distorted by changing demographic characteristics and social, personal and, subsequently, political priorities, will continue. In conclusion of this statistical analysis of the residential property sector, the following, additional observations are made as relevant:

- the effect of the rapidly ageing population and the 'baby boomer' group approaching retirement with different residential property asset demands and likely reconfiguration of their property assets is already having significant effects on regional and general property markets. (Refer also Section 8.7.)
- In a comparable way, the lifestyle and housing choices of contemporary generations X and Y are less than clear but are set demand parameters into the future.
- While having a very localised effect on certain suburbs, the overall impact of foreign investment on Australian residential property markets is very small, overrated and sensationalised by those which may have unrelated motives.
- Using accepted Gini coefficient analysis,[1] there is growing inequality across the Australian community – to some extent in income distribution but more pronounced in wealth distribution. In the later measures, in 2017–2018, the wealthiest 20 per cent of Australian households controlled 63 per cent of household wealth, while the lowest 20 per cent controlled only 1 per cent of that wealth (ABS 2019). Much of that dichotomy is linked to capital investment in principal place of residence but now has the effect of removing an increasing proportion of the community from ever aspiring to home ownership. This may have serious social and political repercussions into the future.
- The low-density urban model, which emerged in Australia post Second World War, was predicated on the use of private motor vehicles which, in turn, relied of access to relatively cheap and plentiful (fossil) fuels (refer also Chapter 4). The effect of ongoing and disproportionate rises in energy costs, combined with growing congestion, increased commute times and environmental impacts now seriously challenge the sustainability of all of those models which will impact both on values in of less convenient locations and on urban renewal and urban design into the future.
- The regional nature of the Australian economy has been emphasised in various Chapters of this text – particularly and Chapters 4 and 9. Of recent years, variations across regional residential property markets highlight the growing disparity between regions. Unlike increases in market activity and values during past cycles, changes in overall levels of value have varied widely. Increases in Sydney and Melbourne and in the inner suburbs of other major cities have often been dramatic. Meanwhile, in many other regions, markets have generally been subdued and, in some locations actually fallen back as a result of falling populations and other economic challenges in those locations. Over time, these differences will almost certainly affect the mobility of population between regions, possibly with both positive and negative outcomes.

- As noted above, arguably the most important demographic change in Australia is the ageing of the population which now sees the mean age at years 37.2 years – a remarkable increase of 7.9 years over the last four decades (Statista 2019). A critical component to this has been the 'baby boomers', whose average age is now 65. The effect of changing locational preferences and on housing stocks demanded by this large cohort (about 22 per cent of the entire population) into the future will be very significant.

Residential rental markets

The operations of residential property markets are discussed in detail in Chapter 13 and also described in the preceding section. However, some commentary on these properties as investment propositions is also warranted and included hereunder.

The investment profile of residential properties in Australia, and therefore those attracted to in it, vary significantly from investors in other types of real property or, indeed other forms of wealth creation. In this case, the vast majority as individuals and other small-scale investment groups, family trusts, et cetera, typically drawn to the sector by:

- low barriers to entry because of the relatively low cost of individual dwellings (compared with other forms of real property)
- a historically sound level of capital growth (like all property investments, this needs to be considered over the longer term and there is certainly no guarantee of capital gains in all cases; however, leverage on borrowed funds and the underlying value of residential property in providing a basic human need provide a high level of security)
- relatively unsophisticated assets that are well known and understood by small-scale investors who dominate this market (i.e. the individual assets appear manageable)
- apparent tax advantages such as deductions for outgoings, interest payments and potentially the negative gearing of losses, which has at least superficial attraction for otherwise pay-as-you-earn taxpayers.

As with any other form of investment, however, residential property, on the face of it, also has a number of issues and potential difficulties. In this sub-sector, these include:

- while longer-term investment in residential property may show attractive capital gains, return on investment (as measured by net rent compared with capital value) is notoriously low in the case of residential investment
- tenancy periods are normally for a shorter term than for commercial properties (typically for six months' renewable, based on gross rental figures [i.e. the tenant not being specifically responsible for outgoings] and normally with no specified rental escalation provisions)
- residential rental properties can be subject to very heavy use and, sometimes, to malicious damage and bad debts
- for a range of political and other reasons, legislation that controls residential tenancies tends to be pro-tenant with eviction procedures and the ability to recover bad debts, et cetera, tending to be protracted and relatively expensive.

Because tenancy legislation, in effect, deals with contract matters and with real property, the relevant laws are typically state-based. Therefore, variations exist from state-to-state but, as well as the observation regarding the emphasis on the protection of tenants' rights, the following general observations should be noted:

- It is essential that the property owner and manager are conversant with, and scrupulously follow, any legislative and regulatory provisions regarding the establishment of tenancies, their administration, extinguishment or other dealings. If not followed precisely, the owner and manager may well find themselves in a legally difficult and compromising situation. Successful property ownership and effective management is based on certainty and legality of process and not on loose or generalised agreements or attempts to take alternative or, perhaps, illegal actions.
- Most state legislation establishes specific tenancy and administrative processes for rental occupancies, typically including processes for holding bonds, mediation and the settlement of disputes. Importantly, most legislation places direct responsibility on the owner to maintain any rented property in a 'habitable' condition regardless of condition at the start of the tenancy or the level of rent paid. Habitability, in this context, is normally construed as meaning safe, sound, waterproof, with all essential services and equipment operational and in a condition considered as reasonable, having regard to that location and comparable properties in the area. It should be noted, however, that these parameters may vary depending on jurisdictions.

8.7 Other urban property sectors – summary observations

The property sectors described above – commercial, retail, industrial, properties used for non–income-producing purposes and residential – represent the mainstream components of urban land uses. In addition to these, there are a diverse range of specialist property types that do not easily align with those generic groupings

Despite their remarkable variety, each of these asset types follows economic, financial and legal parameters outlined in Chapters 2, 3, 6 and 7 of this text. An investigation and assessment of any individual property requires the further consideration of the specific characteristics of that property and its 'context', use and potential – that is to say, its physical environment, the current status and prospects for that type of use and the general state of the regional and wider economy. Together, these will determine highest and best use, functionality and therefore value. Research into such uses and the assessment of individual site and their value is often made more difficult as that use may be uncommon, perhaps unique, in a particular locality and, therefore, have no immediate comparisons are available (API 1996). These matters and the methodologies to be used in assessment are further discussed in Chapter 10.

Based on that general approach, the following comments are made regarding a number of commonly encountered urban land uses.

In this, several initial observations are made. First, and obviously enough, these examples represent only a very small number of the specialist properties that may encounter in dealings within the property sector. Second, even for the cases discussed, the observations are by way of over-view only, setting general guides for the consideration and assessment of that type of property. Much more detailed investigations would be required to obtain a full appreciation of any individual property.

Childcare centres

A significant addition to the urban fabric in Australia has been the development over recent decades of some 15,500 child care centres across the country, accommodating over 1.36 million children through various age groups and providing long- and short-term support (Australian Government 2019).

The growth of the sector reflects the rising proportion of women in the workforce, changing family and social structures and (to some extent) rising affluence.

Given that the safety, welfare and development of young children (principally under five years) are involved, the sector is highly regulated with Commonwealth quality standards applying across all States. Based on those standards and a range of other development and service criteria, each centre requires operational certificates issued through state Child Welfare and Education Departments. Regular compliance inspections and audits from those state departments are also involved.

The property components are specialised but, compared with many freestanding business enterprises, the levels of investment required are comparatively low. Most centres are privately owned and operated, though there are a relatively small number run for not-for-profit community and church groups.

The securing of these highly prescriptive certifications and their maintenance through regular reporting and inspections are not only a legal requirement but are mandatory for parents to receive commonwealth benefits for their child's attendance at that centre. It is important to note that contemporary programmes at such centres does not constitute 'childminding' but rather has significant educational and activity-based components administered by qualified staff.

While government support has varied substantially in the past, it has generally increased over time, now effectively underpinning the growing demand for these services. Since mid-2018, Commonwealth funding arrangements changed significantly now directing support payments for each child to individual centres on that child's behalf. This provided more certainty in cash flow for centre operators but, at the same time, ensured freedom of choice for parents based on the quality of service provided. There remains however a considerable 'gap payment' which households must meet. Consequently demand for places for children within the sector remains very price sensitive.

The corporate model of the sector is unusual. Nationally, ten groups each own 30 centres or more. A small number are publicly listed companies. However over 75 per cent of all centres are owned by small operators holding and/or operating one or two centres only, sometimes co-branded and operating within a defined region where personal goodwill and operational efficiencies can be maximised (Burgess Rawson 2018).

There are economies of scale across these aggregations in the reuse of successful physical centre design, the reuse of programmes developed and approved and the development of staff. Past those advantages, however, advantages from economy to scale may be fairly limited.

In a similar structure to asset investments such as hotels and motels, a two-tier ownership model often applies whereby one party will take a fairly passive role as asset owner with the facility then being leased to an operator under a long-term net-net lease. Such leases will normally be of ten years duration, sometimes more and with a number of options. Rental will often be tied as a percentage of gross income.

For individual centres, the environment is always dynamic with demographic changes in the catchment and as children progress within the age groups in the centre, finally advancing to school. The pursuit of future business (i.e. forward enrolments) is therefore a continual priority.

Goodwill and business success and sustainability over time is normally reinforced by personal referrals (i.e. local 'word of mouth'), based on such criteria as the quality of staff, contemporary premises, quality of programmes and convenience of location. Similar to most other 'service based' enterprises, the most successful centres appear to be run by operators with hands-on experience.

As regards the development process, many local authorities encourage the establishment of such centres in residential areas, given the community service that they provide. The concentration of local traffic and the need to provide safe drop-off areas and off-street parking are important criteria as are fire safety and access requirements. Compared with other business opportunities, the cost of developing such a centre is on the face of it attractive; however, the securing of all approvals can be arduous and 'ramp-up' time to full capacity can take 12 months or more. Further risks can evolve as the demography of the local catchment changes over time and, as noted above, there are few barriers to entry to other competing centres establishing in the same catchment.

Most contemporary centres are purpose built and typically require a 2,000 square metre site in a five room, adaptable configuration with semi-open areas, staff amenities and food preparation areas. The size of centres does vary considerably depending on the circumstances of the case, size of catchment and available built envelope. It is generally considered however that a viable centre will need at least 100 places across a number of age groups and would look to an occupancy when fully operational of about 75–80 per cent.

The two components, ownership and operations, have to be considered in assessing the value of any individual facility. This assessment would be based first on the subsisting lease, its duration and conditions compared with industry benchmarks. The ownership component will be assessed, as with any property investment, on the basis of capitalised rental income and risk profile confirmed by summation of the physical assets. As with any going concern business, earnings before interest and tax (EBIT) and trading trend data over a number of years will be important considerations.

In addition to benchmarking, other criteria important in the assessment of these facilities are:

- location (including growth potential, position related to catchment, accessibility and convenience
- demographic profile and trends
- the built form (including design, structure, layout, condition, approvals, ability to expand)
- current and potential competition
- current certifications and any outstanding government requirements or requisitions.

The viability and success of these facilities rests very heavily on the securing and retention of quality, skilled staff. There are strict staff-to-children ratios prescribed under government regulation. For most centres, a key 'point of difference' is established by the quality of care, programmes offered and the personal goodwill developed by staff.

Additionally, staff need to be able to manage children with special needs, manage emergency situations and illness and be able to provide and maintain detailed records and reports for parents, government supervisors and the centre itself.

Staff costs represent the major outgoing in operations and current viability depends on the maintenance of existing, relatively low, wage rates. This has become a contentious industrial and community issue and its resolution will be important to the future prospects of many centres and the sector as a whole.

Marinas

The assessment of marinas as real property assets brings together components of watercraft, their storage and servicing and in many cases, the opportunity to create additional waterfront land. Through the development phase, this 'creation' of new, waterfront land for sale typically holds more potential for return on investment than the direct, maritime development itself.

There are about 370 such marinas scattered around the Australian coastline with approximately two thirds of those located in New South Wales and Queensland given the suitability of those regions to boating and related activities through most parts of the year (Marine Industries Association of Australia 2010). To be viable, such precincts typically require 100 or more boats, pens or moorings. A number (particularly in large urban or tourist regions) have a capacity of 300 or more.

Most such facilities are privately owned and operated though a number of the larger facilities are owned and operated by non-for-profit yacht clubs and associations. Several remain in public ownership (e.g. through port authorities). In any case, various levels of government will almost certainly be involved – sometimes as a head lessor (of the underlying tenure) and also as the bodies responsible for the issuance of necessary approvals and certifications and to monitor operations.

Commercial maritime activities such as cargo ships, stevedoring, fisheries and major boat building/repairs are typically located separately from these marinas. Many marinas however do include some ship servicing facilities including hardstand areas, crane facilities, workshops and repair yards, et cetera. Typically, these will be operated as freestanding businesses and are of considerable value to the boating community. In practice however, such uses can create a range of industrial hazards and loss of amenity for the locality which are not always easy to address.

There will invariably be environmental considerations in such littoral areas related to the establishment and ongoing management of new marinas. These include the very sensitive issue of the removal of or damage to mangroves, effects on tidal flows, damage to existing eco-systems, dredging and dealing with effluent and the potential for pollution and accidents. Often too, existing foreshore amenity, views and aspects may be adversely affected by the creation of a new shoreline and associated on-water and dryland development. While a significant number of marinas were established, particularly in north and south eastern Queensland and parts of New South Wales and Western Australia over past decades, the level of regulatory control and public reaction now implies that very few such facilities will be further developed into the future, particularly in closer settled areas where demand may well exist.

Arguably, despite the huge expanse of the Australian coastline, the number of sites suitable for new marina developments is surprisingly limited. A successful site must, as noted above, be located close to a large urban centre and have reasonable expectations of securing necessary approvals. It must provide access to a sheltered boating environment and enjoy good geographic and marine conditions together with favourable weather through most seasons. It should also be noted that most suitable regions are already well serviced with such facilities.

This might imply that, with such limited future supply, the unit value of moorings within such facilities will increase over time but this will very much depend on the location and quality of those amenities.

Contrary to public perceptions, most of the boat/yacht owners public are not high-wealth individuals and demand for these facilities is much more price sensitive than might be expected.

Tenure in such littoral zone areas is typically by way of long-term lease from state governments and will normally involve strict environmental conditions. In some juris-dictions, 'inundated freehold' tenures are available to accommodate excavated land now used as the harbour basin. As noted above, major developer interest emerges from the 'waterfront land' so created. High premiums are often attracted to such residential areas given the activity, amenity and aspects that they enjoy, though there may be limitations on height and the maritime influence rapidly falls away with distance from the quay.

Other matters worthy of note here include:

- The construction of marinas and surrounding harbour and residential development is typically very difficult to estimate in advance, particularly as regards dredging and fill characteristics and time for development and the cost of marine construc-tion. High-contingency allowances are therefore essential.
- Given the marine environment, physical deterioration is a particular problem. Much of the infrastructure has a functional life of a maximum of 40–50 years, and continual maintenance of the marinas, channels and other works is obligatory.
- Over time the size of pens and moorings have increased with the optimum size being able to accommodate a 10 metre yacht.
- Numerous examples exist where the marina facilities become progressively congested and the availability of adjoining land for expansion is of considerable advantage.
- On-land commercial opportunities typically exist at such marinas. These may in-clude boat servicing, brokerage, chandlery and refuelling businesses, tourist shops, booking agencies, eateries and associated uses. The viability of any of these will depend, in part, on the scale of the entire facility, levels of passing trade and access available for the general public. (The provision of adequate on-site parking and of public walkways will be important design considerations.)
- In some localities the provision of heavy lift cranes and the rack storage of boats, often within very large specially constructed sheds, have emerged as additional business and use options which maximise available site areas. Their use however has not been widespread, given the construction and management cost, the effects on visual amenity and the resistance by many boat owners to the reduction in availa-bility to those boats.
- Over recent years, further opportunities have been seen to attract super yachts (i.e. vessels of over 25 metres) to base themselves in or regularly visit Australian marinas. Some such vessels do intermittently visit certain parts of Australia, particularly Syd-ney and the Great Barrier Reef; however, the ability to form cluster opportunities around those visits has not been possible and little investor interest in providing such facilities has emerged to date.

As a marina development involves a number of different components, the assessment of functionality and value needs to be considered across those various parts. These matters are summarised in Chapter 10. In these cases, the components to be assessed will include:

- the development of the marine project itself, below high water mark (HWM) to create the marina and boat moorings
- potentially, the dry-land development of apartments, retail and office buildings
- the development of hardstand, boat storage racks and, potentially, fully enclosed boat storage sheds
- boat servicing facilities
- going concern businesses sited in the enclave, including the provision of services and asset management to the various parts of the development.

Clearly, all of these endeavours are freestanding but, at the same time are linked by the marina/maritime environment that has been created.

Retirement/'lifestyle' developments and similar assets

The 'baby boomer generation' (i.e. those born between 1946 and 1964) include 5.5 million people or about 24 per cent of the population. This represents the largest generation in Australia's history and, as they have progressed through various age categories, they have dramatically increased the demand for the built assets and services that typically support that age group. As might have been expected, as the baby boomers have reached or are considering retirement, there is now rapidly increasing demand for and investor interest in retirement accommodation and healthcare.

Commonwealth government policy over several decades has encouraged retirees and the elderly to remain in their own homes and retirement village facilities for as long as possible, only resorting to publicly or privately supplied higher care options as the medical need demands. This, in effect has created two streams of retirement accommodation – independent living and supported living. The latter ranges from low- to high-care facilities. (These supported living facilities are more aligned with hospital and medical services than property developments and are not further discussed here.)

Demographic changes have impacted considerably on the demand for independent living facilities over recent years. A 65-year-old Australian in 2017 (i.e. in the approximate middle of the 'baby boomer' generation) can now expect to live for another 16 years (AIHW 2019b). As well as the increase in total numbers now within that age group and their enhanced life expectancy, this group is significantly healthier than preceding generations. Eighty five per cent of Australians over 65 are capable of living unassisted and are likely to require intensive support in the last year or two of life. All of these changes have continued to produce significant shifts in demand for real property (AIHW 2018).

As regards those able to live independently, options include remaining in the existing principal place of residence (e.g. the family home) or moving into a retirement/lifestyle community which will typically involve smaller living quarters (given particularly that these households will normally consist of only one or two individuals). However, such facilities will supply building and maintenance services, improved security and community activities and other services seen as a priority for that demographic.

These demands of contemporary retirees present quite different criteria from those which produced the low-density and often poorly located 'retirement homes' of the past. Additionally, many of the past occupancy arrangement between centre managers and owners were poorly constructed, misunderstood and/or failed to adequately accommodate the longer-term requirements of those residents. These have become significant community and political issues, though improved regulation, standards and funding models across the Commonwealth and the states have improved these arrangements. Recent positive changes here have included improved transparency in dealings and clear contracts, which identify respective roles and responsibilities, full cost disclosure and entry and exit provisions. Improved dispute resolution mechanisms have also been introduced.

A number of ownership and service models have been developed over time. Despite growth in overall demand, many of the older facilities are proving less than successful given their poor location, low-density and high operating cost together with the need for capital expenditures to upgrade to now required standards. In a number of cases also, those facilities occupy sites that have increased substantially in value and have potential for redevelopment.

Models developed over recent decades have included, in some cases the freehold purchase of units within a complex or the entering into an occupancy agreement with the developer/owner which involves a substantial upfront payment on entry, the regular payment of occupational/operational costs and management and the final return of parts of the earlier deposit, all dependent on the conditions of the original agreement.

Recent innovations and enabling legislation across a number of states now allow new occupancy models whereby a site in an enclave, retirement/lifestyle community can be made available to an individual on the upfront payment of an agreed amount and upon which a residence (often manufactured/pre-fabricated) and provided for or by that individual. An occupational agreement is then signed between the two parties which includes the rights and obligations on both. These include the rights of the occupier to access all services available within the development and to on-sell that interest subject to certain terms and conditions. In return a regular occupancy fee is payable by that individual.

Issues obviously arise in the ability to secure suitable sites and to provide new facilities that are well located as regards other urban amenities. Previous long-standing low-density 'retirement village' concepts required substantial land areas only available in more distant locations. These matters are now addressed through more dense cottage developments (sometimes facilitated under manufactured home legislation discussed above) or in medium to high-rise facilities. While the multi-level centres are becoming more common and can typically provide various levels of care and a range of services, there has been a level of resistance to such configurations from incoming residents more familiar with low-density, suburban environments.

It is important to note however, that while many in this generation have equity in their principal place of residence, for many, other savings such as superannuation are often quite limited. Average wealth of those now entering retirement is about $300,000 plus their interest in the family home (which is an asset that about 75 per cent hold) (Roy Morgan Research 2019). For the majority, this implies that access to welfare support will be necessary from the start of that retirement and fully independent and self-funded retirees represent only a minority of the 15.9 per cent of the Australian population over 65 years old as at 2019. Consequently, the demand for such facilities is very sensitive to entry price, ongoing management fees, arrangements for later on-sale and Commonwealth Government policy and funding, particularly as regards welfare payments.

Given a number of recent controversies, that policy has now been established, in the first instance to keep the aged in their own homes as long as possible. Thereafter, policy seeks to provide government support on an equitable basis, which gives clients a level of choice and control. Given the community and care issues involved, regulation and supervision of the sector have increased and are recognised as a major issue for investors and operators of such facilities.

Given the specialist nature of these facilities and the services to be provided, a number of developers/owners/operators have emerged as sector leaders and demand for such facilities appears to be strong and growing as newly retired households seek alternative and more appropriate accommodation.

Clearly too, some coastal regions providing a range of community and health services and well located relative to large urban areas are increasingly attractive for this type of investment and use.

Service Stations

Service station sites, assets, developments and operations represent another property sub-sector worthy of specific observations. The nature and range of services provided from such assets have fundamentally changed over the last quarter century with the traditional 'fuel sales and mechanical servicing businesses' largely replaced by much larger scale, strategically located facilities, which still provide automotive fuels but now comprise a range of offers including fast foods and groceries and often located within a retail cluster of complimentary uses. This has involved a major rationalisation of sites – from about 20,000 nationwide in the mid-1970s down to about 6,450 sites by 2017 (PCA 2017; ACCC 2018). The remaining centres are typically much larger in scale and have involved very significant new investment from developers/owners.

The sites that remain enjoyed two fundamental advantage – one being the requirement of the owners of all vehicles to visit the site to secure this essential commodity on a regular basis. Second, the remaining sites enjoyed strategic location with access to the large customer base and demand for fuel now required to support such a significant investment. In this regard, Pawley (1998) makes the astute observation that, unlike practically any other form of real property asset, service stations have very little, if any, association with other properties in their immediate locality. Rather they 'connect' to and interact with the entire road network and transport system. Value of a particular site and asset is derived, in no small measure, from that positioning and accessibility related to that network.

Over time and based largely on development control, environmental and traffic management criteria, the securing of quality new sites has become increasingly difficult, thus on the face of it, enhancing the value of existing facilities.

Since the early 1990s, the number of fuel wholesalers and distributors in Australia has also declined. In part this was the result of Commonwealth Government action, which effectively prohibited the vertical integration of supply by oil companies through to retail sales. Their withdrawal provided an opportunity for retailers from other sectors (notably Coles/Myers and Woolworths) to effectively create their own oligopoly reinforced by their existing customer loyalty programmes. For the most part and particularly in the case of green field developments, these corporations tend to avoid direct property development, preferring long-term (20 years plus) leases with the lessee typically taking responsibility for all operational and compliance costs and issues.

Other independent fuel retail companies also exist and, while in aggregate, have secured a market share of about 20 per cent, they have quite limited market influence individually (ACCC 2018). There remain some 4,000 small single-site businesses that follow the traditional service station model (i.e. fuel sales and workshops) though these now represent quite niche, localised businesses.

The rationalisation of sites and the introduction of major corporations at the retail level have provided new, significant development and investment opportunities in the provision of these new, larger facilities. However, the ability to identify, secure and obtain all necessary approvals for a new, large, strategically located site is remarkably challenging and time consuming. For that reason also, investigations often also turn to the full redevelopment of existing, well-located outlets.

From 2003, major retailers in Australia took significant positions in the fuel retailing market and, through attachments to existing brands, rapidly increased market share and secured market leaders in most parts of the country, reinforced by grocery loyalty schemes. Their involvement, though now reducing, renewed investment focus on the property assets supporting the sector. A new investment emerged whereby a third party investor or trust would fund the acquisition of a strategic site (typically located along a major road or highway and full construction to the specification of the end-user fuel retailer). A long-term (15–20 years +) 'net-net' lease would then be established with the property investor then assuming a passive role. Leases with food concession owners/franchisee for the balance of the new facility would then be put in place, typically involving well-recognised brands often with previous association with other tenants.

The small site that remain throughout the country remains under ownership structures much as they always have – sometimes as owner-occupiers, sometimes as sites owned by fuel distributers or other private investors and leased to the individual operator.

In consideration of individual facilities and of the sector more generally, the following observations are relevant:

- Fuel and fuel supplies are an essential and politically sensitive commodity. Involvement therefore has both advantages and challenges. Advantages include the need for regular visitation by customers and, with that, the building up of recognition, consumer habit and goodwill often reinforced by loyalty programmes, et cetera. Challenges arise, however, because of consumer reaction to rising prices and increased government scrutiny and regulation.
- The commodity is of (relatively) low unit cost, heavy, difficult to handle and transport and the facilities require significant and expensive safety and environmental protection installations. Further, retail margins are typically quite low. All of these factors promote economies of scale and have resulted in the continued retail emphasis on large, multi-purpose sites.
- Fuel represents an essential commodity but demand and the composition of the sector itself is remarkably dynamic. Even though the actual number of vehicles continues to increase, each is increasingly efficient and consequently, the link between increasing number of vehicles and increase in the overall demand for fuel

is not as direct as may be imagined. Additionally, the rise of renewable energy vehicles, reactions to the use of fossil fuels and substantial real cost increases in fuel will test the level of inelasticity over time. Such trends may begin to affect strategic decision-making in major investments in the sector into the future.

• In the development and management of these highly specialised assets, particular regard needs to be had to town planning and development controls and conditions, accessibility, contamination and emergency management systems, variations in fuel types and their storage and requirements for truck refuelling, hardstand, et cetera.

For assessment of value and as detailed in Chapter 10, the property investment components of the facility will be assessed, in the first instance, on the basis of feasibility studies and hypothetical development analysis and the later completed project, on the basis of capitalisation of rental income and internal rate of return calculations. The quality of the retailer involve and of the covenant they bring are critical issues here. The valuation of the operating business, again as might be expected, would be principally on the basis of the analysis of comparable business sales and on EBIT.

Hotels (licenced premises)

Following on from English and Celtic traditions, licenced premises have proven in integral part of Australian social and community life. The term itself has a range of quite ill-defined meanings but, in its normal sense it includes premises selling alcohol and proving entertainment and social venues, bars, et cetera, certain clubs and restaurants, resorts and multi-use integrated development precincts. While recognised for their community value, they have also been subject to tight regulation and control as regards the licencing of premises and services that could be provided. There is irony in the prudish regulations that have often been enforced in the past while, at the same time, government has relied heavily on excise and other taxes generated by such activities.

The scale of such establishments has always been diverse, ranging from small country and suburban hotels ('local pubs') through to very large international hotels and resorts in major cities and tourist areas.

Up until the 1980s, the sector was highly structured and vertically integrated with many retail outlets owned or 'tied to' breweries and distribution companies. As part of major corporate upheavals, through that period, many of the traditional breweries became targets for corporate takeovers with the retailer hotels often sold off to trusts and private owners. In a strategy similar to changes in retail fuel sales, large retailing corporations also took advantage of the volatile situation interests in hotel operations particularly related to takeaway sales ('off-license' establishments). During the same period, changing community attitudes to alcohol, evolving work patterns and the strict enforcement of drink-driving laws saw declines in bar patronage and consequent closures of the number of operator hotels across most of Australia. Competition to hotels also emerged from the club sector, which tended to act in a community and social sense.

Through this upheaval however the vast majority of hotel properties (about 90 per cent) remained as simple, small businesses, either as an owner-operator or with a passive landlord and lease to an operator.

The dynamics of the sector changed fundamentally from the early 1990s where gambling and particularly poker machines were allowed in licenced premises, again under strict government control and with considerable revenue flowing back to government, notwithstanding the continuing level of community objection and social issues that can emerge from their use.

What now emerges is often a hotel property that provides three services/incomes streams. The first is a food and beverage offer within the hotel, which while essential to attract patronage and required as part of the hotel licence, is a marginal proposition in many cases. The second is the supply of beers, wines and spirits to be consumed elsewhere. Often these are operated by a third -party retailer. Such activities normally involve that retailer simply taking a long-term lease over vacant space (often little more than a large shed on the hotel site) and taking responsibility for full operations for that part of the business.

Third, the gambling facilities now typically provided the highest return from the entire property and increasingly the layout and function of hotel premises and reflecting that emphasis.

The sector is diverse and, while the above scenario may be common, it is not to imply that all hotels follow that model. Much depends on the catchment population, the location and accessibility of the hotel, and the nature of the facilities and services provided. In regional areas where social cohesion may well be stronger, traditional hotels can still prosper. Some specialise in a quality food offer, others on entertainment or particular theming. Care must be taken in such specialised upfront investment to be confident of market acceptance and the skills required for successful management.

In the assessment of such properties, the physical integrity and style of the property is important as are all necessary licences, certifications, et cetera. Trading figures (including details of costs, volumes and profitability) and trends therein all compared with market expectations and sales analysis will be important matters for investigation as will be changes in local markets and the possible future activities of competitors. For the going concern business EBIT (i.e. Earnings before Interest and Tax) over preceding years will need to be established.

Where the property is on-leased, then the normal capitalisation of net rents into the future and based on market analysis will provide a basic valuation approach. Matters relating to the nature and length of the lease and the quality of the covenant by the lessee will also be highly relevant.

Motels and similar properties

Motel properties together with other 'hospitality related' property types, such as restaurants and hotels, have a fundamental attraction to a range of investors and owner-operators. From the general public's observations, they are well understood and appear to represent both an attractive investment proposition and a lifestyle choice.

While this may be the case, these types of assets belie the need for extreme care in selecting a suitable property, the need for esoteric knowledge of the sector and its operations and the hard work and long hours involved.

There are a number of sub-categories of motels ranging from those servicing the travelling public (typically located on or near major highways), those catering for business travellers, those in resort areas and longer-stay establishments, often in urban areas. Other sub-groups include those attached to particular (tourist) facilities, casinos and entertainment centres, and those with business and conference facilities.

Secondary markets exist in bed and breakfast (B&B) establishments, backpacker accommodation and more recently, through the online rental of single rooms in private residencies, though market evidence to date would indicate that these cater to a some-what different market sector (who may not have been otherwise travelling) and perhaps are not having the impact that may at first be envisaged.

Characteristics such as the type of facility and even its size perhaps detracts from key underlying issues, those being the ability to create revenue, net income and to protect and grow that income stream. In such an enterprise, this is established through developed good will and reputation which, flows through to repeat business and word-of-mouth referrals.

There are typically two levels of investment involved in such ventures. The first is a freehold owner and the second, the operator/lessee. Such leases are normally for over a long period – 25–40 years being common though that may be staged through a number of options. Leases are typically 'net-net' with the owner acting as a passive investor and the lessee responsible for all operational costs, outgoings, and (depending on the lease) certain capital upgrades. Rent reviews are typically based on CPI.

Head leases are often set at about 25 per cent of gross revenue, but that figure can vary significantly on a particular case and circumstances of the lease.

Other matters that will be relevant in the assessment of such a property and an assess-ment of the value of the asset and the business would include the following:

- Strategic location, identity recognition and good exposure are important attributes for motels and similar properties, particularly for an establishment that relies on the travelling public and the 'drive-in trade'. Facing, left hand, corner sites entering a town are normally to be preferred particularly if there is obvious ease of access and parking.
- Location. The ability to easily recognise such a business is critical, particularly for an establishment that relies on the travelling public and the 'drive-in trade'. Left-hand corner sites entering a town are to be preferred particularly if there is obvious easy access and parking.
- As with all going concern businesses, an analysis of audited trading figures includ-ing EBIT must be secured and these should be available over a sufficient number of years to establish trends in trade.
- The conditions of the subsisting lease and the period that it has remaining need to be established. It is in the financial interest of both the owner and the operator for the subsisting lease to be on commercially acceptable terms to both parties and with sufficient time remaining to allow on-sale.
- Difficulties often arise in the provision of on-site restaurants/food preparation/alco-hol sales facilities. Nevertheless, depending on the type of establishment, they may represent a necessary service. In any case, they will require very close management to ensure their viability and profitability
- Such assets are heavily used but must be maintained in fully functional, clean and presentable order to maintain reputation. Consequently, a strong commitment to
- cleanliness, maintenance and investment in renewal and upgrade of facilities is essential.
- Taxation allowances are reasonably generous in these areas and become important in overall assessment.
- The role of both online travel agencies (OTA) and of involvement in motel/hotel groupings are matters of some conjecture. OTA's have over recent years positioned

themselves as a first point of enquiry for many potential clients. Consequently their use by motel owners has been widespread. Once that motel business is successfully established by good reputation and word of mouth, the use of such intermediaries may be more problematic particularly if the individual motel can well position its own internet presence and other advertising and promotion. The value of belonging to a particular motel/hotel grouping depends very much on the type of motel, cliental and services offered by that group.

- Much is made of the 'optimum size' for a particular motel. Clearly that depends on trading figures, competition, the location and region and the capacity of management. Sometimes room numbers between 25 and 40 are suggested as viable and manageable. Such generalisations may however be somewhat misleading as there are a range of other issues involved such as rack rates, actual room rates achieved, vacancy rates and seasonal variations that need to be also considered. As noted above, EBIT and the securing and growth of revenue are key determinates of viability and value.
- A range of home stay, overnight room stays in private residences and other accommodation offers have emerged over recent years, facilitated by the wider use of OTA's.
- As part of the positioning and 'offer' of any motel or similar property, it may be observed that, except for backpacker accommodation, the market is may not be as price sensitive as may appear. Very low-rate motels typically cannot maintain standards and will often fall away in patronage rather than attract larger volumes over time.
- Finally, given the importance of personal service and goodwill, the consistently positive approach by management and staff will be critical to success. These are however business ventures and their success is only achieved through a strategic and business-based approach.

Note

1 Gini Coefficient (or 'Gini Index' or 'Gini ratio') is a form of statistical analysis which measures the distribution (and concentrations) of both income and wealth across a particular population, region or country. In effect, it measures economic inequality, though it may be noted that such inequality is often a critical component of other forms of social dichotomy. A rising index over time identifies growing inequality, which may later cause social, community and political disruption. Further a relatively high index for a population may also reflect higher dependency ratios and reductions in overall productivity and prosperity and social harmony.

Chapter 9

Property sectors
Rural

9.1 Introduction

Earlier chapters, particularly Chapters 4 and 5, provided a general and historical over-view to land use and development patterns in Australia. Additionally, Chapter 10 will summarise a range of methodologies commonly used to assess utility and value of prop-erties within those use categories.

This chapter considers property sectors and land uses in Australia's regional and rural areas. It follows a similar approach to the preceding Chapter 8. In the first instance, it attempts to define those sectors and sub-sectors and observes some of the overall char-acteristics, evolution and current direction of these uses. Second, it aims to provide some current commentary on those activities and how they influence land use capacity and value.

Rural land uses can be defined as those primary activities directly deriving income from plant cultivation and animal husbandry – that is, farming and grazing activities. It is, of course, no single 'rural industry', 'primary production' nor 'rural real estate' as some homogenous grouping. Rural activities in Australia are a conglomeration ('mosaic') of land uses, created from a range of interacting economic, geographic, historic and polit-ical drivers, played out over time and over a huge and diverse land mass.

It needs to be accepted that the definition of what actually constitutes 'rural' or 'primary' production varies depending on the setting and circumstances in which it is used. For example, the term will often be extended to include forestry (native forests or plantations) and seafood production (fish, crustaceans and shell fish, either as wild catch or sea or pond-based aquaculture). (This text does not extend to those activities.)

The nomenclature that surrounds rural pursuits may seem somewhat confusing but, rightfully, it reflects the diversity of activities involved. The following presents an over-view and some general definitions.

The generic term 'agriculture' is taken to mean all knowledge, scientific and prac-tical, investment and activity related to the cultivation plants and growing livestock. Within that, 'grazing' (broadacre or intensive) relates to the breeding, growing and fattening of herbivorous animals for various purposes. Broadacre farming and cropping refers to larger farm areas typically given over to the growing of cereals and fodder crops on sessional rotation, either on an irrigated or dry-farming basis. While much of that production is sold off-farm as a commodity in its own right, a significant proportion, depending on the circumstances, will be used to supplement the feed for livestock on that property.

Horticulture refers to the practice of cultivation, management, harvesting and sale of fruits, vegetables, flowers, herbs, nuts largely for human consumption or processing. The horticulture group is normally further divided into sub-categories of annual/seasonal crops (such as small crops, herbs, flowers and vegetables of various types), orchards (typically involving the growing of perennial trees shrubs and vines and the annual harvesting, principally fruit and nuts) and, third, viniculture – that is, the growing of grapes for use in wine production.

Again, contrary to the 'vision splendid' image often ascribed to rural Australia, contemporary rural enterprises range in size from one hectare or even smaller in peri–urban locations (e.g. market and herb gardens, flower farms and the like) through to remote pastoral holdings traversing hundreds of square kilometres.

As important as the physical attributes including soil type and rainfall are in a particular area, they do not represent a complete analysis of either current or potential land uses. The physical ability to successfully produce a particular commodity from a parcel of rural land does not, of itself, guarantee the initiative's economic value or sustainability. Additionally, a range of service activities (notably health, education and public service departments are typically present in those rural towns and link directly and indirectly with the rural production base). The assessment methodologies regarding physical characteristics, productivity and value are well established (refer Chapter 10), but these only have relevance if they are placed in the context of markets, comparative advantages, regional characteristics and the nature and use capacity of the subject property itself (see Figure 9.1).

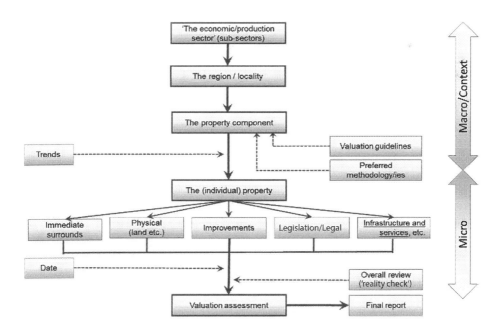

Figure 9.1 Real property assessment – wider context.

Increasingly, many rural producers are open to exploring and pursuing novel opportunities, which can widen the usage of their real property, ranging outside their traditional activities and attempting always to improving their income base, thus enhancing overall sustainability. These are important strategies but, as with all commercial ventures, must be based on rigorous, prior analysis as to both demand and of the capacity for production at competitive prices, quality and scale.

As well as the range of land uses generally defined above, there are a large number of ancillary uses and activities, typically located in regional areas, which are, at the same time, dependent on and are essential to those core, rural activities. These include professional services, scientific research and application, transportation, work contractors, storage and processing and many others.

The primary production activities often have important supply chain links to downstream activities into service and manufacturing sectors. Some of these activities, such as grading and packing, are simply carried out on farm (i.e. 'inside the farm gate'). Many others however, including feedlots, sugar mills, meat works, food processing works, wineries, milk and cheese production are critically important particularly to regional economies, though they are often recognised statistically as part of the manufacturing sector, rather than 'primary production' per se.

Important to this analysis is that a large proportion of those activities are located within the local region, outside the major cities and typically form integral components of those local communities and economies as truly as the on-farm primary production itself. Therefore, in an even more pronounced way than other property sectors, it is essential to consider these rural activities in the context of their region.

Geographic, historic, demographic, community and economic characteristics of that region and the identification of comparative advantages and trends and future prospects will represent a first phase of rural property assessment before the 'finer grained' analysis of any individual property is made. These matters are further examined in Section 9.2.

It is important to appreciate the scale of this overall sector. Rural activities occupy some 58 per cent of Australia's land area. In 2018, there were a total of 85,000 farming enterprises with a total annual production of about $B60. In an annual average from 2014 to 2017, about 70 per cent of total rural production was exported, dominated by grain, beef and veal, wool, wine, dairy and sugar (ABARE 2019, Australian Government 2018a).

Most of that output will be produced 'in field' but there are numerous variations including greenhouse, hydroponics, organically certified environments, finishing feed lots, hatcheries and poultry sheds, piggeries, fish farms and various others. While typically involving larger up-front capital investment and ongoing management, such alternate production systems are becoming more common, providing a more controlled growing environment and higher and more predictable production levels. Because they typically involve much more concentrated and efficient use of land resources, they can often be located much closer to local markets and export ports/airports.

9.2 The nature of regional and rural Australia

A rural asset, region or community is one where the principal activities and income derived are from the land–typically in the form of cultivation, livestock and associated uses and support services.

As it is generally used, the term 'regional Australia' is taken to mean all populations, lands and activities located outside Sydney, Melbourne, Brisbane, Adelaide, Perth, Hobart and Canberra. That description is, however, less than satisfactory – almost defining regional Australia by exception, that is, 'everything else'. While the capital cities have, from first settlement, represented a dominant force in the Australian economic landscape, the regions also have profound geographic, environmental, political and cultural significance to the entire nation. Some 8 million (or about one-third) of Australia's population reside in these areas which generate about 57 per cent of the country's exports and, directly or indirectly, provide significant support for one in three jobs nationally (RAI 2019). Further, as the capital cities now grow to optimum size, much current and future growth is spilling over into surrounding regions particularly if they are within a reasonable distance (perhaps extending to 100–120 km) and enjoy good quality transportation links to that capital city (Polèse 2009).

Eversole (2017) recognises that this nomenclature is more than simply an economic or geographic expression. Certainly, parameters such as a particular geographic catchment, communication routes, political and major topographical boundaries and principal economic activities are important delineators, but the demographic characteristics (including ethnicity, age and educational status) promote a level of social–as well as business – interaction and cohesion that help define that region.

The level of real knowledge by the wider (urban–based) community and business of rural sectors and activities in Australia is quite low–often reduced to quite nebulous concepts, historic images and anecdotes which bear little resemblance to the current realities or future prospects, opportunities and challenges for regions.

In reality, rural and regional Australia presents a quite eclectic picture. There are endemic issues of seasonal variation, under-capitalisation, vast areas of relatively low land productivity, population drift to the large urban areas and the challenges and costs associated with distance. At the same time, there are significant and enduring comparative advantages. The island geography and distance quarantines rural production from many biological threats and the diverse local conditions which emerge across such a vast land mass provides ideal conditions for certain specialised production. Land resources are comparatively cheap and skill levels and infrastructure are generally sound. Again contrary to generic impressions, there are very large areas of the continent, particularly along the eastern and south-eastern seaboard, where high rainfall and land productivity permits high use capacity and diversity.

The key observation is that all of this represents a dynamic environment that continually evolves and changes over time–some regions grow and become prosperous, and others, for a number of reasons and without sound leadership, innovation and capital investment, will slowly decline. None remain the same over time. Second, and of particular importance for researchers and those assessing sectors and individual enterprises, it is essential to investigate past the obvious inherent, seasonal and other volatilities to establish a more holistic view of the region based on trend data extending at least three years but, hopefully, much longer.

The status and future of rural towns and service centres provide an important contemporary example of these evolution processes. The rapid advances in ICT (and specifically in internet usage) and in improvements in road and rail networks, typically radiating from capital cities, have all enhanced conductivity and accessibility from regional and remote consumers to city services and facilities. Given this scenario, significant disruption to the traditional hierarchy of urban centres–from capital to regional centre to local town – has emerged, normally with serious, adverse consequences to those regional and local centres. These trends have been reinforced by the further concentration of government and corporate / financial services and advanced medical and educational facilities in those major cities, often at the expense of the smaller centres.

When considering economic analysis, it should also be obvious that 'micro' and 'macro' components coincide (refer Chapter 2). The economic decision-making at farm level clearly represents single unit decision (micro) making, while, in aggregate, the sectors such as livestock, grain production, horticulture, wine production, et cetera, are largely determined by macroeconomic analysis.

Given the diversity of rural activities, the investigation and analysis of those activities and their regions needs to be based on areas of commonality. These may include:

- by product type
- by location/region
- by supply chain analysis
- by sale – fresh or processed
- principally for local production, national markets or exports, or, more property
- some, appropriate combination of any or all of the above, thus building up a mosaic of the complex interactions, networks, relationships and dealings.

An important early observation is that, for a number of reasons including rainfall variability, low soil fertility, distance and limited infrastructure and development, many Australian regions are limited to only one or, at best, a small number of types of production for which they enjoy comparative advantages over other regions or countries. Examples would typically include sugar, cattle, sheep and grain-and-sheep holdings and regions across much of the country. While this regional specialisation promotes economies of scale and improved skills and know-how in those locations, it also presents serious risks should the market for those (limited number of) regional products experience difficulties. These risks are typically exasperated by high exposure to volatile and often distorted world commodity markets into which much of Australia's rural production is traded.

Some 77 per cent of Australia's primary production is exported and its produce is widely recognised as of high quality, reliable and safe–all increasingly important attributes in the contemporary environment (National Farmers Federation, 2017). Nevertheless, in practically all of those global commodity markets, Australia is a relatively small contributor/producer–and thus very much a 'price taker'. Its high cost base, volatile currency and remote location, particularly for perishable produce, add to the inherent challenges for producers in this sector/sub-sectors.

There are other key issues–some economic, some political–that also need to be emphasised here. First, as in all countries, there is strategic importance in Australia being able to reliably provide quality food, particularly given its largely urban-based population and the country's relatively isolated location. About 93 per cent of Australia's domestic food production is produced within the country (National Farmers' Federation 2017). While not openly discussed, such a high figure would provide considerable confidence for Australian governments in the event of some future emergency.

Second, while the critical importance of supply into 'downstream' processing of rural production has been recognised, it needs to be noted that the significant service sectors in practically all regional towns are directly linked to the success, sustainability and growth of the (typically) small number of rural sub-sectors in the region. A prolonged downturn in externally secured income from that production typically has a profound, negative effect–not only on the various direct contractors supplying those primary produces but also to other local businesses (retailers, banks, hotels, accommodation, et cetera) and on to public services such as schools, health facilities, government offices and the like (Moretti 2013).

It needs to be recognised also that, in any economy or region, a proportion (typically 12–15 per cent) of the workforce rely on sustainable economic growth and new investment to maintain employment. Many of those involved in construction and infrastructure development are in that grouping and, without growth, many will move out of the region, compounding the subsisting lack of growth being experienced. (This again provides an example of the dynamic nature of regions and the inherent difficulty of simply maintaining the 'status quo'.)

This presents a particularly dangerous regional situation where negative multipliers can quickly compound the situation and erode confidence. It will typically be difficult to redress that slide and perhaps take many years to restore services and economic and population growth to previous levels.

The dichotomy between Australia's major cities and regional Australia is widely recognised and yet, at the same time, fundamentally misunderstood–relying largely on perceptions and clichés rather than on facts or evidence. This divergence is increasing and 'geographic judgements' are often made by the majority of urban dwellers stereotyping rural communities. These perceptions range from sympathy for rural producers, given the increasing volatility of climatic conditions and the vagaries of world markets through to an incorrect perception that rural communities, almost by definition, represent dying industries and ageing conservative and reactionary populations (Chan 2018). Those latter categorisations are both incorrect and unhelpful. Regions in Australia are as diverse as its cities. All have progressive elements with optimistic futures or have components that are in decline. As described in detail in Chapter 4, such transitions and evolutions have been a characteristic of Australia's history from the outset and that will continue into the future.

Important to the sustainability of many of rural-based activities is the empathy and support that the wider Australian community and all levels of government exhibit for the production and market difficulties in these sectors (Chan 2018). That strong support is held across both major political parties. While that may be anticipated from the conservative parties, particularly the National Party of Australia, it should also be remembered that the Australian Labor Party was established in rural Australia and many affiliates, including those working in forestry, sugar mills, meat works, shearing sheds, transport and numerous regional and rural activities, are highly unionised and supported by that party. Finally, it should also be noted that such support is common across practically all OECD countries–particularly in Japan, the US and the European Community–leading to major distortions in world commodity markets to the detriment of producers such as Australia.

The total annual Australian rural production as at 2017–18 FY stood at $B66 (ABARE 2018), remembering that these figures are volatile based principally on climatic conditions and prevailing world prices for key commodities during any period. While on-farm workforces continue to decline in numbers, overall productivity in the sector continues to grow remarkably–increasing 34 per cent in real terms in the 20 years from 1998 to 1999, the result of improved breeding, more efficient use of water and fertilisers and, in a number of cases, farm aggregations.

There are about 85,700 farming businesses in Australia (NFF 2017). Ownership structures have changed little in decades with over 99 per cent of farms still family owned and operated. This structure is unlike any other significant economic sector in Australia and an appreciation of that and its ramifications are essential to an understanding of its operations and overall direction.

The level of foreign ownership remains small in overall impact and confined largely to certain major investments in cotton, sugar, feedlots, viticulture and a number of others and to the foreign corporate ownership of certain pastoral holdings–a situation that has existed since the first settlement of those regions. More important in an economic context has been the progressive aggregation of holdings, often by existing farming families to provide greater economies of scale and reliability and diversity of supply.

The concept of 'a living area' can assist in the understanding of individual rural enterprises within their sub-sector and market. A living area in a particular region and for a particular type of rural activity is broadly defined as a holding of sufficient size, to be sustainable as a typical, free standing family farm, through a range of season. As a general measure, such a delineation is of relevance, say, in considering value and suitability of improvements on a particular holding. Like all rules of thumb however, Baxter and Cohen (2009) recognise the major limitations on such analysis particularly in the contemporary environment and that care must be taken in their application given that they cannot reflect the wide range of individual characteristics that each property will exhibit.

An important feature of Australian regions is the social and cultural networks and common understandings that exist uniquely in those centres and regions. Perhaps because of their smaller size and the common issues they typically face, those communities tend to have a more acute, immediate and holistic view of their regions and communities (Chan 2018). It is an interesting observation that in rural land uses and activities, the word 'country' is common in conversational use. Perhaps, reflecting a comparable approach to that of indigenous communities, that word implies a wider affinity to land than the economic criteria ascribed to it in urban areas, where the terms such as 'land' and 'real estate' are more common to the vernacular.

The particular characteristics of Australian regions were presented in Chapter 4. Australia has a national ('macro') economy which provides the basic rules and parameters of economic and business activity, largely established and administered by the Commonwealth government and its various regulatory bodies–notably Treasury, the Reserve Bank. The State governments also have an important but lesser role. However of much greater consequence and underlying those political constructs, the 'Australian economy' is, in reality, made up of about 60 major regions, each with its own comparative and competitive advantages, characteristics and challenges. An appropriate and thorough analysis, both of economic sectors / sub-sectors and the real property within them, must commence with an understanding of that regional context (see Figure 9.2).

The Regional Australian Institute (RAI) (2019) considered the structure of Australian regions and identified four typical activity areas:

Figure 9.2 Australian regions.

- regional cities–typically with a population of 50,000+ and a diverse economy sufficient to generally shape their own future
- connected lifestyle regions–typically developed around an area with particular natural advantages and close to a major metropolitan area
- industry or service hubs–typically with a population of 15,000–50,000 located at some distance from any metropolitan area and with future prospects very much tied to its regional catchment area
- heartland regions not close to metropolitan nor regional cities but where local industry trends and demands for services combined with local ingenuity and enterprise responsible for future success, sustainability and resilience.

Historically, such regions had often experienced waves of settlement and development, initially established on uses such as mining and forestry and then later, depending on the characteristics of that region, dairying and onto other more specialised uses. That evolutionary process continues to the present day as comparative advantages emerge and, overtime, can dissipate.

Over more recent times, the rising dominance of the major cities has seen regional areas in peri-urban locations and those within reasonable travelling distance of and good access to those capitals enjoying significant growth. Other more distant regions typically experience population loss and drift, particularly by young adults into the main cities for education, work and social networks. Many of those areas also have quite narrow economic bases which are vulnerable to volatile world markets and to sparse rainfall becoming increasingly erratic as a result of climate change (Polèse 2009).

In summary, it is erroneous to draw wide observations of rural and regional Australia as if it was a homogenous entity. It is reasonable to observe however that there are a number of endemic issues that, despite the positive action by farmers, rural communities and the government, remain as major challenges. Most important of these are the effects of a rapidly changing climate set against Australia's already fragile ecosystems, seasonal variability and low water availability. Erosion and salination represent immediate sub-sets of this. At production level, food security and traceability require continued vigilance. Demographic challenges include a rapidly ageing population in many rural areas with high-dependency ratios and concerns on succession planning.

More positively, rural production continues to increase overall with improved livestock and plant breeding, more efficient use of water and in many cases, the aggregation of farms into more efficient enterprises. Again, the overall situation is mixed but with those regions with the best natural resources with good access into the major urban areas continues to develop at a much faster rate than more remote regions.

Rural real property is made up of a very large number of sometimes separate, sometimes interdependent sectors based on type of production and influenced significantly by topographical and climatic conditions, location and available infrastructure. This production along with the highest and best use and value is further influenced, in a macroeconomic sense, by market and regional conditions and, at a microeconomic level, the specific characteristics of that particular property and its level of development.

As noted above, ownership structures of Australian farms have changed little over generations. There are some 86,000 farming enterprises in Australia with 99 per cent of those Australian owned and operated. Nevertheless, the farming population continues to age. (The average age of an Australian farmer is currently 52 years, about 12 years older than the average of other sectors (ABARE 2019), though it might be argued that, unlike many other workers, farmers often remain resident on their farms even if they are now only partly involved in those activities.) Perhaps of greater consequence is the lack of succession planning available to many family enterprises with many young adults leaving farms to undertake further education and careers in urban areas.

Equity levels for Australian rural enterprises remain high. About 30 per cent of all farms have no debt at all and, of those who do, a considerable proportion of the debt funding had been obtained for long-term benefit investment such the increase in scale of farms referred to elsewhere in this chapter (ABARE 2014). As at 2018, properties with debt funding had an average liability of a manageable $480,000, comfortably within overall credit limits and with only modest levels of default (ABARE 2018). This reflects long-term, often inter-generational, nature of rural property ownership and the understandable financial conservatism of producers who are well aware of climatic and market volatility. O'Rourke (2018) makes the important observation however that, regardless of gearing, rural enterprises are almost invariably credit dependent because of non-synchronous production outlays vis-a-vis final income payments and, ironically, rising costs (e.g. for feed and water) during difficult trading periods such as droughts where gross farm income is likely to be low.

It needs to be noted also that, in the contemporary environment, individual farm incomes are often supplemented by household members securing income external to that property–be it in rural pursuits elsewhere or employment in nearby towns. Proximity and ease of access to those urban centres has therefore become increasingly important, past traditional supply, educational and community / social reasons.

9.3 Specific rural property sectors – general observations

Given that context and background, there are inherent dangers in becoming overly generalised in considering either individual properties or sub-sectors. There are however a number of larger scale rural production sectors, grazing, dairying, grain production, horticulture and fruit growing and grape and wine production (see Figure 9.3), where certain market observations and the establishment of market parameters can be reasonably made.

Some such observations are set out below, not to provide definitive assessments but to assist in an understanding of how such land uses evolve and some of the current parameters, opportunities and challenges for each. The observations should be read in conjunction with Chapter 10.

Grazing

Mainstream grazing activities in Australia are focussed on either cattle (meat production) or sheep (as lamb or sheep meat and or wool production). The third component, dairying is summarised separately below.

As regards beef production in Australia has about 70,000 producers representing 57 per cent of all rural properties. The national herd is approximately 27.5 million including about 12.5 million in breeding stock, both numbers fluctuating significantly with climatic variations (Australian Government 2018b). Approximately 74 per cent of

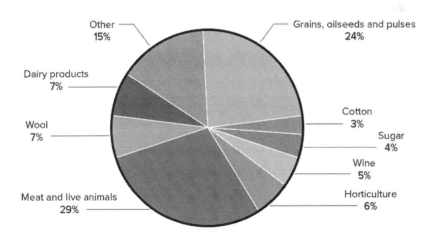

SOURCE: ABARE, Australian Commodities June Quarter, 2017

Figure 9.3 Breakdown of Australian rural production by sector.

final production is exported with about 20 per cent of that number being consigned as live exports.

Depending on country type/soils and availability of water, fodder and grain crops are frequently produced on those properties, grown as stock feed or for separate sale.

Regional and climatic differences are critical to the nature of the herd in any particular region/location. Typically, cattle will be breed and grown (as store cattle) in certain locations and moved to higher quality lands for fattening. Increasingly feed lots are used to finish the cattle prior to processing. Given the need to move cattle efficiently between regions and to feed lots, meat works or export ports, all weather access is critical to operations. The ability to sell and export cattle relates also to certification, traceability and proven quality of that product. In 1989, after a national eradication programme, Australia was declared free of bovine tuberculosis and brucellosis – an essential certification for cattle and bovine meat exports. Over the decades since those matters were successfully addressed, a range of additional food security and environmental issues have become increasingly important and a challenge for producers and processors alike. These have included controversies surrounding genetically modified crops, animal welfare, the usage of water, land clearing and a range of others–all of which appear to continue without final resolution, given some of the extreme views taken by some on both sides of the arguments.

Sheep and wool production in recent years has become more regionalised predominately in New South Wales, Victoria and Western Australia. The national herd stands at about 72 million with about 32,000 producers. As with most other rural sub-sectors, there is high exposure to global markets with some 61 per cent of lamb and 96 per cent of mutton exported.

Wool production in Australia presents a somewhat different scenario from practically any other form of rural production with the country being the world's largest producer of wool and with global recognition for the quality of the clip. Its historic success in Australia is not only due to relatively dry conditions that suit most sheep breeds but because of the fact that bulk wool is such a robust commodity, able to be transported relatively cheaply and able to be held over in storage indefinitely before release to the market.

Herd composition depends on final use with Merino stock still dominating but with other breeds such as Spanish Dorpers now often preferred for both wool and meat production.

Like all rural activities in Australia, sheep and wool production has ongoing challenges with adverse and changing climate and volatile wool prices and exchange rates. In this sub-sector also, challenges arise given the high level of specialisation, competition from man-made fibres. Shearing is labour intensive and difficult work and an emerging problem has been the attraction and retention of skilled labour for these activities. Nevertheless, these sheep and wool markets have proven relatively stable over recent years.

In assessing the highest and best use for the property, regard needs to be had to the mainstream husbandry practised in the region, the immediate location and, historically, on the subject property. On that property, production may be focused on breeding, growing and /or fattening and (in the case of sheep production) whether wool growing and/or lamb and sheep meat production is likely to present optimum returns at least in the medium term. It should be noted that the usage of properties does sometimes evolve over time though major changes, say from sheep to cattle, involves significant infrastructure change and capital investment.

Other critical determinants for any grazing property will include the quality of the herd/flock, breeding rates, certifications and bio-security status, land classification and quality, water availability and reliability, location and accessibility, available supply and production chains, and the type, quality, scale and suitability ('fit for purpose') of improvements. Whether or not the property represents a 'living area' (as defined above) also needs to be considered.

In any of these ventures, the role and acumen of the individual grazier is critical to efficient operations and production and for long-term sustainability. The grazier has to protect and enhance the quality of the base herd / flock over time but, at the same time, minimise the production costs, ensuring that processing is arranged to optimise returns and maintain cash flow. All of this is set against unpredictable seasonal conditions and the actions of other producers also facing similar challenges. The final decision to hold or turn stock off the property will depend on a range of factors including the cost of fodder, seasonal expectations and the market price for animals. Timing, good market intelligence and the experience of that grazier will therefore be critical to success and sustainability, particularly when facing adverse conditions.

Dairying

While one of the oldest rural sectors, dairying arguably remains the most problematic and its long-term sustainability is under a level of threat. Dairying has undergone radical changes over recent decades–transitioning from a large number of small farms and small scale, localised processing to a smaller number of large producers, most servicing the bulk milk market. There are now approximately 8,600 producers in Australia, a reduction of almost 40 per cent since 1990. The national herd comprises some 1.6 million animals and has an annual output of about 4 billion dollars (Australian Government 2018c). Overall however, the sector remains in transaction and dependent on Government subsidies for its continued rationalisation and survival.

The need for very high quality pasture and high demands for water together with the requirements of the bulk handling of such a perishable commodity has seen the sector become increasingly concentrated in parts of Victoria, coastal New South Wales and South East Queensland. Fifty per cent of all milk processing is controlled by two major international companies, and this, combined with oligopoly structures at the retail level, has resulted in severe cost pressures falling back to farmers.

Meanwhile, some 75 per cent of production is exposed to world price variations and even though good opportunities exist for powered milk products and 'high end high quality' and safe products that this country can supply, Australia is typically a relatively small producer in the international market.

A number of opportunities for the sector continue to evolve however. These include the enhanced production of powdered milk and of specialty cheeses and emphasis on organic product. Depending on the location too, regional branding and links with tourism and retail/boutique offerings also present potential.

In considering the assessment of individual farms, the following parameters are of particular interest:

- given the wider challenges to this sector outlined above, contemporary assessment of any farm must look at the entire viability and sustainability of that enterprise – particularly related to its scale, supply contracts and cost structures

- an ageing work force and issues of succession planning
- land quality, seasonal variability and water and energy availability and cost
- herd quality and composition
- improved pasture and ability to grow fodder on farm
- farm management including level of technology and operational efficiency
- ability to diversify and potential alternate uses ('highest and best')
- location and access to processors, urban markets and any opportunities related to regional tourist visitations.

Grain production and fodder crops

The production of grain and, associated with that, fodder crops represent about a quarter of all farming enterprise in Australia with sub-groupings of production including various grain varieties, oilseeds and pulses (i.e. legumes of various types) and representing about 21 per cent of all farm production and about 24 per cent of total rural exports (Australian Government 2018a). Export sales are facilitated by Australia's international trade agreements and arranged through export and grain handling organisations located in major grain growing regions and at major ports (Australian Government 2018a; GRDC 2018). Domestic grain sales are typically arranged through marketing agents. Only 30 per cent of fodder production is traded off-farm – the rest being used for stock feed on the high proportion properties that mix livestock (normally sheep and/or cattle) with the farming enterprise.

Principal grain crops produced are wheat, about 65 per cent, barley about 22 per cent and with sorghum, oats, cotton (linseed), canola and sunflowers grown in significant volumes (Australian Government 2018a). Given variations in soil, water availability, climate and final market (i.e. export, domestic grain or on-farm or domestic fodder), certain crops will tend to dominate in specific regions. This tendency to focus on a small number of crops will be further reinforced given the investment in specialised farm layout and equipment, established production supply chains and the accumulated knowledge of the farmers in those types of production. Nevertheless, the individual farmer will need to take decisions each year on crop selection based on anticipated prices, water availability, the likelihood of summer and winter plantings and a range of other considerations.

Within this broad sector, there are numerous variations and sub-groupings that need to be recognised and taken into account at a regional and farm level before a sound property assessment could be undertaken.

While about 45 per cent of production here is wheat, there are numerous other crops within this group—some for human consumption, the rest for sale for fodder or used as stock feed on the many farms that mix such activities with husbandry. Marketing avenues also depend on the specific commodity and its intended customers.

Principal areas include the central highlands and Darling Downs in Queensland, much of central and southern New South Wales, most of the agricultural areas of Victoria and Tasmania, the south-eastern parts of South Australia including the Eyre Peninsular, and most of south-western Western Australia. Growing areas follow soil types and rainfall with various locations tending to specialise in the sub-groups identified above. Regional seasonal variations, based on prevailing weather conditions are significant.

As noted above, individual regions will tend to specialise in certain grain and fodder crops and produce will be sold into a range of markets–some for export, some locally as grain for human consumption or for animal feed or kept on that property to supplement feed for that stock. The type of produce and the target market will also determine the level of on-site storage (e.g. silos) necessary.

Critically important is the availability of water, whether that water is secured from on-stream pumping, underground aquifers and/or on-farm tanks and dams. Additionally, the reliability and cost of those supplies need to be assessed. Irrigated crops not only provide a high-level of certainty but ensure an appropriate level of soil moisture into the future. Dry farming is obviously problematic depending on local rainfall patterns but provides no guarantee of second (winter) crops. Prolonged droughts may result in no crop at all, year-to-year. Not only is this a serious detriment to farm viability, but it adversely effects the ability to raise debt finance for the enterprise. Changes in climatic patterns and increasing limitations and higher costs associated with irrigation supplies are obvious threats to such farming operations.

As with many other rural sectors, productivity has increased remarkably over recent decades based principally on the higher yields of new seed varieties, more effective and efficient use of water, fertilisers and pesticides. Additionally, the average size of farms has increased significantly over time with broad acre cultivation practised. In this type of production also, the acumen and experience of the individual farmer is critically important in selecting the appropriate crop (having regard to the characteristics of the individual property and likely climatic and market conditions), and during the growing period to apply the right combination of water, fertiliser and weedicide to produce optimum production.

Changing climatic conditions and, with that, uncertainty surrounding future water supplies for irrigation present immediate challenges to the entire sector.

Horticulture and fruit growing

Horticulture (including fruit growing but excluding wine production) earned approximately $B8.7 dollars per annum (Hort Innovation 2019) and is remarkably diverse, producing some 75 significant product lines.

These activities are clustered in a wide variety of locations with production very much determined by local climatic conditions and the availability of water from streams or from storages, size and scale, regional expertise and proximity and linkages to markets.

In practically all of these sub-sectors a number of market and other forces have resulted in significant changes in structure and supply chains. These have included:

- The general increase in size and scale of farming enterprises. While some corporate enterprises have entered a number of these markets, most of the increase in scale has been the result of the significant increase in size from existing, often family-based businesses. As well as reducing the cost base, such aggregation provides those growers with access to improved plant breeding and field and packing/processing technologies and mechanisation.
- The cost and availability of labour with appropriate skills, in particular at peak periods where bottlenecks regularly arise with the need to secure labour, often from overseas, is essential.

- Each business requires quite sophisticated management which recognises the likely demand for particular types of production and, using natural and locational advantages, to secure entry to the market at an opportune time. The farmer need to also recognise production lags and issues of plant selection particularly in orchards where commitments to a particular variety will have long-term implications.
- The role of major retailers in the marketplace and the ability to secure long-term and reliable contracts directly without the necessary involvement of marketing agents and produce markets.
- In seeking higher productivity and returns, farmers now establishing or replanting orchards are typically employing much denser planting layouts and often selecting varieties of plant size that is easier to tend and to harvest.
- Freshness and high quality are important criteria in securing best market price and rapid harvesting, on-farm processing and delivery to the final market and consumer, overnight if possible, are essential to maximum returns.

Grape and wine production

Grape growing and wine production have occurred in Australia since the early 1800s. However, it is only in the last half century that the scale, value and importance of that sector both nationally and internationally have emerged.

There are approximately 65 recognised wine-grape growing regions in Australia with a total area of 170,000 ha under vines in 2017–2018 (Wine Australia 2019) and approximately 2,250 producers (Morrison 2019). The industry has grown remarkably over recent decades with the 2018 vintage crushing about 1.8 million tons of grape, producing about 1.3 billion litres of wine with a retail value of $B 6.25. There is a fairly stable balance between product exported (63 per cent) and domestic consumption of 37 per cent (Wine Australia 2019).

The sector needs to be understood in its component parts. The major grape-growing areas lie in South Australian Riverland area and the Riverina in New South Wales with other significant growing areas in Victoria and south-western Western Australia. Much of the grape production is transported to wineries in other regions and interstate.

In wine production, there are effectively two types of winery. For example, the top five (corporate) producers account for some 85 per cent of total export volume, whereas at the other end, about two-thirds of Australian wineries produce fewer than 5,000 cases per annum (Morrison 2019). The former group supplies not simply to the export market but to major retail outlets across Australia. The latter are small boutique enterprises typically owner operated and relying on cellar door and mail order sales.

Some vineyards incorporate and supply into wineries under the same ownership. Others sell grapes in bulk to wineries elsewhere. The viability of either model depends on seasonal market conditions to sell annual production.

Grapes require particular growing conditions particularly needing well-drained soils, correct aspect and a micro-climate with sufficient cool weather to set the fruit. Vines typically require three to five years to reach maturity but, if properly maintained, can remain in production for decades. Operations are labour intensive throughout. Particular acumen is required throughout the entire process – in selecting varietals that will be well received in markets in the future, in the grape growing and harvesting process and also in wine-making itself.

Many of the smaller wineries rely on the personal goodwill and reputation of that wine maker. In many regions, the winery effectively integrates with the regional tourism offer with the enterprises often including cafes and restaurants and overnight tourist accommodation. The size of such enterprises varies considerably. The capital costs of establishment of vineyards (ground preparation, trellises and plants, and as necessary protective netting) and their associated activities are considerable, and management and operations are labour intensive. Depending on the level of personal involvement of the owners, the size of such enterprises can be quite small–perhaps as low as 5 ha but ranging up to much larger holdings.

Unlike many other rural enterprises, the final product here stores well and can be progressively released over a long period of time and, in fact quality is often improved during that period.

In assessing the viability and sustainability of such enterprises (vineyards and/or wineries), matters for particular consideration:

- scale of operation (whether the size and operations represent a viable commercial enterprise over time or whether, as is sometimes the case, the venture represents a part time interest)
- varieties, quality and condition of vines and expectations for forthcoming vintages
- soils and drainage, micro–climate and aspect
- reputation and wine maker
- on–estate equipment and vine protection
- stored vintages and stock
- supply contracts and outlets
- supplementary/complementary income sources
- regional considerations – reputation, surrounding complementary uses, regional links, proximity to major urban areas, et cetera.

9.4 Rural real property assessment – summary

Chapter 10 considers the various methods of property valuation across a range of sectors, applying a range of assessment methodologies. In conclusion of this chapter on rural real estate some observations are made on their application here.

There are a range of methodologies generally accepted and available in all valuation (Scarratt & Osborne 2014). These are:

- (direct) comparable sales
- capitalisation of (nett) income (nett rent)
- projections of income (D.C.F.), that is, investment analysis
- hypothetical development/hypothetical subdivision
- assessment of going concern operations/businesses
- summation assessment of asset's component parts.

Accepted valuation principles emphasise the primacy of sales evidence provided by comparable transactions. That becomes the basis for assessment of rural properties. An important challenge arises here however, because of the diversity of each rural property (and sale that may be referenced). Each one will have a diverse set of characteristics ranging from size, geography, rainfall and location through to various classifications of land, improvements, livestock, growing crops and a range of other unique features. As important as sales evidence will be, it will be most unlikely that comparisons can be made 'full property to full property' and a further breakdown and analysis of sales will be necessary.

Of other possible assessment methodologies, most, for example, capitalisation of net income, investment analysis and hypothetical development/subdivision, are clearly not applicable. Income and time frame projections are simply too nebulous to provide reliable data. Similarly, even though a rural enterprise can be considered as a 'going concern business', income and costs are impossible to accurately predict over time because of seasonal and market variations. (It may be that some components of the enterprise – for example, a restaurant or accommodation offer at a winery – could reasonably be assessed on a going concern business basis but these are exceptions to the mainstream of a rural undertaking.)

The use of a summation valuation approach, that is, the simple addition of land value plus the depreciated cost of improvements, is recognised as being inferior to other forms of assessment. It is first based on an assessment of cost rather than value and, second, fails to consider the aggregate value of the entire property over and above its component parts. While to be used with caution and never alone, this form of assessment does however have a particular value in rural appraisals in 'adjusting' working calculations on both subject and sale properties to bring them to a basis for close comparison. The general approach is further described below.

As with all property assessments, the valuer is, in effect, constructing 'a narrative' describing the property and its economic and physical characteristics, thereby establishing the highest and best use and, by comparison with the market, an assessed value. As with similar exercises in other sectors, a rural assessment is always to made to a specific date (of either sale or of valuation) and noting the normal criteria that confirm that any sale relied upon represents a legitimate, open market transaction. (These matters are further discussed in Chapter 10.)

In assessing the value of a rural property, it should be noted that a number of country types may be involved. Each type may have differing use potential ('use capacity'), soils and vegetation types, water availability and topography. A mixed grazing property, for example, may have areas which are arable and used for cash or fodder crops (be they irrigated or 'dry farmed'), improved pasture or simply grazing on natural pasture. All of these can be considered and assessed separately but noting the acumen of the expert valuer to also consider the full value as a going concern of the entire holding.

At the risk of over-generalisation, quality of soils and amount and reliability of water supply and rainfall are the primary determents of practically all Australian rural land uses.

In rural and regional settings, sales or components thereof can be drawn from a greater distance than is typically the case in an urban environment, provided that country type, rainfall and water availability are reasonably comparable.

Regards should be had to whether a particular holding reasonably represents a 'living area' (as defined above) or is likely to be best used as part of an aggregation with other lands. This general categorisation may help in the assessment of improvements and their suitability and functionality.

In most regions, rural property sales are less frequent and markets generally less volatile than in urban environments. Therefore, while most recent sales evidence is to be preferred, more dated sales may still have considerable relevance.

In both the analysis of sales and the assessment of individual properties, it needs to be established at an early stage, whether the sale under investigation or valuation assessment is on a 'sold bare' (that is, ex plant and equipment, livestock and growing crops, stored produce and chattels) or on a 'walk in – walk out' basis (that is, sold as a going concern, including all of those items listed above). Particularly in the case of 'walk in – walk out' transactions, the time of year in which the sale or assessment is carried out is highly relevant. That date will often determine the quality, condition and value of any livestock or growing crops to be included in the sale and also, depending on that time of sale, will confirm growing conditions and water availability into the forth coming season.

As with all valuation tasks, the most reliable evidence is provided by the analysis of comparable sales. In the case of rural assessments, such analysis will notionally break down the sale properties into their characteristic of land types, water availability, et cetera, so bringing those components down to a 'unit of measurement' that can be compared between properties. There are two practical components to this:

- To reach a legitimate point of comparison, it is often necessary to nominally adjust the value of certain components of the sale or subject property for anomalies and special features that one may have in comparison with others. For example, one property may have a set of cattle yards suitable and necessary for the running of that property. A second may have no yards but would be well served if such yards existed. To make a comparison between the two possible, the depreciated value of those yards could be nominally deducted from the first property, comparisons then made and, finally that yard cost be once again added to that property. Similar adjustments can be made for other variations in characteristics before a final assessment figure for the subject property is established.

  For rural sales, these 'summation' assessments can be made with considerable accuracy. For example, rural valuers will be aware of the value of stock on farm at any point in time and the value of a standing crop depending on its stage of development and future market expectations: Additionally, the new cost and life expectancy of rural improvements and chattels can be established. To add to this body of knowledge, it was noted above that many rural properties are 'sold bare' and such events give rise to separate clearing sales of plant, equipment and other chattels on properties disposed of in this way. The results of those auction sales are typically monitored by rural and specialist plant and equipment valuers and provide sound evidence of 'as is' value of those items. All of this research and investigations provide further credibility to the valuation adjustments made in the above property sales analysis.

- The final analysed 'unit for comparison' can simply be on a 'per hectare' basis however, depending on the type of production other measures can be used again with the purpose of bringing the various properties to units of comparison requiring limited adjustment. These, depending on property type could include:
  - beast area value
  - dry sheep value
  - irrigable, arable, with water availability, per hectare

- dry farming per hectare
- grazing (various classifications) per hectare
- grazing – valued per hectare, fenced and watered
- for fruit growing and perennial crops, value per tree.

In most cases, it is recommended that as many points of comparison as possible be employed. However, as with any 'rules of thumb' employed as part of a valuation process, extreme care needs to be taken in how they are to be considered and applied (Baxter & Cohen 2009). As has been previously stressed, each rural property is unique with all of its characteristics influenced by a variety of factors. Therefore, while any such general classifications may be of assistance, they need to be applied with a deeper, professional understanding of the region, location and of the particular property being assessed.

The valuation components of this are presented in more detail in Chapter 10.

Chapter 10

Valuation theory and applications

10.1 Introduction

This chapter discusses the role that property valuation plays within the economy and across the wider community. It will provide an overview of the general principles and practice related to the assessment of the market value of real property assets.

Specialist books and papers listed in the reference section of the chapter provide direction for more detailed study and further research.

10.2 The need for established methodologies and professional practice in assessing property value

A large proportion of the resources in Australia are made up of real property assets and, like many other forms of wealth, ownership is scattered across a large number of private owners. There is a huge number of property dealings occurring at any point in time, with demand being the key determinant of both price and level of activity.

Many of these dealings are familiar – such as the buying, selling and renting of residential property and the securing of loans using mortgages on real property as collateral. However, there are many other dealings that occur frequently and are also fundamental to wider property ownership and use. These include the establishment of and dealings with leases, partnership arrangements, personal estates, family law matters, property development and subdivisions, infrastructure agreements, equity establishment and debt raising; compulsory acquisition of privately owned land, Crown and civil legal actions, joint ventures, syndication and securitisation, trust dealings, native title claims, licences, vegetation protection or clearing approvals; and taxation matters among other things – all related to the holding, use and appreciation in value of property assets.

Reliable assessment (valuation) methodologies, standards and protocols for such dealings are essential to maintain business confidence and to help secure much of the personal and corporate wealth of the country.

In many of the dealings and activities listed above, a key task will be to establish the true worth ('value') of the subject property at that point in time. A prospective purchaser, developer or financier, for example, will need to be assured that a particular property is indeed worth the suggested price having regard to all of the nuances of the market and the characteristics of that property.

Sometimes, the transaction or activity coincides with a full sale of the property. This sale, on the face of it, provides evidence of the value of that asset because the vendor and purchaser were willing to actually exchange that amount of money for it. The sale would need to be investigated to confirm that both the vendor and purchaser acted prudently, and that the sale truly reflected the market value for the property. To further support that price, evidence of a number of comparable sales can reinforce the assessment.

In practice, however, there is a large number of property dealings that occur where a price has not been established through a full, 'buy–sell' transaction or where that transaction needs to be verified by wider investigations.

Consider, for example, the following:

- A corporation, individual or bank is investigating the possible financing of a property, investment or development, and seeks independent analysis of the proposal and assessment of the value.
- One firm is taking over or merging with another, or a partnership between individuals is being merged, changed, added to or dissolved, and those dealings involve real property. The part value of these assets will need to be determined and agreed.
- There are off-market property negotiations underway – be they for a proposed sale, lease review, rental or similar – and the parties cannot agree, requiring the opinion of a third-party expert.
- The settlement of an estate or the winding up of a trust and disbursement of its asset to beneficiaries.
- Properties requiring an assessment of value for the payment of stamp duties, rates or other taxes.
- Land is being resumed in whole or part by a government agency/constructing authority and the value and related compensation claims need to be assessed.
- There is argument in a court of law (either a Crown case or tort), involving property and assets and, the court, in making its determination, will need to rely on an expert's opinion (i.e. someone cognised as an 'expert witness', sometimes described in law as 'a friend of the court') as to the true value of the asset in question.

From these examples (and there are many more) it can be seen that the effective operation of economic, commercial and legal systems requires a clear structure and accepted methodologies by which determinations of property values can be made. To achieve this, specialists with independence and a high level of expertise and experience will be required to provide that advice and opinion (Wyatt 2007). Note also that, some valuers build up specialist expertise in certain market sub-sectors (aside from land and building), such as plant and machinery, lease determinations and a range of others.

As well as their role in a specific assessment, property valuers carry a professional responsibility to the wider business and general community. Reich (2018) notes that for any such arrangements to work effectively, the professionals involved need to recognise the concept of a 'common good', that is, obligations that extend well beyond simple compliance with law and regulation. The approach should reflect a positive and supportive attitude to law in creating a stable and predictable environment for business and community activities. Without that, Reich considers the primacy of self-interest in a capitalist system can lead to unnecessary power imbalances and social injustice. While this may be considered as the 'civic duty' for all, a particularly high level of behaviour is required of such professionals, well above personal interests and individual gain or even the risk of penalty.

This does not imply that such professionals should adopt some moralistic nor unnecessarily conservative approach to their role. It does however require that individuals engaged in activities such as independent, property valuations accept wider responsibilities – using their experience, insight and expert observations to support the well-being and advancement of individual clients and, through such ethical behaviour, to reinforce established systems and principles across the community.

Reich notes that the fact that the vast majority of community, legal and business dealings work so smoothly and seamlessly is based on the commitment of professionals, certainly to their clients but also respecting the concept of 'the common good'.

The methods for assessment commonly applied in the determination of value have evolved over the past 100 years and are now at a very mature level. Numerous pieces of Commonwealth and state legislation established and progressively refined those approaches with their importance even recognised in Section 51 (xxxi) of the Australian Constitution (Commonwealth of Australia 1901).

These rules and protocols come together in different ways. First, property valuers are registered in some states. Additionally, through legislation, case law, established methodologies, professional standards and ethics, the behaviour of those professionals is regulated by government and by relevant professional bodies.

An example of this is the publication of the *International Valuation Standards* the latest issued in July 2019 (effective from 31 January 2020) published by the International Valuation Standards Council (IVSC). As property and other investment markets have become increasingly globalised, these standards have gained importance and are now in parallel with the *International Financial Reporting Standards (IFRS)*. Linked to the IVSC publication, relevant professional bodies publish comprehensive standards for their members with adaptations to accommodate their interests. The Royal Institution of Chartered Surveyors (RICS) similarly issues and regularly updates its *Valuation Professional Standards*, known as the 'Red Book', the latest iteration taking effect on 1 July 2017.

The Australian Property Institute publishes the *Australian and New Zealand Valuation and Property Standards* (2012) that likewise sets out the duties, responsibilities and professional standards for its members.

All of these standards are comprehensive and provide detailed direction to respective members relevant to most property valuation assessments.

It should be noted here that, while some valuation experts may practise as 'Registered Valuers' or the Australian Property Institute (API) nomenclature 'Certified Practicing Valuers', many others do not practise directly or overtly. Nevertheless, these professionals will typically use those skills in other property-related areas such as development, asset management, land administration, finance and so on across the public and private sectors.

10.3 Valuation principles

The best evidence of value is derived from the market place through an analysis of transacted sales of comparable properties. This principle is called 'the primacy of sales evidence' which is self-evident and reinforced both in statute and court precedent. In practice, the application of sales evidence can become complex.

As described in Chapter 2, real property markets provide an unusual and complex environment for those who attempt to assess market behaviour or to establish the value of a particular asset. While fundamental economic rules and principles set market parameters, their application to a heterogeneous commodity – where each economic unit (i.e. a parcel of real property) has its own unique physical, legal, financial, ownership and other characteristics – means that final individual assessments and outcomes will vary greatly. This does not imply that the market is chaotic nor that the outcomes are without logic. At the same time however, it is essential that the assessment process recognises and reflects the individuality of each asset; 'one size' certainly does not fit all.

In valuing real property there are a number of terms and concepts that have meaning and implication well beyond their vernacular use. The terms 'cost', 'value', 'worth', 'obsolescence', 'depreciation' and 'time (frame)' have been referred to in various parts of this book but are further elaborated upon here.

Cost is the expenditure incurred in producing a particular good or item. For example, the total cost to a baker in producing a loaf of bread or the total cost for a developer/builder to produce a new house and land package. Cost is therefore a 'supply side' concept as described in general economic theory in Chapter 2. It includes not only direct input costs, but also reasonable margins for the producer (the baker, builder or whomever) as recompense for that person's overheads, skills, coordination, business acumen and so on.

Value is quite a different concept and pertains principally to the 'demand side' of the economic equation. It can be defined as the amount of money that a prudent prospective purchaser would be willing to pay for that item (bread, house and land or whatever) in that market at that point in time (known as 'present value' [PV]). In most markets, there will be a relation between cost and value. For example, for simple commodities in reasonably open markets, if market value significantly exceeds costs, relatively high profits will be available to producers. In that case, new producers will, in the medium term at least, enter the market seeking those attractive margins. The additional production will cause prices to fall back closer to cost – a simple operation of 'the price mechanism'. Irrespective of those factors, however, there are other forces at work.

If cost was the only determinant, the highest value that a particular item could ever achieve would be the amount it cost when new. For some depreciating assets such as motor vehicles, that is a reasonable observation – very few will ever have a market value higher than when they are first 'driven out of the showroom'. Indeed, these goods typically depreciate rapidly over their first few years and, additionally, physically deteriorate with use and mileage. Over time they will also typically suffer from obsolescence as new, more advanced models are brought onto the market by manufacturers keen to have consumers trade-in depreciated, less fashionable assets now at a much reduced market value.

Relevant too, is the concept of 'worth'. In colloquial language, 'worth' and 'value' are often used interchangeably, but there are subtle differences. 'Worth' can often refer to the particular characteristic of an object, asset or even of a person, that produces desirable, useful or valuable outcomes. Worth may or may not be tangible nor have direct money equivalent measures. An important differentiation here is that 'worth' is typically a quite subjective term; the perception of the worth of something by one person may be quite different from another's perspective. It can also reflect on a range of characteristics other than economic impact – as in, a 'worthy person' or 'the worth of an institution' and so on.

'Value', however, presents the opinion or general agreement across a group (i.e. a market) that this object has a certain function, level of scarcity or whatever that can be reflected (normally at least) as a monetary equivalent. The concept of 'value' is more precise and will often be assessable through comparative analysis and in financial terms. Economically, it normally implies that a market exists for that item and, with that, the ability to buy, sell and exchange it. Legal concepts of private ownership and exclusivity will also normally apply.

The assessment of the value of real property assets is a complex matter. Property assets are exposed to physical deterioration and obsolescence (as with the case of the motor vehicle example above); however, there are wider considerations to be taken into account. In acquiring a property asset, the purchaser is securing the use, value and function of that asset over a long time frame (perpetuity in the case of freehold tenure). That ownership allows the potential to alter, rejuvenate and upgrade the built asset progressively over time. Additionally, the asset can benefit over time from the enhancement of its locality and the upgrading of surrounding developments. Consequently, the value of that asset may rise well above original cost (even inflation adjusted) and, because of its unique location or other characteristics, may not be easily replicated (as was the case of the motor vehicle example above).

Of course, the reverse can also be true where a property can fall in value because it now suffers from some form of physical deterioration, obsolescence, or perhaps its locality has lost favour over time or suffered blight. Alternatively, that particular market may become oversupplied, thereby putting downward pressure on prices. The underlying principle here, however, is that there is much more to the establishment of property value than consideration of original cost and physical deterioration.

In investigating this range of value determinants, some influences will be specific to a particular property (i.e. locality), while others may be more generic or systemic, such as interest rates, general levels of investor confidence, taste and fashion, community sentiment and so on. Professional judgement is very important in considering all these factors in a final, assessed value.

'Econometrics' is that branch of economics that uses statistics to model relationships between variables and, therefore, can help predict market behaviour in a more rigorous and structured way. These 'price models' may, for example, analyse sales data to establish the impact on the price (or value) of parcels of residential land, given the distance from transport, available services or any number of other determinants for which data is available. In this way, the price of other land can be better predicted. This analysis, called 'hedonic modelling', is particularly useful in researching property sectors where a wide data base of comparable sales is available, and where a relatively small number of key determinants directly influence final sale prices. Once mathematical relationships are established, the model can be used predicatively to estimate the value of other unsold property.

These techniques, including hedonic modelling outlined above, are known as 'regression analyses' and normally attempt to track the impact of one dependent variable that results in changes in one or more independent variables. As always, assumptions made and analysis undertaken need to be based as much as possible on the interpretation of market conditions – a significant challenge here given that cash flow projections into the future are being assessed. Nevertheless the approach is of particular use in major investment and development projects (refer Chapter 12). In complex projects, by using appropriate software and quality market information, this analysis is valuable in assessing the sensitivity or level of dependency of one component of a property investment or proposal to change in another component (i.e. rapidly responding to 'what if?' enquiries).

10.4 Valuation practice: overview

As with many professional activities in other fields, the analysis and assessment of property value might appear to be a fairly simple and straightforward task. Sometimes it is. Consider, for example, a new residential subdivision, where vacant blocks of land have been selling in a narrow range of values – some blocks that were, say, a little larger or better elevated would sell for a little more than the norm, whereas those with a smaller land area or limited outlook might sell for somewhat less. In this case, it should not be overly difficult to compare another, unsold, block in that locality with a range of recent sales and arrive at an accurate assessment of market value. Hedonistic modelling described above can also be applied to improve the accuracy of the comparison process, particularly where a large number of assessments are made for fairly similar properties.

However, cases as straightforward as that are not common. Often, properties are not physically similar and present a diverse range of improvements, zoning/use controls and other characteristics, making direct comparison much more difficult.

Valuers in effect are asked to interpret the market – using recent sales evidence – to arrive at a value at a specific date (normally present-day value). That assessment will also need to recognise the individual, physical and non-physical aspects of the subject property, its available infrastructure, surrounding development and statutory constraints. These comparisons should also recognise any particular potential of the property, that is, assessment needs to reflect the subject property's highest and best use. Because of the typically large-scale and long-term nature of property, its value is a reflection of what physically exists at the date of valuation and also the potential for a higher return or yield in the future. This could be the result of future capital investment to intensify use and/or further physical development of the property over time.

That potential, and the overriding concept of highest and best use, must be realistic. Any possible potential will vary depending on market conditions, existing improvements, risk factors, time frames and development controls and regulation relevant to that property and location. The assessment of any potential into the future is often made easier where sales evidence is uncovered that is closely comparable (in location, size, types of improvements, statutory controls, et cetera) to the subject property because, imbedded in that sale price, are expectations of any potential into the future.

The observation may well be made that valuation/property assessment is an 'inexact science' – and so it is – though the same observations may well be made of most branches of economics and of town planning, law, medicine and many other disciplines, to a greater or lesser extent. In the case of real property assessment and valuation, the lack of precision is systemic and obvious – each individual parcel or holding is unique and the extent to which each differs (in value and other characteristics) from others will always be open to interpretation. The key principle here, however, is that analysis that includes a sound and comprehensive information base (of the subject property, sales and other market-relevant data) and the selection of appropriate methodology/methodologies, all drawn together through the acumen of a competent and experienced professional, will produce an assessment in which all parties can have confidence.

In valuation terminology, the valuer is attempting to put himself/herself in the position of a 'hypothetical prudent purchaser', that is, to ask: 'what would a person, who is willing to freely buy the property and who has sound knowledge of it and its market at that point of time, pay to secure it?'

The valuer would need to carry out a physical inspection of the property, investigate comparable sales and secure a wide appreciation of overall market conditions at that point in time. Additionally, that expert will need to investigate a range of tangible and non-tangible matters affecting the physical nature, functionality and income potential of the subject property (API 2007b, 2007c). For the recipients of such reports, these observations can be as important as the valuation figure itself. This highlights the important observation, made elsewhere in this text, that a valuation should be seen as much more than a 'snapshot' or 'abstract figure'. Rather, it needs to provide 'a narrative' about that property, setting it in a physical/geographic, economic, sectorial, legal and social context at that point in time. In a relatively simple exercise, such as residential property assessment, such an exercise may be quite straight forward and require little elaboration. In major property investigations, however, this essential contextualisation may require a much more comprehensive and structured process (see Figure 10.1).

Formal valuations, such as those used for statutory, mortgage or court dealings will almost always arrive at a single figure, though there may be caveats on the assessment depending on the circumstances of the case.

It should be observed that a valuation will always be as at a specific date, which may be in the past or the present day. Obviously enough, because of any number of economic, business, legal or political events can happen into the future; a formal valuation cannot be in advance of the date on which the valuation was undertaken.

In some cases, a formal valuation (i.e. arriving at a single assessed figure) may not be appropriate. Rather, using similar analysis, investigations may examine say, property development proposals, asset management options or financial projections more generally. This type of investigation may arrive at a range of values, hopefully in a fairly

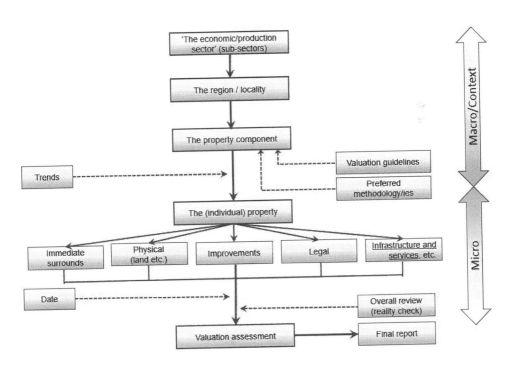

Figure 10.1 Property assessment/valuation – the wider context.

narrow band, and often presented to reflect 'best case', 'worst case' and 'most likely case' scenarios.

As with expert opinion in any profession, two valuers may arrive at different assessments. Given that they should be applying the professional standards described above and be using similar methodologies, variations should not be great, in simple cases at least. However, in major undertakings, variations may be significant. Even quite small differences of opinion as regards, say, development costs, interest rates, development period and/or final achievable capital values or rental levels can have large impacts on the final valuation. These differences in opinion often occur in practice – again reflecting that each property and each transaction is unique and, while sound methodologies will be applied, assessments are largely based on each professional's acumen, judgement and opinion.

Matters of professional disagreement are often addressed at structured conferences between the valuers involved and sometimes require resolution in various tribunals and courts. Arbitration, however, is increasingly seen as a 'last resort'.

In undertaking an assessment of real property value, there are five overall approaches/ methodologies that are generally recognised and in common use. These and the typical areas in which they will be applied are summarised in Table 10.1.

Each has benefits and limitations with some being more applicable in certain situations and for particular tasks. They are not competitive but rather consider the task at hand from a certain viewpoint. In practice, most assessments would involve the use of more than one method, one seen as primary or most applicable, while one or more additional methodologies seen as confirming or checking those results.

10.5 Valuation practice: *use of comparable sales*

Given the primacy placed on recent, comparable sales in valuation theory, it should not be surprising that references back to market evidence (either direct or analysed) permeates most of the approaches summarised in the above table. Each completed sale represents the financial affirmation of that purchaser and vendor's judgement of the market at that time. This tenet is based on the landmark High Court decision *Spenser v Commonwealth of Australia (1907)* (Ref: HCA 82; 5 CLR 418; 14 ALR 253) and repeatedly cited property court determinations since that time. It is also an important premise of the various valuation guidelines identified in Sections 10.3 and 10.4.

In a simple example, such as residential land in a subdivision described earlier, direct, full comparison ('block-to-block') can be made. The more developed and complex the property, the more difficult such simple comparison becomes. In considering the assessment of, say, improved residential properties, the valuer is still able to compare one property with another but the assessment will now involve much more significant price adjustments than those, described earlier, between vacant residential allotments. The valuer will also need to ensure that sales used are as comparable to the subject property as possible (e.g. in the same general locality, of comparable size and with comparable improvements, condition, design, potential and so forth).

The manner in which comparable sales are to be analysed and interpreted now becomes more complex with the process based on established professional guidelines and case law noted previously. Suffice to say, not all sales represent evidence of open-market

Table *10.1* Summary of valuation methodologies

Methodology	Process	Typical applications	Observations
1. Comparable sales (direct market evidence)	Using established analysis procedures, evidence of value is obtained by comparing the subject property with comparable recent sales.	Fundamental evidence of value applicable across all sectors and applications – either 'whole property to whole property' or comparison of component parts.	Primary methodology and the basis/ reference point to all other methodologies.
2. Capitalisation of net rental income (traditional investment approach)	Value is assessed by first establishing the net income derived from a property per annum (typically gross rent minus outgoings). That net figure is multiplied by a factor known as *Years Purchase* – that is, 100 divided by an acceptable rate of return deduced from the analysis of comparable sales.	Applicable to the assessment of all income-producing properties, particularly where a rental stream and outgoings are known. All protected cash flows and anticipated returns on investment to be linked with market comparison.	Makes the critical link between income and capital value and risk. Accuracy depends on the reliability of market information. Is the basis for more advanced investment analysis (below) but lacks the ability to include the level of detail of those approaches. Is relatively easy to understand and apply.
3. Investment analysis (acquisition or new development)	Using spreadsheet analysis, track the positive and negative cash flows of some investment or development project over its life so arriving at either a present-day value or as estimate of realisable value for that undertaking. Accurate use requires referencing to various market components throughout the process	Investigations of investments for existing, income-producing properties and the proposed development of new investment properties and subdivisions. Establishes *Net Present Value* (NPV) and *Internal Rate of Return* (IRR) viability of a project and in comparison, with other projects.	Accommodates complex cash flows through the life of the project, well recognising the time value of money. Allows exploration of 'what if?' options. Issues arise from the need to make a range of assumptions regarding income and costs well into the future but where accurate assessment is not possible. Widely accepted and used by the financial sector but limitations are recognised in valuation practice and court decisions.

(Continued)

Methodology	Process	Typical applications	Observations
4. Summation (cash-based assessment)	An assessment based on the sum of land value (normally deduced from sales evidence) plus the depreciated value of any improvements.	Not normally used as a primary valuation methodology but has value to check more sophisticated approaches. Has use in replacement assessments (e.g. insurance) and in the analysis of certain sales evidence (e.g. nominal value of rural improvements).	Pertains to cost considerations much more than market-oriented value. Is quick and fairly easy to apply though is not to be singularly relied upon.
5. Going concern business	Investigation of a business enterprise having regards property and non-tangible components typically using the assessment of asset value (i.e. assets – liabilities), direct market comparisons and/ or assessment of income projections.	Assessment of businesses – retail, manufacturing, services, et cetera	Such assessments will include real property considerations – leases, function, et cetera, but will also consider a range of other trading issues (e.g. net earnings, goodwill, trends, management, et cetera)

dealings and a proportion of sales will need to be dismissed or used with caution because of particular characteristics. These include:

- sales well before the valuation date
- sales subsequent to the valuation date
- sales between related parties
- sales involving a number of properties or other dealings
- distressed or forced sales (e.g. mortgagee-in-possession sales, winding up of trusts or estates)
- transfers involving compulsory acquisition
- part sales
- sales to adjoining owners or purchasers with unique interests
- sales on unusual terms or conditions
- demonstrably ill-informed ('imprudent') vendors or purchasers and/or
- sales that are significantly different from the subject, in size, location, physical characteristics, overall scale/capital value, zoning, highest and best use, available services, et cetera.

It is not implied that these types of sales are always excluded. In some cases, there may be little or no quality sales evidence and any market activity may be able to give some guide. The point is, however, that these sales are, on the face of it, suspect to a greater or lesser extent and open to challenge. If considered at all, they must be used with caution.

Likewise, given both the unique nature of each property and the variations that can exist in sales evidence, it is not acceptable and, in fact, erroneous to apply the average of a number of sales prices.

An important underlying concept is that of 'comparability'. The word means able to be compared or similar enough to make comparisons. In simple cases, such as the residential example given above, closely comparable sales may exist, making the task relatively straightforward. This is not always the case, but the search needs to be extended until some basis of comparative sales can be found. Again, it is left to the acumen and the professional judgement of the valuer to make necessary allowances and mount a logical argument for an assessed value and for the use of particular sales in arriving at that assessment.

Sometimes, markets for certain types of property may be regional or even national, in which case sales may be drawn from a considerable distance with necessary adjustments made. There are many land uses where only one such activity may, for any number of reasons, exist in a particular locality. Examples here might include a marina, a quarry, a racecourse, a funeral home or cemetery, a golf or bowls club, a motel in a small town – and numerous others. Location ('position') is an important criterion in assessing values but where a particular property has no comparable sales nearby, then it is in order to widen the investigation. For example, characteristics and elements that will be uncovered in, say, the sale of a motel in one provincial town could, on the face of it, provide evidence of value for a subject assessment of a motel in another, comparable town in the general region.

Similarly, the valuation of regional and sub-regional shopping centres, the sites for such developments and for certain national chain outlets provides further examples of such markets. Here, prospective developers, final owners and major retail traders will have similar national project criteria regarding catchment, turnover and profitability. Thus, sales evidence for different properties, physically remote from each other, may still be highly relevant and comparable.

In any case, the basis of comparable sales, or evidence analysed from those sales, is preferred to other methodologies. Note that, to be accepted as value, a sale must have actually been transacted and completed. In considering such matters, courts have been unwilling to accept options to purchase, failed auctions or collapsed or other incomplete property dealings as definitive valuation evidence – past perhaps, and depending on the circumstances of the case, providing some indicator of overall market conditions.

Once investigations extend to large, highly complex properties – for example, a large office building, shopping centre, or rural holding – comparisons on 'entire-property-to-entire-property' basis can become almost meaningless given the unique physical, economic, financial characteristics of each. Even within the same asset type, significant differences in total capital value can have major effects on likely purchasers of that asset and their priorities in determining value. By way of example here, consider the investment market for a small group of local shops compared with that of a major shopping centre. Certainly, they are both 'retail property investments' and will share general criteria and priorities for property investment but financing, management, leasing arrangements and performance expectations would be so different as to make close comparison impossible.

For a whole range of reasons, the capital value of sales may vary by very large amounts and the 'positioning' of the subject property within such a diverse range would easily lead to wildly inaccurate results. Overall, however, if the valuer is to assess a particular complex property relative to others (i.e. by sales evidence), then sales need to be reduced or analysed to a point where they may be reasonably compared. As well as these underlying principles, their application requires specific knowledge of that sector, the property's function and the business activities undertaken (API 1996). Established valuation methodologies make this possible.

The following sections provide examples of the diverse application of sales/market analysis to a range of sectors and assessment tasks.

10.6 Valuation practice: *income-producing properties*

A significant proportion of real property is held for the express purpose of earning income. This income is typically derived in one of two ways: directly through renting the property to a tenant or, second, as an input to the wider commercial activities of the owner on that site. In either case, the actual or equivalent right to receive that income stream into the future provides a basis for comparison and valuation.

The principles behind the capitalisation of net income as a method of valuation assessment were introduced in Chapter 5.

Except for owner-occupied residential properties and the assets owned by government and certain not-for-profit organisations, practically all property is held for income-producing purposes in one way or another. That includes a huge variety of property types, all with their individual physical, financial and legal characteristics that must be considered in arriving at an accurate assessment. All are purchased with the expectation of income/cash flow into the future. In each case, the purchaser (investor) has accepted a certain anticipated return (i.e. financial reward) for the risk taken. The purchaser (like any investor) also realises that the financial returns may take some years to materialise, for which the purchaser as an investor will expect financial recognition and reward – the 'time value of money' concept discussed previously.

The key observation is that, in valuation practice, the use of sales evidence is not confined to the comparison between entire, freestanding properties. Under the (income) capitalisation approach, the comparison between the subject and sale properties can successfully be undertaken on the basis of their respective ability to earn net income and the risks that need to be taken to secure those returns (API 2007c). This relationship is shown in the fundamental 'capitalisation' equation which holds true for many investment, not only real property.

$$\text{Capital value} = \frac{\left[\text{Gross income} - \text{Outgoings}\right] \times 100}{\text{Acceptable rate of return}}$$

The investigation of the current and future value and the income streams of both the subject and sale properties will need to take into account not simply physical and geographic characteristics but also non-tangible parameters and issues that would affect the certainty of that income stream now and into the future. These would include the commercial and legal structure of subsisting leases and their conditions, whether the current rental represents true market and, particularly, whether appropriate rent review mechanisms are in place. Also important are the quality and security provided by the tenant's (known as the 'value of the covenant') and the Weighted Average Lease Expiry (WALE).

WALE is the average time in years or months until current leases on that property expire – and when, potentially, the income stream could cease. A consideration of possible vacancy rates, rental levels and market conditions at that date would also be relevant, for both the subject property and sales.

These matters are further discussed in Chapters 5 and 13.

From an investigation and analysis of sales, net income per annum can be established by deducting the annual outgoings for which that owner is responsible from the gross income being received from the property in that year. By comparing that net income with the purchase price, the level of return (capitalisation rate) that that purchaser accepted in acquiring that investment can be calculated.

Note that, because this rate of return is a quotient in the capitalisation equation, an inverse relation exists between capital value and acceptable return. That is, a property that shows a return of, say, 6 per cent represents a more secure investment proposition (and, therefore, can demand a higher capital value) than, say, a higher-risk investment showing 6.5 or 7 per cent.

The acceptable rate of return will also typically include what the purchaser sees as the potential for capital growth of the property into the future. If, for example, the property's current use is below 'highest and best' or the purchaser considers that market values are likely to rise into the future for other reasons, that purchaser may be willing to accept a lower rate of return (and, by implication, pay a higher capital value) to access that potential. By its nature, such possible potential is uncertain and nebulous and a clear differentiation must be drawn between realistic expectations and simple speculation.

In these considerations, there will often be something of a trade-off reflected in the final capitalisation rate deduced from that sale. This considers the security of the net income, on the one hand, and the potential for rental capital growth, on the other, each having its own risk profile of likely outcomes over time. In capitalisation assessment, it is not really possible to differentiate between these two elements – that is, current rental return and longer-term, less predictable rental and capital growth; therefore, the basis of assessment here rightfully focusses on the more certain rights to receive the net income stream.

Based on that approach, the analysis of a number of comparable sales will normally show that capitalisation rates cluster quite closely and exhibit reasonable predictability. Again, the valuer will need to use professional skill and experience in considering the reasons behind any capitalisation rate variations. In building that market evidence across a range of sales in a particular location or property sector, the valuer is in effect assembling a detailed picture of the market at that point in time within which the property to be assessed can be positioned. Such assessments/valuations are not abstract exercises but rather must reflect the manifestations of forces, both positive and negative, influencing that property and its functionality – including physical, legal and economic characteristics, its surrounds, history and trends and reasonable expectations into the future.

The assessed capitalisation rate can then be incorporated together with the net annual income from the subject property in the capitalisation formula above to arrive at a capital value for the subject property.

In practice, software-based spreadsheets are often used to reflect cash flows, time preference and time value of monetary consideration; however, the fundamental principle remains in the capitalisation formula above.

In terms of a wider analysis, it also needs to be noted that any investor ready to enter a market has a range of investment opportunities, not only in that location nor in a single sector (such as property), but potentially in many others across an increasingly globalised market. As discussed in Chapter 4, property has a number of specific characters that may act as either attractors or, sometimes, disincentives to investment.

Consequently, it is not only the comparison of return on investment on comparable properties assets that are of overall importance. Also relevant are the market conditions of competing investment types and other economic externalities that affect buyer sentiment and levels of activity.

These external factors will typically include overall levels of economic confidence, interest rates, inflation, exchange rates and their prospects into the future, stock market conditions, the political environment and government policy and regulations and a range of others.

Other commonly used assessment methodologies, which also reflect positive and negative cash flows and risk profiles, are called 'hypothetical developments' and 'hypothetical subdivisions'. In these cases, the valuer is attempting to make a professional assessment on the viability (or otherwise) of a particular future development of a property, be it a building development (such as offices, retail centres or others) or the subdivision of land. Again here, any assessment needs to be analytical, considering how the 'prudent hypothetical purchaser/investor' would assess both the potential and the risk in proceeding with that particular project. These assessments could be used to determine the initial price that should be paid for the en globo site or, alternatively, as the basis of a feasibility study into the project to calculate an anticipated final project value.

Such analysis can be undertaken in either direction, that is, a study from a known land purchase price or hypothetically proceeding through the development process to arrive at a potential future value for the project. Often, the assessment commences with a considered market value for the final product, based either on anticipated sales or on a future rental return. From that, development and holding costs, interest and the developer's profit and risk are hypothetically removed to arrive at a residual value for the land in situ. These matters are further discussed in Chapter 12.

This is complex analysis and extreme care must be taken to ensure that a range of variables, many of which may be identified as 'project assumptions', are properly investigated, assessed and included. Variations in construction costs, holding charges, final project timings, interest rates and final realisable value will have significant impact on the overall quality of the analysis and therefore the final investment decisions made. The risk profile of such assumptions can only be lowered by detailed investigations, effectively 'unbundling' the project into its component parts and testing each of these in their application to the case.

These forms of assessments are in common use and, when properly applied, provide a detailed analysis of project/investment breakdown and risk profile. Overall, they can establish whether a particular project meets the hurdle or internal rate of return required by a particular investor or, alternatively, is attractive and realistic compared with the development market as a whole at that point in time.

10.7 Valuation practice: *investment analysis*

In this chapter and Chapter 11, the assessment of the viability of existing or proposed income-producing buildings or land development projects are examined. Examples of such investigations might include proposals to invest (in an equity or a debt funding arrangement) in the purchase an existing office, retail or industrial building, or perhaps to build a new asset or extend or refurbish an existing one. Other common examples based on similar principles would include the assessment of land subdivisions or similar developments.

Several important observations on these methodologies need to be made.

First, this analysis is complex not simply because of the numerous financial transactions involved (both as expenditures and income or returns) but also because these dealings may occur over an extended period affecting the overall risk profile and rendering 'time value of money' considerations highly relevant (see below). Because of this complexity, such assessments substantially involve mathematical calculations and modelling based on market evidence and reasonable expectations.

A second key parameter is the need for consistency in application particularly in a common understanding of terminology used. The colloquial language used across the Australian property sector is notoriously imprecise. Key terms such as 'rent', 'yield', 'outgoings' and many others in common usage typically have a range of variants that require precise definitions for use in any particular case. Given that such assessments are normally comparative (i.e. assessing one project against others), it is not only important that the terms used are both acutely defined and used consistently.

There are several sources by which this accuracy can be maintained. In such investigations, there will frequently be existing legal documents – subsisting leases, development contracts and the like – that will include key definitions of words and terms used. Additionally, the Property Council of Australia (PCA), 2019, has issued a Glossary of Terms which, while generic, assist greatly in a consistent approach across the sector. Table 10.2 is extract drawn principally from that document and includes a number of terms particularly relevant to income investment analysis.

Table 10.3 contains extracts from the same document but here providing definitions pertaining to investment analysis and yield.

It will be noted that these assessments are largely based on established economic and valuation principles previously discussed, including the concept of 'time value of money' and 'the primacy of sales/market evidence'. There is however, a fine but important distinction between a project analysis and formal valuations. A valuation will attempt to provide an evidence-based assessment of the market worth of a property at a particular date; project analysis uses similar information but attempts to build up a financial model as to how a particular investment proposal can be structured (sometimes considering several options or scenarios). This latter approach typically attempts to establish whether that particular proposal is worthy of investment – often comparing it with other opportunities. Considerable case law exists that reinforces these distinctions and the different roles that each approach fulfils.

Table 10.2 Lease and rental terms – adapted from Property Council Australia, Glossary of Terms, 2019

Gross lease

One in which all operating costs on the property are included in the rental charged rather than charged as a separate amount. The landlord generally pays for all base year repairs, taxes and operating expenses incurred through ownership. It is the opposite of a net lease in which these costs are borne by the lessee either in whole or in part depending on the terms of the relevant lease document.

Outgoings

The expenses incurred in generating income. In real estate, these expenses include but are not necessarily limited to property rates, insurance, repairs and maintenance and management fees. Operating expenses when subtracted from gross income equal net operating income.

Rent

A payment made periodically by a lessee to a lessor for the use of premises. The term 'Rent' is often associated with a variety of other terms outlined below:

- **Gross**: In a gross lease, all operating costs on the property (excluding direct tenancy expenses) are included in the rental.
- **Net**: In a net lease the owner recovers outgoings from the tenant on a pro-rata basis (where applicable)
- **Face**: The rent shown on a lease document which may or may not include incentives and may or may not include outgoings.
- **Effective**: The actual liability for rent and outgoings after adjustments for any incentives to the face rent are taken into account.
- **Passing (or contract)**: The rent specified by a given lease agreement; although a given contract rent may equate to the Market Rent, in practice they may differ substantially, particularly for older leases with fixed rental terms. The term contract rent is North American usage; passing rent is Commonwealth usage (IVSC).
- **Equivalent**: Equivalent refers to the rent being adjusted for the effects of any market rent reviews that will occur in the period of consideration.
- **Base**: The minimum acceptable rental provided in a lease. In retail leases the base rent generally refers to the commencing rent which is supplemented with a 'percentage rent' based on the tenants turnover.
- **Break-Even**: The point at which a tenant's base rent is equal to an agreed level of sales above which percentage rent takes effect.
- **Concessionary**: A discounted rent, usually during the initial lease term.
- **Market**: The estimated amount for which a property, or space within a property, should lease on the date of valuation between a willing lessor and a willing lessee on appropriate terms in an arm's-length transaction, after proper marketing wherein the parties had each acted knowledgeably, prudently and without compulsion. Whenever Market Rent is provided, the 'appropriate lease terms' which it reflects should also be stated.
- **Turnover/Percentage/Participation Rent**: Any form of lease rental arrangement in which the lessor receives a form of rental that is based upon the sales of the lessee. Percentage rent is an example of a turnover rent (IVSC).
- **Overage**: US terminology.
- **Peppercorn**: A term used where it is desired to reserve only a nominal rent for any period. A minimal rent which is below market values.
- **Rack**: (a) Market Rent (b) Advertised rate for hotel rooms.

Table 10.3 Investment analysis and yield terms – adapted from Property Council Australia

Discount rate
The interest rate used to discount future cash flows to determine Present Value.

Internal rate of return (IRR)
The percentage interest rate at which the net present value of all the cash flows (positive and negative) from a project or investment equals zero. It is typically used in comparative analysis between possible projects or investments. All other things being equal, the higher the IRR, the more attractive the proposal. Project IRR will often be benchmarked against the required return for that investor (known as the 'hurdle rate').

Net present value (NPV)/Present value (PV)
The measure of the difference between the discounted revenues, or inflows, and the costs, or outflows, in a discounted cash flow (DCF) analysis. In a valuation that is done to arrive at Market Value, where discounted inflows and outflows and the discount rate are market derived, the resulting present value should be indicative of the Market Value by the income approach.

Yield
The derived percentage return of a property assessed from the net income and the market value or price. It is calculated by dividing the net income by the opening market value or price.
There are several specific yield types depending on the nature of the net income. Terms commonly used in Australia are:

    Effective yield
    The effective yield is the percentage return on value or price derived from the current net income after adjusting for rent incentives or impending vacancies.

    Equity yield rate
    The equity yield rate is the percentage return on the equity portion of a property investment. It is the net income after deduction of the annual debt service divided by the equity portion of the asset value.

    Initial (passing) yield
    The initial or passing yield is the percentage return on value or price derived from the current net passing income. No allowance is made for any future rent growth.

    Market yield
    (a) The market yield is the percentage return on value or price derived from net income that reflects current market rent levels. If the current income from a property is at market level, then the market yield is the same as the initial (passing) yield.
    (b) An average yield identified by analysts for different classes of buildings (e.g. you may see an analyst quote a yield for Premium and Grade A office stock in a specific central business district [CBD]). The yield quoted is an estimate.

    Reversionary yield
    The reversionary yield is the percentage return on current value or price derived when the current market rentals are payable. This yield relates a future net income to a current value or price and it is normally quoted together with the date from which it will apply. To calculate a reversionary yield, one would determine those leases that are subject to a market rent review within the period of consideration and adjust the income from those leases for the effect of the reviews (it should not be confused with the reversion yield) (see Terminal yield).

    Terminal yield (capitalisation rate) (also exit yield or reversion yield)
    The terminal yield (rightfully a capitalisation rate) is the percentage return applied to the expected net income following a hypothetical sale at the end of the cash flow period. It is a capitalisation rate used to determine the terminal value in a discounted cash flow exercise.

(Continued)

Yields based on income over several years:
Annualised yield
The annualised yield is the total holding period return on the net income of a
property expressed as an annual compound rate.
Equated yield (refer Whipple 1995, pp 335–336)
The equated yield is an annualised yield that is derived from the current
net income and future changes to the net income over time with specific
consideration of future rental growth. It is the rate of return over a specific
time period that has been adjusted for rental growth.
Equivalent yield
The equivalent yield is an annualised yield that is derived from the current net
income and future changes to the net income over time but no allowance is
made for future rental growth. It is the rate of return of a net income stream
over a specific period of time that reflects current actual rents and costs and
current levels of rental values.
*It is recommended that other yield terms are not used in Australia without an associated
definition.*
Yield to maturity
The percentage rate of return paid on a fixed security such as a bond if it is
bought and held to its maturity date. To calculate the yield to maturity, the
investor must take into account the coupon rate, time to maturity and the
market price at which the security was purchased.

Another important distinction is that, if properly conducted by a suitably qualified
expert, an assessed valuation typically arrives at a single figure or perhaps in certain
circumstances within a narrow range of values, which should hold true for any party
considering that property. As noted earlier in this chapter, that is not to suggest that
valuation is an exact science nor that valuations are invariably correct; however, given
the typical purposes of formal valuations, they often find their way into financial and
legal dealings and, in some cases, onto the public record.

A project analysis, however, will present a financial dissection of that particular project. Even if, as would be hoped, the input variables (costs, interest rates, time and so
forth) are sound, this assessment aims primarily at providing a synopsis of anticipated
returns from that development scenario.

Once the initial assessment is made, the project analysis typically becomes more subjective to the potential investor and their particular circumstances. For example, one
investor may believe that the proposed project does not reach the rate of return (also
called the 'hurdle rate' or 'target rate') that that investor seeks and, therefore, may not be
willing to accept that particular risk and responsibility for the assessed return. Another
investor may consider the same analysis and believe that, through their experience,
gearing, access to funds or the like, they can make more of the project than the analysis
suggests and, therefore, consider the investment acceptable.

Project analysis can obviously be applied to a wide range of business ventures, not
only to property development or investment investigations. In each case, the investment is considered in terms of cash flow out (e.g. original cost of acquisition and/or
construction plus operational costs, holding charges and interest) and income back to
the project, up to and including disposal or final sale of the asset. These factors occur at
different times during the life of the project. To allow for a realistic comparison, all costs
and income need to be brought back to the same point in time – normally the present

day – often called 'present value' (PV) or 'time zero'. To account for the time differences involved, a discount (called the 'discount rate' or 'discounting factor') will need to be applied to reduce all cash flows back to the present-day equivalent.

The actual percentage to be applied is a matter that needs close consideration, and the principles of the time value of money discussed earlier apply. That is, the investor will require adequate compensation for the fact that he/she must wait an extended period – and carry the risks associated with that – before the income progressively returns to the project in the form of rent or eventual sale. This percentage must reasonably reflect the risks involved and the length of time during which these risks are encountered. Consistency in the use of discount rates across a project and, indeed in comparing projects, is essential, so that the 'like-with-like' maxim be maintained.

The two most common tools used in this type of analysis are Net Present Value (NPV) and Internal Rate of Return (IRR). Both provide insights into the financial performance of a proposal. They may be particularly useful when comparing and ranking projects under consideration; however, as noted below, all applications and comparisons need to be carried out carefully and with a wide appreciation of all prevailing influences relevant to each case.

NPV is simply the final quantum (in dollars) of the cash flow analysis described above. IRR (also known as the 'investment yield') is a percentage showing the project's ability to repay the capital invested in that project (i.e. its profitability). It is, in effect, the discount rate that makes the NPV of all cash flows for a project equal to zero. All other things being equal, the higher a project's IRR, the more attractive an investment it will represent. It can provide a basis for comparing the attractiveness of a number of investment/project options. Further, a particular investor or corporation may have established a particular IRR (called a 'hurdle rate') required for investment and against which all project proposals can be assessed.

With a major project or investment under consideration, a huge number of individual calculations are required to reduce the various cash flows back to present-day values. The use of Microsoft Excel, or a comparable specialist software package, provides a platform to quickly undertake such calculations relatively. They typically incorporate a range of functions and formulae required to investigate various options. Furthermore, such software allows for different scenarios to be quickly constructed, effectively asking 'what if?' questions through various parts of the project and considering the sensitivity of change in one project element to other variables.

Most project analysis evaluations would start with a 'base case' scenario where all owners' equity is used before any debt or interest calculations are introduced. The inclusion of debt may well improve the results of the IRR and NPV calculation, following the maxim that 'debt is cheaper than equity'.

In this analysis, the impact that a change in one of the components in the project (called the 'independent variable') has on other components (dependent variables) is called a 'sensitivity analysis'. This analysis can be undertaken on the basis of bilateral considerations or the impact of change between a number of components (called 'multiple analysis'). These techniques are very useful in considering, for example, the effect a change in interest rates on other external or internal factors would have on the other components (i.e. costs and income), on other parts of the project or the viability of the project as a whole.

These software analyses are of significant value in quickly addressing complex calculations. However, the apparent sophistication of computer-based modelling and spreadsheets produced can be deceptive. The reliability of these forecasts depends very much on the accuracy of information used to set the parameters for the calculations: assumptions regarding time preferences, discounting rate used and predictions of the quantum of both costs and revenue. Sometimes quite small adjustments in any of those inputs or assumptions can have dramatic effects on final assessments.

Software tools, therefore, need to be applied with care, and always with professional judgement and reference to current realities and future market prospects for the project/investment. A range of court decisions and property-related litigation have noted these potential risks and reinforced the observations above. Post the global financial crisis 2008, lending institutions typically take a much more careful and conservative approach to software analysis and modelling. Ultimately, it is the underpinning market reality that will determine project success.

Rowland (2010) recognises the value of the range of forecasting tools and assessment options now available, but he notes that in selecting the appropriate methodology in any particular case a number of critical questions need to be asked: Is the methodology sound and proven for these cases? How accurate and stable are the data sets used, both historically and to the present time? Third, are these approaches likely to remain robust and meaningful into the future?

These matters are further discussed in Chapter 12 as part of wider considerations of land subdivision and property development projects.

Soundly based and well-applied modelling to establish IRR and NPV assessments can also provide decision-makers a sensitivity (regression) analysis to review projected outcomes against a checklist of expectations for a project.

For valuers, analysts and their clients, the construction of models and projections of how financial or other events might evolve into the future and impact the project is an important and legitimate consideration. These deliberations will be enhanced by good quality data and sophisticated statistical techniques. A key point, however, is that they can never replicate or mimic the complexity of real life, particularly in multi-faceted areas such as real property. Statisticians are the first to recognise that any model is presented with the parameters, boundaries and limitations for that analysis. They aim to help explain (components of) reality, but never claim to be reality. Difficulties invariably arise where this premise is not understood or truly appreciated. Much of the underlying failings of the 2007–2008 global financial crisis can be traced to a basic misunderstanding of this and the consequential poor management of risk, particularly related to debt levels (Reinhart & Rogoff 2009; Ferguson 2013). In valuation practice, modelling is essential to accommodate large volumes of data, its analysis and testing. Nevertheless, that practice requires experience and an acceptance that there are elements of art as well as science in all of this (Isaac & O'Leary 2012).

In this regard, the eminent English–American statistician George E. P. Box wryly observed that 'all models are wrong, but some are useful' (quoted in Fung 2010).

10.8 Valuation methodologies applied: rural properties

Valuations of this sector will be based on market analysis, referencing comparable sales and, as much as possible, the valuer would select sales within the same district or region and with comparable location, access, country type and soils, rainfall, surface and underground water and so forth. However, the nature of rural holdings in Australia requires that the comparative process undertaken be adjusted to reflect the complexity of these land uses and their sectors.

It may be argued that, given that these are business enterprises aimed principally at deriving income, assessment should parallel that of assessment of other income-producing real property – that is, based on the capitalisation of actual or nominal rents, productivity and risk and risk management. While there is some rationality to that argument, the volatility of income returns and of cash flow, and the vagaries of the Australian climate, making accurate predictions near impossible and the use of such methodologies here would provide unreliable assessments. In any case, a counter argument would hold that all of these considerations, both positive and negative, form part of the deliberations of the 'prudent purchaser' and are therefore already embedded in the price that that person paid to secure that (comparable) property.

Particularly for high-use intensity rural enterprises, such as dairies, orchards and small crops, and feed lots, sales may well be for the entire enterprise, as a going concern – 'walk-in-walk-out' – and include production allocations or assignments, water allocations, et cetera (Baxter & Cohen 2009). More generally, however, if sound, comparable sales are found, then a comparison may be drawn on a per-hectare basis for the various types of country that may be identified. Adjustments can then be made to recognise the particular characteristics of both the subject and the sale properties.

Almost invariably, other factors will emerge that complicate direct comparison. For example, one property may include a homestead, stock yards and ramps and equipment; dips or spraying equipment, some may include stock or a growing crop; another may be 'sold bare, excluding stock'. A range of adjustments will, therefore, need to be made to nominally adjust these factors to a 'common denominator', so that the range of sales and the subject property can be compared, like-with-like. However, this dissection is not without risk, and issues such as the overall viability of the property (as a 'living area'), the quality of farm management and the valuation ascribed to assets, such as water rights and licences, both now and into the future, are all matters that may be difficult to quantify. Further complications in investigations can arise because the turnover (i.e. number of sales) of rural properties is often quite low, particularly in more remote, pastoral regions.

To establish a platform for later comparison and adjustment, the valuer can consider reducing all sales to some relevant 'unit of comparison', say, for farmland, their value 'per irrigated hectare ex value of the growing crop' or, for grazing lands, value 'per-hectare fenced and watered, but with no stock'. From those benchmarks, the value of other improvements could be added or subtracted as required. For broadacre grazing or pastoral holdings, other measures such as 'beast area value' or 'dry sheep equivalent' are also sometime employed as the basis for comparison (Baxter & Cohen 2009). These concepts were presented in more detail in Chapter 9.

Similar analysis is possible for a range of rural pursuits. In undertaking these adjustments in rural or any other property type, the values to be ascribed to various improvements need close consideration. Those values need to reflect the contribution that a particular improvement or asset adds to the overall function of the entire property (i.e. 'added value'). That may or may not be the same as the original, replacement or even depreciated cost of a particular improvement.

Rural property assessment in the Australian context is a remarkably complex and multi-faceted task. Many of the vagaries in these assessments relate to seasonal uncertainties, the fragility of ecosystems and, generally, poor fertility and low rainfall compared with those encountered in other countries. Most production is sold into highly volatile global commodity markets, adding to inherent uncertainty to the overall instability of the sector. Finally, an additional important parameter of the markets for Australian rural properties is the culture, ethos and demographic profile of the Australian farming community.

The comparison of a subject rural property with analysed sales and perhaps also using an animal-area-value approach is justified and generally accepted; however, a wider, business/enterprise-based analysis over the longer time frames involved should be increasingly considered in such assessments (Eves 2004). This change in emphasis would also better reflect major structural changes occurring across many parts of the contemporary rural sector. These include the increasing economies of scale being experienced throughout most rural sub-sectors and fundamental production and supply chain changes resulting from evolving buyer preferences, food security, quality and animal rights issues and the rise of live livestock exports (Hefferan 2015).

An elaboration of regional and rural property and land uses is contained in Chapter 9.

10.9 Valuation practice: *summation*

There is another simple methodology called 'summation', which is sometimes applied to property assessment exercises. The summation method, as the name implies, simply adds together the value of the land and the depreciated value of any improvements that attach to that land. This approach is relatively easy to apply: land value, from known sales of unimproved and lightly improved land, is added to the replacement costs of the improvements, less an allowance for depreciation given their current age and state of repair. This second component can be established using quantity surveyors' building cost schedules, life cycle analysis and depreciation allowances.

However, this approach again highlights the issue of confusing cost (even depreciated cost) with the concept of value. Clearly, a summation approach should be used only as a 'rough guide' or check measure and has fundamental and conceptual flaws. Except for the land component, it has no reference to market-sale evidence and does not address fundamental valuation considerations including income-generating capacity, risk profile, functionality or obsolescence. The inability of the summation approach to detect either undercapitalisation or overcapitalisation of a particular asset also represents a significant issue (API 2007b, 2007c).

A quite useful role for this methodology lies in the component of rural valuation practice outlined in Section 10.8. Therein, it was noted that, to better establish 'like with like' comparisons between sales and the subject property, positive or negative adjustments would often be required to account for particular improvements across those properties. In that way, a consistent basis for comparison could be established and applied. Many of those adjustments involve the application of 'an assumed value' for those components. In these rural applications, the methodology often has enhanced validity as the 'as is' value for second-hand plant and equipment, chattels, et cetera, can be ascertained from the results of clearing sales common in rural areas.

Overall while being of assistance in situations such as the examples above, summation is demonstrably an inferior method of valuation assessment when compared against the analysis of comparable sales and other assessment methods that derive from them.

10.10 Valuation practice: *Going concern businesses*

Except in the case of owner-occupied residential properties and public and community-owned assets, real property derives its capital value by income through function – providing a platform for business by an owner occupier or for a second party (a tenant or lessee) paying rent to that owner. The ability of that property to support the business activity is a major determinant of the value of both the real property component and, second, the sustainability of the business itself.

Even in cases where the business owner also is the freehold owner of the real property, it is common practice that the two components – the capital asset represented by the land and buildings and second the 'going concern' business – are identified as separate activities. Often the real property assets will be held under a separate name (e.g. a holding company or trust) to reinforce this delineation. Such structures can represent sound business practice as regards the raising of debt, taxation administration and overall risk management. All of that notwithstanding, and whether the property is leased or owned outright, there is a symbiotic relationship between the real property and the business it accommodates. (Where the real property and the operating business are, in effect, in related ownership, the 'marriage value' that is created may be further secured by that link.)

A business that has been operating successfully for a considerable period of time, perhaps for many years, clearly has achieved sustainability and a record of success that enhances its capital value. The difference between a 'going concern' business and one which is 'opening its doors' on its first day, is known as 'goodwill'. Goodwill can have several aspects and components which need to be understood and assessed in each particular case.

The nature and effect of goodwill and the strength of the relationship between that type of business and the real property that supports it vary considerably with the activities undertaken. In some cases, the business can only exist on that particular parcel of land – a farm, a quarry, a mine or a port facility – would be some obvious examples of that type of relationship. In other cases, the business remains heavily dependent on its location though it could potentially exist elsewhere in close proximity.

Dependency on a particular location for a business is known as 'site goodwill'. Service stations and local schools, hotels, child care facilities and convenience stores provide examples here. In other types of business undertaking, customers are largely attracted and retained by some particular goods, professional service, skill or reputation that it enjoys. That component is known as 'personal goodwill'.

For the majority of businesses, location is an important consideration; however, as shown in the above examples, the impact and value of security of tenure in a particular location varies significantly, depending on the type of business or service.

Most businesses have at least some elements of both site and personal goodwill. In general terms, 'site goodwill', by its nature, is saleable with the business and will have recognised value in the market. The market value of personal goodwill is more problematic and, by its nature, less likely to be tradable. Its value will depend on the specific case and whether any of this benefit can be retained after the current owner is no longer involved. In practice, there are a range of business arrangements and systems all aimed at protecting and enhancing the site and personal goodwill of that enterprise. These may include longer-term leases or options, franchise and exclusive supply agreements and lease conditions that give to, say, a tenant in a shopping centre, specific and exclusive rights to use.

Goodwill is dynamic and, in any situation, will change over time. For example, in the contemporary environment, the impact of product commoditisation, new distribution arrangements and internet-based transactions on business and retail dealings are disruptive to existing business arrangements. These provide an increasing challenge to the 'time and place parameters' at the customer-to-business (retail) level and for business-to-business ('B-to-B') transactions and can also question the primacy of one physical location over another. New functional relationships between virtual and physical environments are profoundly changing business operations and practices and their relationship with and use of real property assets. While such trends have been evolving for several decades, the current COVID 19 pandemic presents a new urgency and obligations to ensure that these new, hybrid relationships both satisfy the contemporary demands of all stakeholders (including safety and security) while also maximising the potential of the physical assets, new technologies, globalisation and new communication platforms. The form that all of that will have and the effect on business location decisions and on business networks and customer relationships (including concepts and value of goodwill) will take time to mature (Siggelkow & Terwiesch 2019). These matters are further discussed in Chapters 4 and 5 of this text.

In the determination of market value, the real property assessment and the valuation of the going concern business must be assessed separately, though there is obvious overlap in the future projections and risk profile related to both.

Consider, for example, a fashion/clothing shop located in the main street of a town. Here the 'goodwill' will take several forms. The shop's location, exposure, passing trade, complimentary businesses nearby, convenience, the 'image' and size of the shop, and so forth, all represent influences on site goodwill. Within the store, the range, quality and brands carried, the knowledge and level of service provided by staff and personal rapport established with customers are all examples of personal goodwill attaching to that business.

In considering the assessment of such businesses four key observations are made from the outset:

- The separate valuation of the going concern businesses is a specialised activity drawing together a range of disciplines. As well as essential property professional opinion, the skills of taxation and management accountants, analysts with esoteric business expertise in that sub–sector and, potentially, legal opinions will be needed to provide comprehensive analysis and an accurate, overall assessment.
- Second, the purchase of a business is, in effect, the acquisition of a right to secure an income stream for some period of time into the future. Consequently, access to exact, audited operating figures for the business (profit and loss, balance sheet and cash flow) will be essential. Figures over at least the past three (and preferably five) years will be required to establish overall viability and also trend data. Consistency in analysis is essential and a valuable comparative measure here is 'EBIT', (net) earnings before interest and payment of income tax. (These figures should be reasonably easy to secure from the financial statements above. The level of gearing – and therefore interest payments – and income tax liability are excluded as they are esoteric to that individual owner at that point in time and not germane the property/business itself.)

- Third, such assessments must not be considered as an abstract nor an isolated exercise but rather be considered in the context of prevailing economic, sectoral and local conditions at that point in time. This generally follows the process summarised in Figure 10.1.
- Finally, as with asset assessments, including real property, there is rarely one methodology or approach that fully reflects all aspects of such complex matters. Financial analysis will be essential but, as outlined below, there a number of methodologies that may be relevant and helpful in investigations. All of these should be considered, to a greater or lesser extent, as contributing to as comprehensive understanding as possible.

As with the assessment of any enterprise, capital value is, fundamentally, a function of the ability to earn net income over time, the risks that will be encountered in securing that income and risk management techniques available to mitigate adverse effects. The property component is in effect providing a 'platform for business' which needs to be recognised both for the costs involved (through the payment of rent, et cetera) and the overall contribution made to provide not simply physical 'space' but also 'place' – in the form of identity, functionality and goodwill.

The 'going concern' business component can be assessed in a number of ways and, as with all valuation undertakings, accuracy is normally served by the use of range of methodologies – though in many cases, one method will prove more relevant and the others providing support. All analysis must be grounded in market investigations and evidence and in these cases too, any additional benefit derived from 'marriage value' needs to be considered. Finally, any such assessment must contemplate the unique nature, not only of the real property but also of that business and the goodwill that attaches to that business. The application of generic or sector-wide observations is therefore accommodated with inherent risks.

Three methodologies in common use in such business assessments are:

1 'Net' worth of the business (asset value):
   This is a simple measure of the difference between what a business owns ('its assets') less its financial obligations ('its liabilities'). On the face of it, this represents a simple 'snapshot' of the enterprise and may include useful trend data from previous periods. However, the valuation of intangible assets such as goodwill and longer-term business prospects and risks here are much more problematic to assess. Its overall value, therefore may be fairly limited.

2 Direct market comparisons:
   This involves directly comparing the completed sales of comparable businesses to the subject property – adjusting for the specific characteristics of each. Because of the wide diversity that might emerge, 'whole business to whole business' comparisons may only provide a fairly general but often useful guide. Analysis using EBIT may also be valuable here and sales can be further refined to points of comparability after considering the additional matters identified below.

3 The assessment of income projections (and potentially the use of 'business multipliers'):
   This process involves determining as accurately as possible the financial state of the business, normally measured as net income before interest and tax are considered. Second, on available evidence and using professional judgement and experience, it is

ascertained as to whether that level of performance appears likely into the foreseeable future. As noted above, the input of a number of experts will typically be required to establish if that level of income can be sustained into the foreseeable future value to an incoming operator. (Note that a range of ancillary matters – such as the value of existing stock, plant and equipment and similar business assets – may have to be taken into account before a final value is confirmed.)

The final step in this analysis is to consider the present-day (date of assessment) value of that right to receive that income stream. The selection of an appropriate discounting rate for that cash flow will be necessary to reflect risk and the time value of money (TVM) overall.

Comparison with other potential investment returns may provide some general parameters but the specialist expertise and analysis of assessors with experience in that type of enterprise will be essential.

To assist, financial assessment tools (such as CAPM – Capital Asset Pricing Models) can be employed. These can more accurately establish discounting rates by considering the particular risk premium associated with that business, sector and/or location over and above a risk-free investment.

A widely employed, simpler but much less accurate variation to this general approach involves the use of 'business multipliers'. Under this approach, the initial date collection and sales comparison steps are undertaken as outlined above. From that information, the relationship between the price achieved from the sale of those businesses and their past earning performance can be established. That numeric relationship, repeated across a number of comparable business, provides some simple benchmarks for assessing the value of other going concern businesses.

That multiplier can take the form of a single number. For example, the analysis of a number of a certain type of retail shop may establish a typical 'business multiplier' of, say, 1.5 – indicating that this type of business might sell for about one and a half times the average net income of the enterprise. (Note that there may also be a range of adjustments that will need to be made to that overall figure to accommodate the particular characteristics of this business. Some of these are summarised below.)

While such simple analysis may well provide a general guide they need to be applied with particular caution. As with any property and business asset, each business represents the unique fusion of many real property, economic, locational and other characteristics that simply cannot be reflected in generalised or averaged figures. This is particularly the case in contemporary environments where rapid business change and shortening business cycles result in highly volatile environments.

As well as the assessment of the going concern business itself, such analysis will almost invariably encounter particular aspects or components of that business that will also need to be assessed and the value of the business adjusted positively or negatively, depending on the issue involved. Some of these might include:

*   Details of leases
    In most cases security of tenure for the business will be provided by a commercial lease. Detail of the nature and operation of such leases is included in Chapters 7 and 13. Over time subsisting leases can evolve to be at variance with market conditions. In those cases a 'profit rent' can occur either in favour of the landlord or

tenant for as long as that anomaly exists. Such variations must be taken into account in considering the overall value of that going concern business.

- The nature and transferability of goodwill involved

    As noted above 'goodwill' can take a number of forms notably 'site', and 'personal'. Additionally, practically any business will have a range of licences, regulations and agreements to enable the business to maintain operations. In assessing business value, care needs to be taken to ensure the transferability of all of these essential components to the incoming owner. (To limit the impact of continuing personal goodwill, it is not unusual to include in the contract of sale of the business a 'non-competition' clause which will restrict the vendor from re-establishing a comparable business in that geographic area for some specified period of time into the future.)

- Plant, equipment and chattels

    Most going concern businesses will involve the use of plant, equipment and chattels of various types. In any assessments here, items for inclusion in any valuation or sale must be clearly identified. Many such sales will be on a 'walk-in, walk out' basis and therefore many such items will need to be recognised in sales analysis. Ownership (be it outright, leased or involving debt funding) needs to be clarified before value is ascribed. Specialist plant and equipment valuers exist to carry out assessments where substantial and specialised equipment is involved. In any case however, value must be established as regards type, age, condition, tax depreciation arrangements and, most importantly the functionality of that item and the added value (or otherwise) that it brings to the business.

- Stock, supply contracts, stockpiles and advanced orders

    At the date of valuation or of transfer to an incoming owner, going concern business will have a range of activities underway and assets in transition that need to be included in assessment. Inventory/stock held for that trading period should be included at its wholesale value. Supply agreements with other business raw materials, services, et cetera, and forward orders will need to be considered. Such arrangements often provide additional security for the business into the foreseeable future but each case is different. Each must be considered on the basis of contractual obligation and also an assessment as to whether those agreements have additional value (or indeed represent a liability) for the businesses compared with current market rates.

- Franchise agreements

    A proportion of going concern businesses operate under franchise agreements with other, 'upstream' and/or larger corporations. There are numerous examples across retailing, vehicle sales, fuel distribution and many others. Along with the actual sales and distribution of goods and services, such arrangements often involve corporate identity branding and marketing, the use of patented equipment and know-how, business systems, et cetera. In assessing these as part of a going concern business the exact details need to be thoroughly investigated (including corporate history, costs, obligations, duration of the contract, and the ability to transfer such arrangement and on what terms). While such arrangements are often essential to that business, this should not be taken for granted and the merits of otherwise of such structures need to be carefully considered on an individual case-by-case basis.

- Business management

    Management at both a management and strategic level represents a key component of any going concern business. At the strategic level, this can include overall production, sales and profit figures over time so establishing trend data. At an operational level important aspects here would include relationships with suppliers and customers, staff satisfaction, training and turnover, safety and environmental performance and suitability and effectiveness of plant and production systems. Location and real property suitability should also be considered here.

Though a very generalised observation, it is often noted that the 'simple look' of a business – the manner in which it is presented and operates, obvious organisation, cleanliness and the positive attitude of staff is a remarkably accurate indicators of wider success factors.

Finally, it should be noted that current poor management may not always adversely affect sale price as much as might be expected given that an incoming purchaser may well see an opportunity to rebuild the business into the future.

10.11 Summary

This chapter has outlined the role and importance of a well-structured and applied assessment of value of real property assets for business and financial dealings – and across the wider community within a developed capitalist county such as Australia.

It noted that, even though undertaking a real property valuation appears quite straight forward, it is normally a very complex task – reflecting the remarkable diversity of the property assets themselves and the variety of dealings and interest that might be involved. Most of the chapter has attempted to summarise assessment methodologies in common use for the various real property asset classes encountered, noting that the purpose was to provide a general understanding of each method, rather than an in-depth study, which can be pursued through specialist valuation texts included in the chapter's references.

To provide a concise summary of the valuation methodologies, it is worth noting the clarification provided by Scarrett and Osborne (2014), which itself reflects RICS guidelines.

In the overall sense, they recognise five fundamental methods of valuation assessment. Again, it is not to be construed that these approaches are in some way competitive or that one is, by necessity, superior to others. Rather, it reinforces the observations that different asset types and activities lend themselves to certain situations – all have positive characteristics and all have limitations of which the valuer and, indeed, the final recipient of the assessment need to be aware.

Paralleling much of the preceding chapter, Scarratt and Osborne summarise the categories as follows:

- various methods of direct comparison
- the traditional approach to investment – related particularly to capitalisation of rental income
- investment analysis involving the investigation of the parameters of that particular project and its funding and income profile, including discounted cash flow techniques and, typically, arriving at its final realisable project value or underlying land value
- the assessment of value of going concern business
- cash-based (summation-based) assessments.

All of those, as noted previously in this chapter, need to be considered within the wider economic, financial and legal context of that property, its location and that property sub-sector.

Chapter 11

Compulsory acquisition

11.1 Introduction

In the operations of a contemporary society and community, situations will arise where some public good may need to take precedence or override the rights of the individual. Some such cases can involve the taking of that individual's subsisting rights to real property. Examples of this may arise where some necessary public work – be it for a road, right of way access for infrastructure, public building, dam, environmental protection areas or so forth – can only be secured using land currently held in private ownership.

In those cases, there needs to be a process whereby a particular public authority, with suitable legislative power and following strict procedures, has the right to take back that land or part thereof from the private owner. This process is called compulsory acquisition. (This process is colloquially known as 'resumption' and, in other jurisdictions, such as the US, 'eminent domain'.)

This chapter explores the structure, parameters and procedures involved in such activities. Compulsory acquisition matters are, by their very nature, most serious. Enabling legislation strictly limits the rights of government agencies (in these cases called 'constructing authorities') in their ability to acquire and deal with private real property. Further, such legislation will typically be prescriptive regarding the process to be followed, the dispossessed owner's rights to claim compensation and the reference to third-party, independent bodies to determine compensation disputes.

With certain exceptions regarding the rights of the Commonwealth Government (refer Section 11.2), land law pertaining to compulsory acquisition is the domain of the individual Australian States. The legislation in each jurisdiction is based on consistent principles, outlined in the following sections. It must be recognised that these are complex legal and valuation matters, and this chapter aims to provide a general understanding of the subject and an overall working knowledge of its operations.

It is essential that, before applying any of these principles to specific cases, practitioners are familiar with the specific requirements and nuances of the enabling legislation and regulations and with precedent court decisions. The bibliography to this chapter identifies further resources and the work of Brown (2009) and Hyam (2014) together with the specific legislation from each State and the Commonwealth will be particularly relevant. As much as possible, observations in this chapter have been kept generic however, where specific examples have been necessary, the Commonwealth Land Acquisition Act (1989) and the Queensland Acquisition of Land Act (1997) have been used.

As well as general or underlining principles outlined particularly in Section 11.3, a number of important court decisions are identified throughout this chapter. These are used to identify some notable aspects and decisions. It is stressed that these examples only represent a small, albeit important, proportion of cases in this complex area. Wider observations made in this chapter are generally based on court precedent though, in the interests of conciseness, are not individually identified. These can be further studied through, among others, the Brown and Hyam legal texts.

11.2 General approach and processes

The basis of any compulsory acquisition lies in specific legislation enacted in each Australian state and, separately, by the Commonwealth for matters prescribed under Section 51 (xxxi) of the Australian Constitution.

The general principles that underlie these matters are quite consistent across these jurisdictions, based originally on English law and precedent. That notwithstanding, it is again stressed that there are important differences across the various pieces of legislation and regulations which, while outside the parameters of this text, need to be recognised and applied in individual cases.

Given the importance and potential complexity of these activities, it may seem surprising that, within those pieces of legislation, the core elements of compulsory acquisition are quite concisely established and only briefly defined. It might be surmised that lawmakers recognised the importance of setting key directions and parameters (in legislation and regulations) but also realised that there were inherent difficulties in becoming too prescriptive or detailed in their application to the almost limitless number of acquisitions, scenarios and real property parcels all with diverse characteristics. Legislators therefore typically and appropriately leave the actual process and methodologies for assessment to established professional practice supported by precedent court decisions and providing, as necessary, access to litigation for any claims that cannot be resolved through negotiation.

The entire processes of compulsory acquisition is, by nature, sensitive and sometimes controversial causing disruption to the normal rights of 'quiet peace and enjoyment' of disposed owners. It would be observed that the legislation and processes that operate in Australia have been refined over many decades and represent a fair and equitable approach balancing the need for necessary public works to be carried out in an efficient and procedurally correct way while also ensuring fair and wide compensation to disposed owners.

There is particular reliance on the adherence to due process together with considerable procedural and administrative obligations placed on the Constructing Authority. The knowledge and professional and ethical behaviour of those representing both sides are also prerequisites to resolving such matters expeditiously and fairly. In this environment, the precedent and guidance provided by the courts and other bodies previously adjudicating on matters in dispute take on a highly significant role (see Section 11.3).

Any compulsory acquisition activity in Australia is only possible under enabling legislation from the Commonwealth or from the various State governments. In the case of the Commonwealth the current legislation is the Land Acquisition Act 1989. The justification for that Act flowing from Section 51 (xxxi) of the Australian Constitution. These powers are limited to matters relating to 'purposes in respect of which the Commonwealth Parliament has powers to make law and must be undertaken and acquired on just terms' (note, however, that the expression, 'just terms' is not further defined or qualified in the Constitution).

In practice, the number of Commonwealth compulsory acquisitions are very few – confined principally to defence, aviation/airport facilities and telecommunication dealings. The activities of State governments and their associated bodies – statutory corporations, local councils et cetera – typically have a much larger number of dealings as they carry out their duties in the provision of public utilities (electricity, water, sewerage and drainage) and public services (schools, public buildings, transportation and roads and the like).

Such acquisitions can involve the taking of the whole of the property, parts of a property or, in the case of utilities such as electricity and water, the acquisition of a 'right of way' or easement rights to cross privately owned land.

The acquisition powers of the States are established through specific 'acquisition of lands' acts in each jurisdiction. While the details vary, common themes of these pieces of legislation are typically:

- for the application of compulsory legislation and for compensation claims to be possible, land must actually be taken. This may seem obvious, but it establishes a very important maxim that, regardless of the impact that nearby public works may have on a property or locality, no claim can be entertained from any party who has not lost, in whole or in part, land or some equitable interest in that land. As with any legal description, the term 'land' is taken to include improvements attached or forming part of that land
- the acquisition must be for a 'public purpose' (i.e. those purposes are often listed within that relevant legislation)
- a specific constructing authority is identified (typically a government department, statutory authority or similar though the level of delegation will vary)
- the acquisition is instigated by specific intent by the constructing authority, setting off a process and series of steps. This includes the identification of and advice to any party with an interest or potential interest in the property (normally through the issue of a Notice of Intention to Resume ['NIR'] issued to them). The owners are advised of any rights to object to the proposal and of the specific rights granted to the construction authority to undertake investigations necessary to verify the feasibility of the project
- the land to be acquired is identified, for example, through cadastral survey
- if after due process and consideration, the Constructing Authority decides to proceed with the acquisition, the gazettal (government proclamation) that the land has been acquired, as at a specific date, is made. This date is of critical importance as it marks the loss to the Crown of all other interests from any other party – whether identified or not and including potential native title claims. At that point any such existing equitable rights, in effect, become rights to claim compensation
- in certain jurisdictions, the obligation lies with the claimant to submit a detailed and quantified claim. Time limits for such claims (typically three years) may apply
- such claims are frequently settled through 'without prejudice' negotiations or other Alternative Dispute Resolutions (ADRs); however parties have a right of redress to arbitration, typically to a state Land Court or Commonwealth Administrative Tribunal.

The compulsory acquisition process is complex and varies depending upon the jurisdiction and the legislation involved. Figure 11.1 shows the generic steps that typically apply.

11.3 Underlining principles

Inherent difficulties typically arise in compulsory acquisition dealings flowing from the unique and complex nature of impacts on each of those properties affected. Consequently, the precedent decisions of courts that have addressed comparable matters in the past will be of assistance in future cases, at both negotiation and litigation stages.

Court decisions are usually handed down with a description of the case and the reasoning behind the final determinations. Over time, a very large number of such decisions are accumulated with some having greater importance and relevance to a particular case than others. Further to the legal environment generally described in Chapter 6, some of the determining factors will be:

*   the legislation to which the case and the determination refers. As noted above, there exists significant variation in the legislation across jurisdictions that may have well influenced that decision
*   the status/seniority of the court involved
*   whether the case is Australian, English or from another jurisdiction – noting that jurisprudence has evolved separately in the Australian context particularly over recent decades
*   the date of the decision, noting that with the passage of time, the relevance of certain determinations will become superseded or be influenced by subsequent decisions.

As important as precedent decisions may be, courts are careful to focus on the particular piece of legislation and its application to the case at hand and not feel unnecessarily inhibited by precedent. Achieving this balance is succinctly described in the following observations:

> In actual practice, two cases are rarely if ever alike…thus a judge may be given a wide discretion in deciding to follow either precedent A or precedent B both of which seem to have considerable relevance on this case but which, unfortunately are completely contradictory to one another.
>
> (Carr 1942)

> Even more wryly, 'When the ghosts of the past stand in the path of justice clanking their medieval chains, the proper course for the judge is to pass through them undeterred'
>
> (Lord Atkin in United Australia Limited the Barclays Bank 1941)

Over time and in specific areas of the law however, some precedent decisions will be seen as of fundamental importance and widespread relevance. Progressively subsequent decisions will refer to that original determination and a basic principle emerges – often bearing the name of that original case. As with any precedent, the establishment and recognition of such a principle does not imply that it must be followed in every individual case though, as a principle grows in importance, variations from it becomes rare. It would be observed too that alignment with established principles typically creates points of commonality in negotiations and potentially later in arbitration.

A number of such principles exist and are widely practised in the area of compulsory acquisition and the assessment of compensation related thereto. In summary, some of the more important principles are as follows:

a   Spencer Principle (Spencer v Commonwealth [1907] 5 CLR 418). This is a general principle wider than simple compulsory acquisition. Critically, it defines 'value of land' through a comparative analysis of the actions of a prudent purchaser willing but not obligated to secure the property. This is typically best assessed by the analysis of comparable sales.

b   Raja Principle (Sri Raja Vyricherla Narayana Gajapatiraju Baradur Guru v Revenue Divisional Officer, Vizagapatam [1939] 2 All ER 317). This principle recognises that a compulsory acquisition has a number of complex components and includes special interest value and potential. The assessment needs to be a logical quantification of all of these issues encapsulated in that property and that claim.

c   Rosenbaum Principle (Rosenbaum v Minister for Public Works [1965] 114 CLR 424). As regards claims for compensation in such matters, any claimant has the right to have their claim considered on its own merits and free standing from any other claim that happens to be forthcoming from that compulsory acquisition.

d   Pastoral Finance Principle (Pastoral Finance Association v Minister [1914] AC 1083; 15 SR [NSW] 535). In compulsory acquisition dealings, this represents a fundamental tenet, whereby it is recognised that the compensation payable should be more than simply land value (as established in Spencer) but needs to also reflect any special value that the property has to that dispossessed owner, thus, on the face of it allowing claims for severance, injurious affection, disturbance, et cetera.

e   Turner Principle (Turner v Minister of Public Instruction [1956] 95 CLR 245). This principle established that the dispossessed owner is entitled to the full monetary equivalent of the value to that person which includes the realistic expectations as to the probable uses in the foreseeable future. This principle also links to Horn v Sunderland Corporation (1941) All ER 408, which confirms that the land is to be valued at its highest and best use but, once that is established, claims can only be accepted if relevant to that highest and best use. (To use the vernacular, this principle prohibits the claimant from 'having it both ways' – for example, making claims based on existing usage while also claiming compensation for some other, higher valued use.)

f   Executor Trustee Principle (Commissioner of Succession Duties South Australia v the Executor Trustee and Agency Company South Australia Ltd [1947] 74 CLR 358). This principle recognises that the assessment of compensation in compulsory acquisition cases is rarely clear cut and, given the nature and impost of compulsory acquisition, a reasonable generous approach to assessment should be adopted, giving some benefit of doubt to the dispossessed owner. However, this was not to imply payments should be overly generous, and, as much as possible, must be based on evidence and normal valuation guidelines

g   Marshall Principle (Marshall v Department of Transport [Qld] [2001] 205 CLR 603). This principle applies to injurious affection to the residual area following partial resumption (see also Section 11.4). It held that the assessment of compensation should consider the whole 'scheme of works' being undertaken by the Construction Authority, not only the activities of that authority on the resumed land.

This Australian case superseded more restrictive assessment of such compensation in older, English cases (known as the Edwards Principle) and exemplifies the independent course now followed in Australian judgements.

h   Pointe Gourde Principle (Pointe Gourde Quarrying and Transport Co. v Sub-Intendent of Crown Lands [Trinidad] [1947] A C 565). This principle establishes that the actual construction activities and final improvements undertaken by the constructing authority on the acquired land are to be completely disregarded in assessing compensation – either as damage or enhancement ('as if they had never existed').

i   San Sebastian Principle (The Housing Commission New South Wales v San Sebastian Pty Ltd [1978] 140 CLR 196). This principle held that no prior action by the constructing authority or others – either by way of physical works, rezoning or other activities in anticipation of that compulsory acquisition – can be taken into account into the assessment of land value or compensation.

j   Milledge Principle (Commonwealth v Milledge [1953] 90 CLR 150). This principle reaffirms the importance of valuing and assessing compensation to the property's highest and best use and that claims for compensation, particularly for disturbance, can only be sustained when considered as relevant to that highest and best use.

k   The 'Before and After' Principle. This overarching tenet is referred to in many compensation cases, though without particular reference to any particular judgement. It establishes that, notwithstanding the various claims in any particular case, the overall objective of compensation is to financially re-establish (i.e. as much as money will allow) the claimant/dispossessed owner in the same position after the acquisition matter is completed as before. Obviously, with land taken, the situation can never be exactly the same nor can full reinstatement (e.g. of a business) necessarily occur. The task in such dealings is therefore the establishment of the monetary equivalent of the loss.

11.4 Heads of compensation

The concept of 'compensation' in compulsory acquisition that is to remunerate or make good for some loss, to neutralise the effect of a loss or to counter-balance a loss with some monetary equivalent. This generally implies that compensation will extend 'as much as money will allow'. Consequently, while emotional attachment and sentiment will often affect dispossessed owners, they do not represent grounds for financial payment.

Across the various pieces of compulsory legislation in Australia there are variations in the heads of claim that may apply and even the definition of those terms varies considerably. This reinforces the importance of a close understanding of the enabling legislation before such assessments are made. Nevertheless, the Pastoral Finance Principle (i.e. land value plus special value to the owner) and the overarching 'Before and After Principle' apply to all circumstances. Generally following the approach by Brown (2009), and depending on the relevant legislation, items that may be claimable include:

- market value of land and improvements taken
- special value (i.e. to the owner) – which may include other claims below
- disturbance losses
- other losses, expenses and costs stemming from the resumption
- items of function, utility and therefore value to the dispossessed owner even though that value may not be recognised in the wider market

- in the case of partial resumptions, severance loss and injurious affection
- solatium payments.

Several of these terms commonly encountered require closer definition.

A claim for 'severance' arises in the case of a partial compulsory acquisition of real property and refers to any damage which the retained land has suffered as a result of the acquisition of the resumed area. The term 'severance' may be somewhat misleading. While, for example, a road resumption may 'sever' the original block into a number of residual pieces, the term may also apply to any case where only part of the parent block is required – for example, along one side. A severance claim may be extended across several parcels of land, whether contiguous or not, provided that they are in similar ownership and are used in aggregation.

The quantum of compensation for severance will vary depending on the physical and functional effect on the property in that particular case. For example, where a small rear section of a large rural holding is acquired, special value based on 'severance' may be on little financial consequence – though by definition it would still exist as would other claims for the value of the land and improvements taken. At the other extreme, in the case of the compulsory acquisition, of, say, a major road (cutting in two) a small urban land parcel, severance damage would be extreme. (In such cases, it is common that the constructing authority would acquire the entire property.)

A claim for injurious affection can arise in the case of the partial resumption. It relates to the damage done to residual lands (i.e. lands remaining after a compulsory acquisition has taken place) as a result of the actions and activities of the construction authority and the subsequent use of public purposes. These adverse impacts may be in addition to the simple severance effect outlined above. The close association of the two types of claim is noted and, in some jurisdictions, injurious affection is simply seen as a sub-set of severance. In any case, the different character of this type of claim needs to be recognised, as does the relevance of the Marshall Principle identified above. (Marshall establishes that the whole impact of the 'scheme of works by the constructing authority needed to be taken into account, not simply that part of the works carried out on the land acquired from that owner.)

A 'solatium' payment is where a general or additional payment is made by the constructing authority to the dispossessed owner simply to reflect the inconvenience, general disruption and personal distress that such a compulsory acquisition may cause to that claimant, over and above specific claims identified above. The term is derived from the word 'solace' which gives an indication of intent of such payments – that is 'to console' the affected party, though it is recognised that monetary measurement and payment cannot always equate to such intangible and non-economic impacts. Such payments are mandated in some Australian legislation – sometimes as a lump sum or an (additional) percentage on top of the aggregate of the other components of the claim. In some cases, the ability to claim a solatium payment is restricted to owner occupiers. Other legislation is silent on the matter and therefore such claims could not be entertained.

11.5 Assessment

The manner in which the quantum of compensation payable is to be assessed is a matter of valuation practice and standards, undertaken within the parameters of the legislation of the specific case. The assessment is also guided by any relevant precedent court determinations. The task at hand will always be the establishment of the monetary equivalent of the losses incurred under a specific claim.

As with any valuation exercise, the first step will be to establish the highest and best use of the property and to assess value on that basis. This 'highest and best' assessment will have regard to the physical nature and location of the land and the type and function of the improvements thereon and also to the nature of surrounding development, services, patterns of development and so forth. A critical determinant will also be 'legal parameters' controlling current and future use – such as land use zoning, development density and servicing requirements for current and future potential uses. Such potential must represent a reasonable expectation, achievable in a commercially acceptable timeframe and with market viable return. (Note that with many parcels of land and under contemporary planning controls, there may be a wide range of potential uses. It is, however, another thing altogether for any of those uses to represent realistic potential and thereby have a positive effect on the land over and above its present usage and state of development.)

Such considerations are of interest in the compulsory acquisition processes. Once the highest and best use is determined, claims can only be entertained if they are relevant to that best use. This will impact the compensation payable for any existing uses (not relevant to highest and best use) and may also limit claims for disturbance to and losses related to existing activities. (These observations form part of the Milledge Principle referred to above. The maxim of 'equivalent financial position after the resumption as before' is also relevant here.)

On occasions, those losses can be mitigated by the swapping of land or other improvements/works or 'trade-offs' that may be possible in that particular project. While that may prove opportune for all parties in some cases, courts have noted that such dealings can become very complex and less than certain or fair. Consequently, settlement on money equivalents or on the maxim 'as much as money will allow' are normally to be preferred.

The approach, as always in such matters, is based on the prime evidence provided by the recent sale of comparable properties and the analysis of those sales to ensure the value comparison is on a realistic basis (the Spencer Principle). In the cases of compulsory acquisition, it is highly unlikely that comparison can be drawn from sale properties that involve compulsory dealings. Consequently, and as highlighted under the Raja Principle, the final assessment will need to be built up by the logical assessment of the component parts.

At the beginning of the process for a particular assessment, a checklist should be established which identifies the enabling legislation, its requirements and direction, the date of the resumption, the land (including area and dimensions), that has been taken through this due process, the establishment of highest and best use, the categories (types) of land resumed and the improvements thereupon.

Building upon evidence derived from market transactions, normal valuation methodologies may be applied as relevant to that particular case. These could include, for investment properties, the capitalisation of net rents, investment analysis involving the present-day value and internal rates of return in building development or subdivision projects, the assessment of 'going concern businesses' and/or a cost-based summation assessment. All of that represents typical valuation practice to which is then added the 'special value to the owner' assessments and claims for disturbance and costs as prescribed under the relevant legislation. In application, and in accord with the Executor Trustees Principle, a reasonably liberal (but not unjustified) assessment of the claims in favour of the dispossessed owner should be made.

The aggregation of all of these components represents the assessment of final compensation. In a final review the 'before and after' maxim, as explained earlier, would need to be considered.

Where compensation for the securing of a right of way, 'engross' easement is involved, the assessment will include only a percentage of the value of the land affected. Again, aligning with normal valuation practice, that percentage will vary according to the level of impact that the activities and installations undertaken by the constructing authority have in that particular case.

Even with the best of intentions to finalise matters without delay, compulsory acquisition processes can become protracted, particularly if litigation becomes necessary. Given this scenario, legislation will typically include provisions to protect the interests of the dispossessed owners. These may include:

- payment to the dispossessed owner of advances of settlement amounts up to, for example, the compensation offer made by the Constructing Authority
- 'hardship provisions' whereby, if the dispossessed owner finds themselves financially distressed because of the compulsory acquisition, part of the compensation payments can be 'fast tracked' to relieve that distress
- interest at an appropriate rate being added to the compensation payment for the period of delay. (On this matter, interest will typically only accrue if the dispossessed owner has in fact lost the ability to derive income or benefit from the property during that period.)

Compulsory acquisitions may involve the dislocation or loss of businesses, either in whole or in part and such losses will, in all probability, be claimable by those with an equitable interest. Such 'disturbance' claims would be predicated on the legality of that activity and its ability to trade profitably into the future if not for that compulsory acquisition. In some cases, particularly where site goodwill dominates, the business may be lost completely. In other cases, compensation may involve the temporary loss of business plus the costs of re-establishment and other disturbances that the business endures. On the face of it, full reinstatement and re-establishment elsewhere may sometimes appear to be the most equitable solution. However, in practice, such outcomes are often extremely difficult to achieve and risk unfair 'new for old' compensation claims.

Expenditures by way of costs incurred by the dispossessed owner in legal, valuation and other professional assistance in preparing that claim would normally be met by the Constructing Authority, though reasonableness and proof of expenditure will be required.

In the case of the partial compulsory acquisition of a property, the balance (or 'residual') area may be enhanced in value as a result of the entire scheme of works undertaken by the Constructing Authority. This is known as 'betterment' or 'enhancement'. (This is separate from, and not affected by, the Pointe Gourde Principle discussed above which refers only to actual construction works.) Some legislation include the assessment for betterment which can be offset in favour of the constructing authority against claims for compensation by the dispossessed owner. In cases where such provisions exist, the onus of proof of betterment typically lies with the constructing authority and this counter-claim cannot exceed the claim from the dispossessed owner. In practice and given the overall parameters and sensitivities of compulsory acquisition dealings, a constructing authority would probably only make such counter-claims where the betterment was both substantial and easily demonstrated.

As regards assessment of compensation also, two other concepts require comment. Typically, major public works such as a new major road scheme, dam or urban redevelopment, et cetera, are announced well in advance of compulsory acquisition and construction activity. In the intervening period, the ongoing threat of such works and disruption will often discourage normal investment and other dealings in that locality and, with that, the level of business and investment activity and comparative property values may fall. Such a general effect on a locality is known as 'planning blight'. While its negative impacts can be significant, there is (on the face of it) no claim possible for compensation, given that no land has been resumed at that time.

A variation exists in this regard under what is known as the San Sebastian Principle (Housing Commission New South Wales v San Sebastian Pty Ltd [1978] 140 CLR 196). This principle holds that any specific action taken by the constructing authority in anticipation of a later compulsory acquisition (e.g. a 'down-zoning' of the land to facilitate later acquisition) is to be disregarded in any later assessment of compensation.

The second matter refers to the personal or sentimental losses that individuals or groups may experience because of the loss of a particular property. The depth of this emotional attachment is not doubted, particularly where individuals or family groups are long-term owners and occupiers of houses, businesses, rural holdings, et cetera, that are now to be lost, in whole or in part, to some public use. However, the overarching compensation approach here is based on payment 'as much as money will allow' and therefore claims based on sentimental attachment are generally not allowed. It might be observed that solatium payments (if available under the relevant legislation) may go some way to addressing such effects as might the Executor Trustee Principle discussed above. A generally sympathetic and diplomatic approach by the constructing authority and its expeditious dealing with wider claims may assist in lessening such impacts.

11.6 Due Process and the critical role of the Constructing Authority

The strict adherence to due process has been emphasised in the preceding sections. That process will be specifically defined and closely controlled through the legislation and regulations under which that particular acquisition is being undertaken. Figure 11.1 outlines the typical, main steps in the process, noting that, in following the requirements of particular legislation and regulations there will almost invariably be additional steps and more detailed activities, investigations and processes required.

Compulsory acquisition in Australia is closely controlled through legislation, regulation and precedent because of the important underlying issues involved. The constructing authority under that relevant legislation is central to that, instigating the process in the first instance and, as later steps, securing the land, paying compensation and undertaking the public works and activities involved.

While public works will typically involve very large capital expenditures, the property acquisition components and physical and community impacts are, in the contemporary

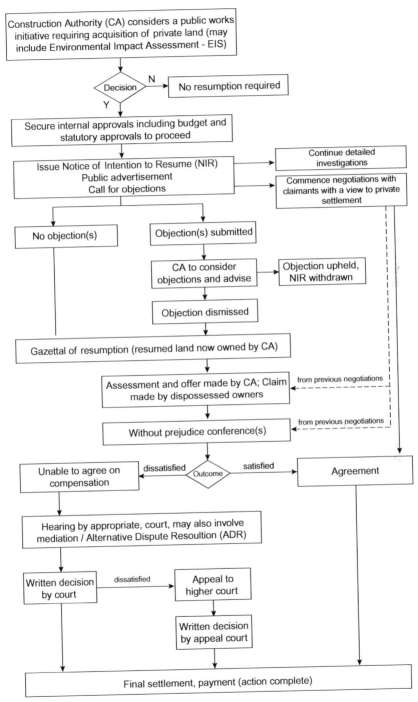

Figure 11.1 Typical steps in compulsory acquisition (variations exist depending on jurisdiction).

environment, as likely to be key determinants of the location and overall feasibility as construction engineering design and construction.

For the constructing authority, the strict adherence to due process, professional, fair and courteous behaviour with dispossessed owners and good relations and communication with the wider community will all be essential to ensure property acquisitions and later construction proceeds expeditiously. Any suggestion of acting outside due process ('ultra vires') or acting in anything less than good faith in what will already be a sensitive matter, will severely compromise the entire dealing and project.

The constructing authority is well advised to ensure that it has full access to the facts of the matter and be reasonably sure of the optimum course of action before making public their intentions. Thereafter, Notices of Intention to Resume (NIRs.) will typically be served on all owners and those with other interests in that property. If, after any objections are considered, the matter is to proceed, the process will move to gazettal and the actual acquisition is completed, normally with the negotiations to settle compensation to follow. In all legislation, appeal provisions for arbitration of compensation payments that remain in dispute. The vast majority of claims are settled by negotiation, using without prejudice conferences and other forms of Alternate Dispute Resolution (ADRs) with only a very small number finally requiring court determination.

Again, to maintain good faith and to limit distress, the constructing authority should move through these phases expeditiously, avoiding procrastination and unnecessary delays.

The compulsory acquisition processes are normally completed well in advance of construction activity and public use. In such circumstances, it is not uncommon that the dispossessed owner may remain in occupation for some time after the acquisition is complete. Such a practice can, in fact, benefit both parties. It may give the dispossessed owner more time to vacate and reinstate elsewhere. It can also act as a sign of good faith by the constructing authority and also provides a level of ongoing security to the property. It should be noted however, that such arrangements do not change the full property rights now held by the constructing authority nor does it normally imply a residential tenancy arrangement. Eventually the property will be required for use and the constructing authority will have full rights to take possession.

In practice too, such constructing authorities schedule such works well in advance of requirements and, where appropriate, will attempt to buy up required properties, simply via private negotiations or as those properties become available for sale on the open market. Given that compulsorily acquisition processes take a period of time, a point will be reached where those processes must be instigated to ensure that construction schedules are met.

Even then, purchase negotiations with affected owners will normally proceed in an ongoing attempt to secure the required property, thus avoiding litigation.

Through all of these various dealings, the constructing authority must be careful to take a reasonable and balanced approach. On the one hand, it will need to ensure that, in line with the 'Executor Trustee' and 'Pastoral Finance' principles (outlined above) the dispossessed owner is provided with a 'reasonable benefit of doubt' in negotiations. At the same time however, the authority will be well aware that it has accountability for the expenditure of public funds in a legitimate and justified manner. Quite pragmatically also, many projects involve the acquisition of a considerable number of properties and the constructing authority needs to be careful that it does not establish precedence nor encourage unrealistic expectations resulting from overly generous or unjustified payments.

On occasions, planned projects do not proceed because of technical, financial or political reasons. On some occasions too where a compulsory acquisition has been completed there may be residual land that is not used or required. In such cases the constructing authority may wish to dispose of those lands for which there is no further public use, either by full sale or amalgamation into adjoining lands by way of boundary correction. Many pieces of legislation contain provisions that foreshadow such circumstances. These may prescribe an initial offer back to the dispossessed owner and the manner in which value is to be determined in such circumstances. These matters can often be sensitive in that particular locality and the constructing authority needs to strictly adhere to those parameters.

Chapter 12

Land subdivision and property development

12.1 Introduction

'Land' or 'real estate', in its natural state, is rarely, if ever, at its highest potential (i.e. 'highest and best use'). Typically, it requires the input of labour, capital (in various forms) and entrepreneurial skill to bring it into economic production.

Particularly for urban land, this will commonly involve one of two processes: the subdivision of large parcels of land into smaller lots (with infrastructure and services suitable for more intensive uses), and/or the physical development of improvements (buildings) on subdivided lots.

This chapter summarises some of the key components, processes and issues involved in both those generic activities.

12.2 The nature of development in the property sector

The owner of a property asset can typically enhance the value of that asset by investing additional capital in the form of development works of one sort or another.

Regardless of the complexity of actually doing this, the underlying economic proposition is both simple and consistent in all cases. The owner seeks to make a profit in terms of increased capital value, rental income and/or increased productivity and functionality when that further investment is complete. This is sometimes called the 'value add' or 'developer's profit and risk'. In economic terms, this would be described as the 'marginal profit', not referring to the quantum of that profit, but because it represents the differences between the marginal (or additional) revenue that can now be secured from market dealings and the additional cost expended affecting those changes/additional works.

On the 'supply side', the timing of cash flows in and out of the project will be critical, given the time value of money considerations previously discussed in Chapters 5 and 10. On the surface, the capital costs, including holding costs, interest and so on, should be reasonably easy to measure. The increase in value may be a little more problematic given that, for this exercise, it is the added value (i.e. over and above what the property was already worth) that needs to be assessed at the date when those new development works are completed and available for use. Feasibility analysis and Internal Rate of Return (IRR) calculations will be undertaken from the start of the proposal to test project viability and will be frequently revisited and upgraded during the project as more accurate information becomes available. (Again, these approaches are further discussed in Chapter 10.)

The added value is determined not as the actual cost incurred, but rather by what the market is willing to pay for the finished product. This may be in terms of additional capital value at the date of completion, or the owner's right to receive increased rental income streams or additional production from that asset. As explained in Chapter 10 also, 'cost' and 'value' are two separate concepts; while they may have some relation, there is a range of reasons why they are unlikely to coincide in quantum.

Property assets include many examples of 'development works'. In the rural sector, development works may involve new fencing, yards or timber treatment; in the mining sector, they might include the development or expansion of a mine, minerals processing or related infrastructure; in the tourist sector, they might take the form of the development or expansion of a resort or hotel; and in retailing, the development, extension or refurbishment of a shopping centre or similar facility.

The role of project home builders in the residential sector also represents another common example. These builders can be large-scale, national organisations, or smaller, individual building contractors operating within a certain locality. Regardless of scale, all are involved in the primary economic activity of using additional capital, labour and their own management skill to (hopefully) 'value add' to a real property (land) asset.

In the property sector the term 'development' has a wide and generic meaning, though it has more specific definitions in various pieces of legislation. In general terms and as described in various planning and development regulations, it is taken to mean the physical carrying out of building works, plumbing and drainage and the installation of other services, operational activities, the reconfiguring of a parcel of land (i.e. subdivision) or making a material change of use to that property (i.e. more intensive or specific use).

One of two basic activities are therefore involved. The first is 'land subdivision' – that is, the division of a parcel of land and the on-sale of the individual, now subdivided, smaller lots. The second is colloquially known as 'property development' – that is, the securing of a parcel of land and, thereafter, the construction of a new building, for example, a multi-residential, office or industrial building or retail centre. A variation on the latter category would be the extension, refitting or refurbishment of an existing building that, in its current condition, is no longer at its highest and best use. Where a bare parcel of land is secured and a whole new asset is built, the project is often referred to as a 'greenfield' development. Where additional investment and work or changes to an existing built asset is involved, it is commonly referred to as a 'brownfield' project.

These activities effectively produce new or improved assets for business, the wider community and the public sector, and are vital to overall economic stability and growth, particularly at a regional or local level. Because of the time taken to bring these projects to completion, phases of transitory shortages and oversupply are not uncommon (known as 'production lags', which could extend from six months to several years).

Land subdivision and property development projects are remarkably complex and have become increasingly so, given the variety of demands of the contemporary urban environment and its end users. These include the difficulties and costs involved in securing approvals and providing infrastructure, including roads, water, electricity, sewage, drainage, ICT, et cetera, together with the coordination of a huge range of construction and other activities that will see the timely delivery of a product that market will accept.

Some of these activities will involve physical works, others will relate to the payment of agreed amounts to the local authority or to public utility providers to carry out to affect those required works themselves.

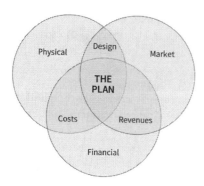

Figure 12.1 Components of a project plan for development.

The links between the property, construction and finance sectors are obvious and the developer needs to successfully bring the diverse economic (i.e. market), financial and physical components together into a project plan and, thereafter, to deliver it according to that plan. The graphical representation of these components and their interrelations is shown in Figure 12.1.

Each project will be unique in scale, location, opportunities and risks. Success is typically based on market analysis, meticulous planning, sound risk management and excellent project management and coordination. Some of the largest corporations in Australia operate in this sector and, invariably, these components – strategic under-standing of the market, planning, risk management and coordination in delivery – represent the common elements of their success.

As with considerations in the financial sector, it is not suggested that any property professional in general practice – be they a valuer, asset manager, financier or other – could have a detailed, day-to-day knowledge of all components of these complex activities. There exists, however, a number of important development market characteristics and principles that are reasonably easy to comprehend. These are essential knowledge for all property professionals given that, to lesser or greater extent, they all have exposure to the development sector.

The balance of this chapter will identify and briefly discuss the important characteristics and components of this sector – focussing particularly on 'land subdivision' and 'property development' projects defined above. It will be noted that while they in effect produce 'different commodities' – one being subdivided allotments and the other a building – there are, understandably, many common characteristics.

12.3 Common characteristics and issues – land subdivision and (commercial) property development

In essence, any land subdivision or property development project has the same core elements, many paralleling the components of any other business venture. Primarily, the developer must have the confidence in his/her ability to be competitive and to deliver the required 'time, cost and quality' into that market. The second is the necessary skill level to carry out all those functions in a way that manages risk and has good prospects of securing an attractive and timely profit (Jowsey 2011).

In property projects, these issues are particularly important because of the scale of the projects and the very long time frames, often extending over several years. The management of costs and cash flow, the securing of the necessary statutory approvals, finance, contract and construction management, provision for insurances and adequate contingences are all vital for successful delivery.

Additionally, the developer is creating and/or adding to fixed assets. This implies that once the commitment is made to proceed, it is, to all intents and purposes, irrevocable. This may seem a trite observation but is fundamentally important. For example, if an investor had, say, $10 million in equity to invest, one option may be to simply buy a 'near cash' share or bond portfolio. In that case, decisions made can be incremental and are certainly, as long as the investor is willing to accept the going-market price, near-instantaneously reversible. Alternatively, that investor may have used the $10 million as equity, perhaps leveraged it with, say, another $20 million debt funding, to develop a residential subdivision or commercial building. Once those key decisions are made, the land is bought, approvals acquired and construction contracts executed. The project and the financial exposure that goes with it, now $30 million, will continue for some years. That process cannot be stopped or reversed without probable disastrous financial effects on the project and the security of the original investment and debt commitments. Overall, their potential for very significant return is presented in such ventures but, with it are accompanying serious risks that must be identified, qualified and quantified and managed.

With all of that in mind, some fundamental principles become apparent.

An understanding of the market and of demand

Any project will be a success only if, at practical completion, there is a final consumer/consumer group for that product, whether that person is a final buyer of a residential lot or a commercial building, and/or someone who will commit to lease and use the space created. It follows that, before any significant commitment is made to embark on a development project, a thorough understanding of the final market for the product is essential. This includes considering issues such as:

- market–acceptable final price/rental values
- overall condition of the market and, particularly, the likely future condition of the market when the product finally comes on stream
- any unique, unusual or previously untried features of this project and what investigations have been made into their feasibility, risk and acceptability to the target market
- what the particular/typical requirements are for that type of property asset, that is, demand priorities
- historical and likely future absorption/take-up rates
- emerging issues including energy-saving sustainability, accessibility and/or transport
- locational issues and likely changes
- actions/likely actions of competitors
- legal/regulatory requirements and likely changes
- likely financial conditions, for both funding the project and final purchases
- the reliability/level of certainty of all of the above forecasts – including adequate provision for contingencies.

Many of these matters may appear obvious and simply common sense. It is remarkable, however, that in the enthusiasm and urgency of an apparent opportunity, some of these necessary investigations are bypassed or approached with unsubstantiated optimism even, at times, by larger, experienced developers.

Successful developers often spend much more time in these research and design/pre-design activities than they do in actual construction. They do this on the principle that one can undertake the best and most efficient construction possible but, if the product is not what is required or is poorly located for that purpose, it is most unlikely to succeed. Moreover, research needs to be done in a structured way using expert advice. Hearsay, hunches or unsubstantiated rumours can be misleading and dangerous. Sometimes too, an investor/developer can become captivated by a novel concept or project and badly read the 'market realities and acceptability' of the proposal. This is not to say that developers, designers and constructors are not innovative; they certainly are and must be to provide an attractive product. However, the successful ones, and their financiers, need to be assured that there is a final 'take-out' market for that end product.

Careful selection of a suitable site represents a vital component of this research. An inappropriate site for the development represents a flaw for which no subsequent design or management will normally remedy. Depending on the development project involved, site acquisition may account for only 15 or 20 per cent of the final developed value of that project. However, its impact on the final value of the completed development, either positive or negative, can be far greater. The purchase of a site that is poorly located or that suffers from major infrastructure or headworks charges, lengthy approval times or construction problems, will have fundamental, negative ramifications for the capital value and success of the final product, even if the site initially appears to be inexpensive.

Competencies and scale

Successful developers are almost invariably well experienced. They understand markets, approval regimes and project and construction issues within their sub-sector and/or locality. They will often have strategic alliances in place with their financiers, principal contractors and marketers and will have assembled, or be able to quickly assemble, a team of professionals and consultants from in-house and external sources.

In Australia, developers will often specialise in certain types of development and, except for major corporations, often tend to concentrate on a particular region where they already have detailed knowledge, experience and networks and where their set-up costs can be minimised.

Despite the large scale of many of these projects, the management approach needs to emphasise fine detail and coordination. Certainly, large-scale projects require a strategic vision and overall project leadership and direction. However, all such projects are made up of a huge number of small-scale activities, sub-contracts and packages, and it is typically at that level where many of the innovations, efficiencies and savings in time and cost can be made. It is at that 'fine grained' project management level where emphasis and resources must also be placed. Time and cost issues are central to the majority of failures of development projects and, in turn, many of those issues can be traced to problems in a comparatively small number of components and their compounding effect on the project.

Additionally, the development needs to be of a scale that can be adequately managed and supported by that developer, even under 'worst case' scenarios. The level of capital investment, project duration and gearing (ratio of debt-to-equity funding) typically increases with the scale of the project, together with the overall risk profile. In all of this too, cash flow throughout the project is critical. Project failure is often not related to the soundness of concept or quality of design or construction; but rather, as a result of unforeseen issues, the project can simply run out of cash and is either liquidated or becomes a forced sale.

Risk management

Any type of property development project will be complex, requiring specialist skills and effort. By nature, it will be relatively high risk. That is why the reward/return sought for such ventures (known as the 'developer's profit and risk') will be much higher than the return for many other forms of investment – often 20 to 30 per cent or higher, depending on the project. Contrary to some public perceptions, property developers and their financiers tend to be cautious and conservative in their approach.

Risk can be understood and managed; in fact, the real competitive advantage of successful developers lies in their high level of research into all aspects of the proposed venture and, based on that and their own acumen, to understand and better manage risk than their competition. This superior knowledge and planning will reduce risk exposure and provide a more competitive product without the risk aversion strategies that less knowledgeable competitors will be forced to adopt.

As discussed below, some characteristics of property investment and development require particular attention and risk management. These are typically large-scale and multi-faceted undertakings, extending over quite long time frames. A particular risk and challenge here lies in the need to integrate a diverse range of diverse financial, construction, legal and market elements within the development process. That notwithstanding, Sutton (2005), Malkiel and Ellis (2010) and Mendleson (2009) identify the fundamental concepts of risk and its key indicator levels that are fairly generic across all commercial ventures.

Sutton considers the following fundamentals:

- The need for decision-makers to truly understand the concept of the 'optimum' scale[1] of a venture – established at the point where maximum profits are likely to be secured within acceptable risk parameters. This point may have little relation to maximum production levels possible and, indeed, the singular pursuit of higher volumes or increased scale has inherent corporate dangers, particularly if a primary aim is to 'trade-out' of difficulties or losses. The business/project must be stable and viable before expansion is contemplated.[2]
- Debt funding represents a key component of most ventures and, properly managed and controlled, it may well be essential to project success. However, that potential also comes with a risk that grows significantly as the debt-to-equity ratio ('gearing') increases. If a project or other commercial venture encounters difficulties – for example, because of unforeseen time delays or changes in the market or general economic conditions – good management may make adjustments to lessen the effect. However, debt repayments and financing costs continue regardless and, depending on prevailing monetary and lending policies, may indeed become worse. Highly geared projects (i.e. with a high debt-to-equity ratio) are particularly vulnerable in this scenario.
- While there will always be vagaries and risks, particularly in future revenues and profits, there is a great deal that management can and must do to lower risk. As

noted above, detailed current market research is essential throughout the venture, as is the close control of both capital and operational costs. Cost control should not act to undercapitalise the venture, but the overall objective is profit maximisation – that is, ensuring the widest difference between revenues and costs. Therefore, the astute management of costs has a similar 'net' effect to increasing sales and, at the same time, has positive effects on cash flow and the lowering of risk profile overall.

- Time and timing will always be critical to any commercial venture at a number of levels, and are particularly important to property projects, given the relatively long time frames and levels of exposure typically encountered.
- Linking all these factors is the composition, skill and experience of a team of individuals who run the ventures and their ability to properly communicate and report. They not only need a project plan, but also must adhere to it. Their efforts and skills need always to focus on the real outcomes and success factors established from the outset (Mendleson 2009).

In the first instance, key decisions must be made in a timely way. Detailed research will take time and important decisions cannot be rushed. However, procrastination will lead to lost opportunities and an extended period of exposure.

It may be the opinion of some that property developments or investments are somehow so different in character that they are protected from commercial realities.

These misconceptions are dangerous. Certainly each asset class, including property, has its unique characteristics that provide both opportunities and particular risks, but the fundamental rules of economics, investment and risk remain unchanged.

Risk obviously occurs at various times throughout any project and the developer must recognise and qualify those various key risk areas and reduce exposure (called 'lowering the risk profile'). Risks within a project can be defined in qualitative and quantitative terms. Using a range of management tools and modelling, they can be specified as to size, impact, frequency and duration, and from that, appropriate risk management techniques can put into place (Dobbie 2007).

Risk falls into two basic categories. One is known as 'systemic risk', which relates to the general risk encountered by all within that economy, sector, region and so on, at any point in time. This includes variations in factors such as interest rates, market confidence, taxation, cost of living, building cost and inflation. The other is 'non-systemic risk', which are the risks internal to the project or reasonably under the developer's management control. This includes, for example, the securing of approvals, financial arrangements, delivery systems, project costs, time frames and project management. Significant too are risks related to time delays and the ability to secure and hold prospective final purchasers and/or tenants.

As risks are shared – for example, with other suppliers through construction contracts and sub-contracts – then, it must be reasonably expected that a commensurate part of the overall anticipated profits from the project will need to be shared with those others now contracted to take that part of the risk.

Examples of risk management techniques typically employed within development projects include:

- use of milestones or 'gates' (i.e. not proceeding with a project until certain pre-conditions, such as certain levels of pre-sale or pre-leasing are achieved)

- inviting an equity or joint venture partner into the project who brings new skills, resources or capacities to the project
- tracking and responding to market changes in the interim, especially for non-committed components of the project or its final use
- securing an option to buy a property, say, subject to development approval rather than buying the asset outright in the first instance
- securing a fixed-price construction contract and nominated construction period
- allowing for sufficient contingencies to accommodate unforeseen circumstances and delays, particularly on unfamiliar or difficult projects
- developing the project in stages; often, not possible in a building/construction project, but used in subdivisional projects, where the sale of initial stages can provide cash flow for the development of later stages
- securing fixed interest rates during the duration of the project
- taking out suitable insurances for accident, public liability, loss of profits or wet weather, et cetera
- timing construction so that it is likely to avoid seasonal wet-weather delays
- employing sophisticated project management and coordination techniques to avoid delays in construction and to coordinate, integrate and overlap activities, where possible, to avoid unnecessary time exposure
- ensuring sound construction management and labour/union agreements that provide a safe, efficient and cost-effective site and project.

The risk profile changes through various stages of a project and with various activities. The highest risk and highest cost components in practically any property project are:

- The initial land acquisition cost – not simply because of its considerable capital value, but also because it is a cost normally incurred very early and, therefore, needs to be carried – including interest, rates, land tax and related costs – throughout the full life of the project.
- Holding and financial charges and approvals costs; again because these are progressively held throughout the life of the project and, depending on financial arrangements, may vary over time and from original budgets.
- Actual construction/infrastructure/development costs; typically, these represent the highest cost component and may be subject to substantial variation and unforeseen additional costs. Note that, depending on the project, certain activities may present greater risks because of the potential that unforeseen issues, costs and time delays may occur. Examples here might include issues with the securing of approvals, geotechnical and excavation problems and the installation of unusual or complex services and equipment. Detailed prior research and close management during those phases should assist in reducing the risk profile.
- Time delays at the end of the project in leasing or final sale. These are particularly important because, by that stage, the total capital cost of the project has been expended and is exposed to financial charges on that full amount. Any unexpected delays through that last stage will prove very damaging to final profits and, potentially, to project viability.

Given that the majority of projects will exhibit this typical profile, most risk management activity should be concentrated on these areas of highest exposure.

12.4 The development process

In any location or region, there is a process through which property development, either subdivision or new building construction, progresses. It is a blend of business process and project and risk management, combined with specific town planning and building requirements. Figure 12.2 outlines the generic steps involved which, depending on the scale and complexity of the particular development, could involve time frames of several months but extending up to five years or more for major, staged developments.

Typically, and as discussed in more detail below, the process involves the knowledge, investigation and analysis of:

- the wider economic environment and projections
- the region and the vicinity as regards growth pattern, change, infrastructure and government policy
- site-specific physical and legal considerations
- demand analysis, market/marketing considerations and targeting
- concept development design and testing
- potential activities by competitors
- financing options/vehicles
- construction and delivery of the project.

The fundamental steps are fairly similar for either subdivision or building development, though certain variations are noted in Sections 12.5 and 12.6. However, note that the details of the process will vary considerably, depending on the type of project, its scale and complexity and the statutory planning and other laws and regulations that manage these processes in specific jurisdictions.

Further observations on each of these stages (as per Figure 12.2) are as follows:

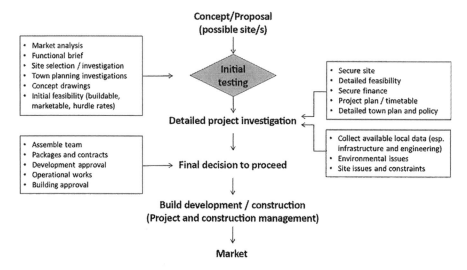

Figure 12.2 General outline of typical development process, property building development or subdivision projects.

Concept/proposal

The original concept or proposal can come from various sources but is most often brought together or instigated by the principal/developer.

It may be that the developer has identified an upcoming potential demand or shortage in a particular area. Perhaps the developer has become aware of an opportunistic purchase of a particular site, or a pre-existing concept that has not yet proceeded. Perhaps the developer has uncovered a particular parcel of land available for sale. Initial ideas can emerge from a range of observations or 'leads'.

In any case, and, sometimes by drawing on the advice of others, an initial concept of the project is developed. The developer's previous experience in the sector and/or locality will be of advantage as a functional brief – often simply a few pages and concept drawings. Critically in this phase, and indeed throughout the entire project, the focus must be on the final marketability and competitiveness of the project at completion as presented to final consumers: that is, the project's required outcomes, not simply what physical characteristics or activities are required to get to that end point.

It needs to be noted that any developer will typically explore a large range of possible projects, sometimes over a long period of time, before deciding whether to proceed with any more formal, costly and time-consuming, investigations.

These considerations pertain to the business basics of the rate of return sought (the 'hurdle rate'), risk profile, and time and cash flow parameters. At this stage an initial feasibility study, a functional brief and some early concept drawings of 'what the thing might look like' may be undertaken together with some investigations as to the market acceptability of the project. A preliminary analysis of development controls, infrastructure and services under the local town plan, will also be carried out at that point.

Early investigations aim to identify fundamental problems or an unacceptable risk profile. Typically, very few concepts will proceed past this point.

Detailed investigations

Where a project is identified as having good development potential, more detailed planning and investigation is undertaken. This stage which will include the undertaking of feasibility as IRR modelling is vital to the final success of the project, as fundamental errors in concept design, costing or targeting can later cause serious delays and cost overruns and, perhaps, for the project to fail altogether.

Activities include site acquisition negotiations – to at least hold the land through an option to purchase – geotechnical and related studies and advancing concepts towards detailed design stages. Also included will be the completion of a comprehensive feasibility study specifying costs, income streams, anticipated time frames, holding charges, infrastructure and contingencies over the life of the project. The investigation process is assisted by the large amounts of geotechnical, environmental, hydraulic and service supply information typically now available on public websites. Further relevant information can be obtained from other, local town planning applications that are already lodged and are now in the public domain.

Using such innovative investigations, a comprehensive analysis of the proposal can often be secured without, at that stage at least, expensive primary data collection. Through this phase, applications for development approval (identified in some jurisdictions as 'material change of use' and/or 'reconfiguration of a lot') will be made and design, infrastructure and other charges negotiated with the local authority. Again, the review of recent decisions by the relevant local authority of comparable projects and as available, decisions of the planning and environment courts, can assist in ascertaining the likely success of project applications under submission here and the conditions and costs that are likely to apply to that application.

It is important to note that, through this phase, there will normally be no absolute commitment to proceed with the project, or even, if possible, to have settled on the purchase of the site. This is typically a high-risk period given the level of effort and cost applied to the proposal and, yet, it may still be abandoned if costs, market investigations and / or dealings with local authorities and financiers result in the project becoming, in the opinion of the developer, unacceptable. As a general rule, it is much better to abandon the proposal at that point and absorb the costs rather than continue in the hope that the prognosis will improve as the project advances.

Even where a development approval (DA) has been secured, the developer may still decide not to proceed. Subdivision and/or the DA attach to the land for a particular period of time, typically between four and six years depending on the jurisdiction. It is not unusual that the original developer/applicant will on-sell the property with an enhanced value from the attached DA without that party proceeding to physical construction.

Final decision to proceed

All of that information and the results of market analysis, dealings with financiers and the local authority eventually come together and, if the project still appears viable, then earthworks, infrastructure, building and trade packages will be assembled and tenders called, particularly on major items.

With that completed and funding in place, the developer needs to make the final decision whether to proceed which, depending on the individual project, may start with the acquisition of the identified site, followed by securing final approvals for building and operational works from the local authority, a design freeze and, finally, the letting of contracts. The project is, from that stage, virtually unstoppable without very significant financial and other losses.

The apparent high returns for developers on successful projects need to be considered in this context – not only the risks involved with the projects that proceed, but also, in part, by way of recognition of the time, cost and effort involved in investigating the many potential projects that never proceed.

12.5 Particular considerations – land subdivision

As noted previously, both types of property developments – subdivision and building construction – share many generic characteristics (refer Section 12.3). In a project involving the subdivision of land into individual new titles/lots to be sold, a number of other unique characteristics need to be noted.

Typically, subdivision projects contain a high proportion of civil construction work as the landform is altered and as infrastructure and other services are provided to end users as prescribed under the relevant town plan.

Infrastructure will typically include roads, water, sewerage, stormwater, drainage, telecommunications, electricity, earthworks, community facilities, pathways, bikeways, parks and so on. For development projects, however, 'infrastructure' relates principally to physical works. The definition of this term is rapidly evolving and increasingly applied to much less tangible components of the development, such as 'social infrastructure'. Social infrastructure relates to the community services and networks that underpin how the neighbourhood actually operates and evolves over time. Examples might include libraries, civic places and, potentially, other community benefit initiatives.

'Infrastructure' in the form of larger mains and services constructed externally that reach to the boundary of the land to be subdivided or developed are called 'headworks'. The distributed service lines within the new subdivision/estate are normally referred to as 'reticulation'. Within the subdivided blocks, the services (conduits, pipes and wires) from the boundary to the buildings are called 'service lines' and are normally the responsibility of that individual lot owner, or body corporate in the case of a community title development.

Under various pieces of state legislation across Australia, subdivision is effectively controlled through the issuing of Certificates of Title: the legal, registered document formally representing that subdivided parcel, its ownership and record of other dealings. While some variations exist, it is generally illegal to contract to sell or attempt to transfer land that is not legally subdivided. That is, the legal precept is that each parcel of land must have its own certificate of title and, therefore, have a separate legal identity before its sale and transfer can be recorded. Certificates of Title can be issued only by the appropriate arm of the state government, normally a Titles Office, Department of Environment and Resource Management or similar, and that only occurs subsequent to the registration of the final subdivision survey. In turn, these subdivision plans will not be accepted for registration unless they are also sealed, that is, signed off as approved, by the relevant local and/or state authority. (This represents an important 'fail safe' on the entire system. It is impossible to transact in land and have that dealing legally registered unless the land itself has separate, registered title and recognised, separate ownership.)

The relevant local authority normally has operational control of the local town plan, the issue of development approvals and, under legislation, also has the power to apply 'reasonable and relevant' conditions upon that particular development prior to, or concurrent with, sealing the plan that finally approves the subdivision.

Those conditions can be quite comprehensive and include lot size and dimensions, specifications for roads, water, other infrastructure, environmental protection, appropriate provisions for drainage, stormwater, sewerage et cetera, and provision of certain community services and the like. The system is based on a 'user pays' approach whereby the developer, representing the individuals who will buy and enjoy the subdivided lots, is required to pay those costs rather than the local authority or the wider community.

The design, engineering specifications and other typical requirements for headworks and development will normally be laid down in the local authority's subdivision and development regulations (also known as 'by laws' or 'ordinances' in some jurisdictions).

On that basis, the local authority can require the developer to carry out certain works themselves or make contributions, under 'deed and bond agreements' with the local authority who, under those circumstances, will carry out those tasks. In a major subdivision, these conditions and costs can be extensive and, in some cases, what exactly constitutes 'reasonable and relevant' and the quantum of those costs is a matter of conjecture, debate and, sometimes, litigation.

Urban areas in Australia are typically surrounded by large tracts of unused or lightly used lands. Often these are zoned (reserved) for 'future urban' purposes or, in more recent times in some regions, are identified as 'within the urban footprint' of that region. This identification and allocation allow for the progressive expansion of the urban area as demand requires. The objective here is to help coordinate and sequence that urban expansion, ensuring that expensive infrastructure and trunk mains are efficiently used and that any adverse ecological impacts are kept to a minimum.

Land holdings on the fringe of existing development with potential for subdivision, are sometimes referred to as 'accommodation land' or 'en globo land'. How valuable or immediate that potential might be depends on its location, levels of current and future demand and the extent to which services and infrastructure are available.

It should not be surprising that englobo land often has a relatively low value per hectare in its undeveloped state. Most of it will require very large investment in external infrastructure for which, depending on sequencing, the local authority may be quite reluctant to provide assistance. Demand for land can be uncertain and development time frames protracted. Furthermore, because of the time frames and risks involved, the potential additional value, may be very modest and the land value little higher than rural use.

The general process of land subdivision (highly simplified) is shown in Figure 12.3.

From the 1960s through to the present time, the vast majority of urban subdivision has produced low-density, single-unit dwellings based firmly on the principles of conventional subdivision emanating from the US (i.e. a strict hierarchy of roads and segregated land use, all enforced by town plans and the segregation of residential, commercial, retail and industrial areas). These urban forms demanded high levels of household mobility and the use of private motor vehicles.

Because of urbanisation pressures in many regions, the high cost of infrastructure provision and the inefficiencies of these low-density uses, alternative new subdivisional

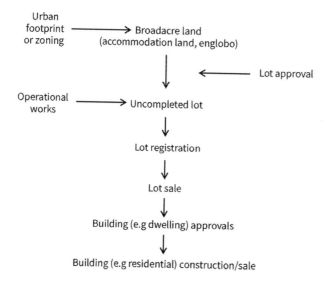

Figure 12.3 Typical land development process (basic steps only).

models have rapidly emerged. These include Master Planned Communities (MPCs), Transit Oriented Developments (TODs), urban villages, in-fill densification projects and so on. At this stage, all of these represent a relatively small proportion of the market. However, given that these models are, on the face of it, more economically and environmentally sustainable and can better address significant demographic change and rapidly increasing energy and transport costs, it is anticipated they will become more common (Ehrenhalt 2013; Kelly & Donegan 2015). These changes are further discussed in Chapter 4.

Another important subdivision trend in most parts of Australia is the increase in the scale of development and subdivision projects. While there are still smaller scale, in-fill densification projects, the bulk of subdivision development is now carried out by large corporations–often working under master plans for entire suburbs or communities with a time frame from five to ten years.

This trend reflects the high–risk and large capital expenditures required, particularly in upfront infrastructure provision, as well as the ability for larger developments to reach the necessary economies of scale to implement a more holistic approach to issues such as community, social networks, sustainability and sound environmental practice. It could be argued also that these developments generally provide wider options for housing types and densities. Additionally, they can facilitate better integration of public housing and other facilities and hold improved potential to create commercial and other uses within their precincts. This can include the provision of employment nodes/clusters, thus avoiding the need for long commutes. The physical development and the subsequent evolution of community may take a decade or more to mature. Therefore, it remains open as to whether these new, large-scale forms of development will produce the sustainable integrated urban form their proponents envisaged (Wardner 2013).

The move in many regions to open large englobo land areas for development now limits opportunities for small-scale ventures. Lots produced must meet a particular standard with provision in infrastructure mains for power, water, telecommunications, roads to the site, and stormwater drainage and sewerage. This may well require new mains and conduits, substations, pumping stations, treatment works and a range of other capital works that must be provided before the first block is available for sale. This, in turn, may demand the expenditure of some millions of dollars in physical works or in work contributions to the local authority.

Such payments can take several forms including specific payments for carrying out the works immediately, the provision of bonds/guarantees for their future payment of upcoming works or, finally contributions retrospectively to works that have already been undertaken by council in anticipation of the development. In all of these cases, the very large upfront capital costs can be accommodated only if they are amortised over the large number of blocks being subdivided. Under these parameters, small-scale subdivisions present an increasingly marginal proposition.

As with all development projects, cash flow is critical and the initial cost of acquisition and early headwork contributions weigh heavily on the entire venture. Unlike single building development projects, however, subdivisions have the ability to be staged and progressively brought to market at appropriate times. For example, once a commitment is made to, say, a multi-level commercial building, financial expenditures must continue and no return will be secured until the entire building reaches practical completion.

In the case of the subdivision, however, the developer will need to finance the initial land acquisition, the cost of headworks, holding charges and the initial development of, say, 20 to 30 lots. Thereafter, the proceeds of the sale of lots in early stages can help fund the subdivision of subsequent stages. Furthermore, those stages can be brought forward in time frames that optimise market opportunities. These opportunities and options clearly lower the project's overall risk profile.

It needs to be appreciated that developments, even when staged and pre-approved, will be subject to significant 'supply lags', that is, the physical design, approval and construction of these works inevitably take a considerable amount of time to complete. These time delays ('lags') between recognition of a market opportunity and the delivery of physical product, be it in the form of subdivided allotments or a complete building, add to the inherent risk of the development sector. A key role of the developer and project management is to take all reasonable steps to reduce time lags and, therefore, reduce exposure to further systemic risk.

A contemporary example of this lies in the online access to a vast array of information available.

12.6 Particular considerations – (building) development projects

This type of project involves the securing of a divided parcel of land and the construction ('development') of a new building. Alternatively, the project may involve the acquisition or ownership of land that includes an existing building and, thereafter, its refurbishment, upgrade or extension. Examples include the development or refurbishment of an office building, industrial premises, hotel, tourist facility, retail centre, apartment building or a range of others. All of the general comments noted in Sections 12.3 and 12.4 apply here. In regard to building development, some further observations need to be made.

In most cases, the development sequence, assuming successful financial analysis and internal approvals, will involve:

- the securing of a site, if not already owned by the developer
- the establishment of a functional brief (i.e. what the asset is meant to achieve or the definition / specification of demands to be met)
- the development and acceptance of a design brief (i.e. what the asset will look like and the components to be included)
- the securing through the local authority of a development approval (DA) and, progressively, other necessary approvals
- the securing of a building approval (BA)–in some jurisdictions, through an external consultant called a 'building certifier' who has a working knowledge of all of the necessary requirements of the relevant building codes and aligning with the necessary statutory approvals for structure, services and finishes
- the letting of tenders and actual construction work through to practical completion (i.e. the handover of the finished asset back to the owner). As part of this process, a point will be reached where the design and delivery process are agreed and these become the basis for further detailed contracts
- the commissioning of the building to operational standard
- the leasing, use and (where required) on-sale of the building to a final investor.

As part of the risk management strategy for a particular project, a level of pre-commitment (i.e. advanced contracts to future tenants or purchasers) to the final product will probably be required before a decision to proceed with construction is made. This may, for example, take the form of a percentage of lease pre-commitments of a proposed office building or a number of units in a proposed multi-unit residential development sold 'off the plan', that is, contracted for sale when the development is complete and a separate (community) title is available. Such pre-leasing/pre sales may be a requirement for securing debt funding. In some cases too, conditional contracts to a third-party investor may be in place for the on-sale of the entire project, once completed.

Depending on the scale of the project, time frames could extend to three years or more. These time frames again emphasise the level of risk exposure potentially involved and also, as noted above, the need for careful pre-planning, research and sophisticated project management to reduce time exposures to an absolute minimum. Both short-term development finance and longer-term investment finance to buy out the final product will almost certainly be involved (Jowsey 2011).

Regarding site acquisition, a 'call option' in favour of the developer represents a way in which those major costs can be deferred until approvals are in place or a higher level of certainty is secured. Again, while this is a legitimate and familiar technique, it may come at a cost. The existing owner will almost invariably require a higher price or other conditions to compensate for the delay in settlement and for the risks of tainting that property if this development approval is not finally forthcoming. Furthermore, even if an option is secured, it will eventually 'sunset' (i.e. come to an end). Projects can sometimes be lost because negotiations for approvals, headworks and the like become protracted and are not finalised by the end of the option period.

Investigations into these projects will normally involve some form of feasibility study. A feasibility study is the notional progressing through the development project and building a time and cost schedule to prove or disprove project viability. While this represents a valuable decision-making tool, it must be recognised that there will always be a number of systemic and non-systemic risks that, at best, can only be estimated. The usefulness of these studies will depend on the quality of the research and, particularly, the assumptions made as to acquisition and construction costs, time, acceptable profit, risk, financing and other charges, together with the estimated final capital or rental value of the finished product.

A reasonably conservative approach needs to be adopted and suitable contingencies provided, particularly in unusual or unfamiliar projects. Contingencies are additional sums, normally included as a percentage, set aside in the event of encountering unforeseen issues in design or construction. Some projects are more susceptible to cost variations than others. These include instances where foundation quality is uncertain, where wet weather is likely to be encountered at critical times or the refurbishment of old or heritage buildings is involved. Novel or unfamiliar design may represent another case where contingency provisions may have to be increased. Practically all projects will provide a small allowance for contingencies, perhaps 4 to 6 per cent – but the rate considered depends on the circumstances of that case.

Overall too, it should be noted that contingency allowances that are too high may well reflect an overly conservative approach being taken, with projects thereby assessed as unviable. An overly high contingency allowance may well indicate that additional research and investigation is required. Furthermore, it needs to be remembered that, when completed, the final product must be market competitive and a balance struck between optimising market position while at the same time avoiding unnecessary risk. Simply inflating costs and contingencies will result in a non-competitive project.

Feasibility study analysis can be undertaken by going either forward or in reverse. Going forward is when a hypothetical development notionally commences with land price and adds estimated construction and other costs to arrive, with profit and risk added, at a final capital cost of the completed project. This final figure and risk profile can then be compared with the estimated market value of the completed project at the anticipated date of completion to determine if the project appears viable, given the risk profile involved and compared with other opportunities.

Alternatively, this analysis can be undertaken in reverse, where the future market or capitalised rental value at project completion is estimated and, from that, the profit, risk and all development and other costs are notionally deducted to arrive at a residual land value that could be paid and still result in a financially viable project. These matters are further discussed in Chapter 10.

To a large extent, development will always follow a sequential process. The land needs to be secured, approvals acquired, construction and finally occupation, leasing and/or sale. Time, however, is of the essence and in many cases, it is possible to overlap these activities at least to some extent. There are numerous examples of time-saving opportunities. These include the commencement of functional and design briefs and the engagement of professional consultants as early as possible, commencing preliminary discussions with the local authority early in the project, and ensuring that marketing programmes are underway well before completion of the construction phase.

Notes

1 Optimum scale implies a very close knowledge of the market and market trends. As Malkiel and Ellis (2013) observe, no one has better knowledge of a product than the market for that product, be it property or anything else, and the close analysis and reading of that represents the closest approximation to truth that can be found in any venture.
2 These observations are in fact simply the practical application of economic marginal analysis theory discussed in Chapter 2.

Chapter 13

Property management and facilities management

13.1 Introduction

Chapter 1 recognised a number of key characteristics of property assets, highlighting the complex and multi-faceted nature of its development, function and value.

Like any commercial undertaking, success is measured in the competitiveness and marketability (i.e. value) of that investment over time. Specifically for real property, this is reflected in the ability to generate satisfactory levels of net income and capital growth, considered against the level of risk involved and the ability to manage that risk. Securing those outcomes involves a wide range of activities and skills – from the owner and portfolio manager through to those involved in property and facilities management and service providers.

While this indicates some sort of hierarchy or ranking in importance, it needs to be recognised that, for complex property assets, each of these disciplines and activities needs to be performing efficiently if all the interests of stakeholders – owners, financiers, tenants, occupants, managers and the wider community – are to be satisfied.

Pivotal in that success are those involved in property management and facilities management. They provide the critical link between the owners, their corporate objectives and the asset itself in its day-to-day ability to earn income and operate effectively and efficiently.

This chapter describes this interface and the manner in which these duties are performed in a contemporary environment. It also particularly considers the management issues for various real property asset types.

13.2 Initial observations

Property management and facilities management roles were defined in general terms in Chapter 5 as follows:

- Property management is the 'translation' of the required objectives set by the asset owner through to the day-to-day operations and functioning of the building. Typically, it will include, in whole or in part, such aspects as leasing and tenancy management, budgetary cost management, operational management, contracting out and reporting, under the legal concepts of agency, to the owner as principal.
- Facilities management refers to strategy, technical and operational matters and activities aimed at securing the optimum performance from a particular building

asset. It will include, in whole or in part, such matters as services, layout, workplace planning, optimisation of expenditures, environmental health and so on, and the overall creation of effective and efficient workplaces and assets.

Care needs to be taken to avoid becoming overly definitive or involved in the semantics of the various definitions and terms used, and there are, in practice, obvious overlaps in some activities.

The definitions would correctly imply that a property manager acts as the agent for the owner, and also undertakes professional 'duty of care' responsibilities to other stake-holders, particularly tenants and occupants (see Section 13.3). The property manager's first area of responsibility is tenancy management: that is, securing income through tenancy dealings, lease management and so forth.

The second fundamental role is to manage costs and operations, in particular, budg-eting, financial operations, managing contracts and overseeing the work of other op-erational staff and service providers. Third, they must provide professional reports and advice back to their principal, the owner of the asset.

From the general definitions above, it will be seen that a facilities manager's role, at the risk of over-simplification, relates more to physical issues pertaining to the build-ing and its occupants. These include the operational efficiency and effectiveness of the building in use; its physical components; and how it is to be operated and used in the short and longer term to maximise its functional performance for tenants, occupants and owners alike.

The activities of these managers are, perhaps, not given the emphasis and profile they deserve. Certain, high-profile events – original acquisition/construction, major refurbish-ments or final sale – always appear as the dominant life cycle of any real property asset. However, in reality, it is the long-term management and operation of the 'asset-in-use' – the domain of property and facilities managers – that represents the real functional value and, therefore, income potential of the property. These long-term, lower profile components of the life cycle of a building actually reflect more of what the building aims to be and achieve, rather than simply its creation, design or image (Edwards & Ellison 2003).

Before further consideration of the roles and responsibilities of property managers and facilities managers, it is opportune to recognise some of the legal and other param-eters that underpin these relationships.

Some of those involved in these relationships will be the direct ('in-house' or 'day labour') employees of the asset owner. In managing complex property assets however, it may not be physically, nor sometimes legally, possible for the owner or the owner's direct employees to undertake the full range of activities required. Consequently, many skills and services will have to be secured from other external organisations.

Some will be quite specific tasks, for example, to retain a marketing agent to sell an asset; engage a builder to carry out certain building improvements; or retain a con-tractor to paint a building and so on. However, some of the services required will be much less specific and long term: retaining a company to provide property management services over, say, three or five years would be an example. In all of these cases, the arrangement represents a contract with all of the legal ramifications that such arrange-ments apply. These matters were summarised in Chapter 7.

In property dealings, the specific word 'contract' is sometimes used, such as a 'service contract' to maintain a particular property, or even a 'building contract'. At times, other words are used, such as 'management agreement or licence'. Whatever terms are used, they are nevertheless, contracts, normally with specific parameters and other statutory obligations particular to dealings in property.

While the works involved will vary case-to-case, uniformity, predictability, consistency and certainty in such agreements and documentation are fundamental to any well-managed and high-performing property asset or portfolio. To that end, it is important to use standard and consistent documents in leases and service contracts and, as much as possible, use recognised industry standards, specifications, systems, procedures, audits and reporting protocols throughout. Fortunately, because these activities are frequently encountered by all property owners and managers, there is a range of standard forms for contracts and agreements, as well as templates for audits, assessments, reports and so on. Some of these are available through various government agencies or industry associations such as the Property Council of Australia and professional groups such as state law societies. These 'standard' documents have been in use over a long period of time. They will, in many cases, be based on widely accepted technical and performance standards, the majority emanating from Standards Australia, the national peak body in these matters.

Additionally, there is a range of generic and specialist software packages and management systems that are commercially available for use. Given the diverse nature of property assets, it may be necessary to make certain adjustments to accommodate the idiosyncrasies of particular portfolios or property assets. The key point however is that those standards and industry-accepted documents, specifications, legal agreements, templates and the like should be a starting point in the development of any management systems and varied only for sound and considered reasons. To attempt to develop these complex documents from first principles would be time-consuming, expensive and represents unnecessary legal and operational risk.

For some decades, the property and development sectors have proven themselves to be early adopters of both new technologies, in building components and building monitoring and service controls, and systems and software for asset, facilities and tenancy management.

Sometimes, this is based on the need for regulatory compliance – in areas such as fire services, alarms and communications, environmental and indoor air-quality (IAQ) monitoring and a range of others. Often, it was the large capital cost of assets and their components, together with the volume and complexity of operational and income cash flows, that encouraged investment in innovations and systems development that enhanced performance and competitiveness.

Finally, commercial asset owners, their managers and professional advisors, consultants and contractors are typically large, long-term businesses willing to invest in innovation in ways not always available to smaller-scale corporations.

In design and construction, computer-assisted design (CAD) systems had developed over some decades to very advanced and sophisticated levels. As regards systems and management across construction, building services assets and facilities management, a comprehensive computer-based process and information base called 'Building Information Modelling' (BIM) is now widely applied. First conceived in the 1970s, but subject to remarkable development and advances to the present day, BIM effectively manages the physical and functional components and characteristics from the initial construction/development project through to its continued use as a live asset.

The ability to represent these components graphically and with supporting data on a single, reliable integrated database is of immense value through both construction and operational stages. As discussed in Chapters 1 and 5, the property and development sector has systemic challenges in the coordination of the various disciplines and activities involved in solving complex problems. Typically, those problems need to be addressed within the confines of a construction site or building and, often, under very tight budgetary and time constraints. In this environment, incremental and sometimes major changes to design and plans need to be made to accommodate emerging problems. They will almost invariably affect other parts of the building and its operations. In the past, these alternations were often made on site during construction and not well recorded in the 'as built' drawings, resulting in all manner of later maintenance and refurbishment challenges and costs.

As well as issues of currency and completeness of data, one inherent problem with conventional two-dimensional plans and elevations was the difficulty in visualising all spatial concepts and implications of the design, and design changes, together with the successful integration of various component parts. There were also common problems in translating ideas, changes and instructions between the various disciplines and individuals involved – particularly for those unfamiliar with the interpretation of relevant plans and documents.

BIM, particularly in its advanced forms, addressed many of these issues including, for example, visualisations of the building/project in three dimensions, transparently showing service lines, layouts, structural components, aspects and perspective views, et cetera. Of great value is the ability of these systems, as they parallel financial modelling (discussed in Chapter 10) to pose, through the design and construction stages, 'what if' questions, that is, say, to provide a quick and accurate investigation of options to address particular problems, opportunities or possible innovations.

Overall, BIM systems provide a level of cohesion across the broad range of design, construction and management tasks that surround a particular asset, with costing, financial, procurement and scheduling modules able to support the delivery and ongoing operations of the asset in its entirety.

13.3 Agency and responsibility

Each task delegated or distributed to others will clearly carry with it different types and levels of responsibility. Some will be fairly straightforward, such as retaining a trade contractor to carry out very specific work. Others will be more extensive and sometimes less able to be closely specified; some of these will, in effect, ask a third party to act as the representative of the owner in particular cases. This concept is known as 'agency'. In either case, however, there will be an established obligation to undertake the works prescribed; but, as well as those direct and primary responsibilities to the principal under the contract (normally the owner), there will be what is known as a 'duty of care' to others. Property tenants, occupants and people visiting the building can reasonably expect that the building will provide what is required of that type of asset – be it residential, retail, commercial or other – in a safe and functional way and to an acceptable standard.

Despite the fact that property and facilities managers do not have a direct or formal contract with many of those parties, there is still a serious obligation ('duty of care') that is being increasingly reinforced by specific legislation. The concept of agency needs some further explanation, particularly in the case of property managers. A certain level of delegation of responsibilities from the owner to the property manager will be inevitable to allow for the day-to-day coordinated and efficient running of the asset.

The concept of 'agency' is well recognised in law as well as in commercial activities and across the wider community. Agency effectively means 'to act on behalf of another person' or to 'represent that other person in terms of a pre-existing agreement or delegation'. Asset managers, solicitors, accountants, tax agents and real estate agents are common examples of this arrangement. Trust arrangements, that is the establishment of a power of attorney by one person to look after certain affairs of another, provide another common example as does the role of a 'Clerk of Works' as the owner's representative on a development or building site.

The acceptance of the role of 'agent' comes with a range of contractual and legal obligations. Some are explicit in the formal agreement setting up the agency. However, some obligations of an agent, such a property manager, are also implicit; that is, they exist, whether specifically mentioned in the agency agreement or not, and created by the very establishment of that agency arrangement (Duncan 2006, 2008).

The effective operations of most business and legal dealings in a country like Australia depends heavily on the effective operations of agency, trust, power of attorney and similar arrangements, and the responsibilities they create cannot be taken lightly. Courts typically take a very stern view of any illegal action in this regard, identified as a 'breach of trust' (Scarrett 1995).

The legal responsibilities of an agent such as a property manager are complex, but, in summary, they include:

- a duty to perform specified actions in a timely manner
- to undertake those duties with the professional skill required, and with a duty of care to all those directly and indirectly involved
- to offer correct professional advice to the best of the agent's knowledge, whether that advice is asked for or not
- not to delegate authority without the principal's express prior permission
- to act, in the interest of the principal, in good faith
- a duty not to act on his/her own behalf or on behalf of his nominee except with prior exposure to and agreement from the principal
- a duty not to take secret commissions or otherwise profit from the arrangement
- a duty of care to keep the principal's business confidential
- a duty of care to maintain accounts and keep the principal informed at appropriate intervals.

The establishment of an agency agreement, such as a property management agreement, should be in writing. Again, industry pro forma and standards should be the basis of these agreements. Typically, the agency (management) contract will include (Duncan 2008):

- nominations of principal and agent, such as a property manager
- nomination of the personnel and representatives actually involved in the project/activity
- identification of any particular qualifications the agent/their employees must possess; again, these may well also be required by statute for certain activities

- definition of key words used
- identification of the property to which the agreement refers
- duration of agreement and options and option terms (if applicable)
- specific identification of duties, including downstream levels of responsibility
- provisions for securing services (e.g. for service contracts, how quotations are to be called and tenders managed – if that is involved) and nomination of any preferred contractors
- access to, and use and return of records
- payment, including bonuses and penalties
- level of financial delegation and involvement in decision-making and recommendations
- special emergency provisions
- confidentiality
- reporting and provision of information and recommendations
- insurances (if any) that the agent is required to carry
- administrative processes
- penalties
- termination of agreement
- provisions for variations
- dispute resolution procedures.

While the above generic clauses may be commonly found in such agreements, it is the specific and exact requirements of the particular agreement that must be adhered to by both parties. Whatever the good intentions, it is legally and financially dangerous for the property manager to undertake wider duties than those sanctioned under the management/agency agreement without the prior consent of the principal (i.e. owner).

13.4 The role and activities of a property manager

A general definition of the role of a property manager and of agency arrangements was provided in Sections 13.2 and 13.3. Both are relevant to the further details provided below.

Because agency arrangements are set up bilaterally between the owner and the property manager, the scope, range of activities and delegation levels can be established in any variation or combination that the two parties agree on, providing, of course, the agreement is within all statutory requirements.

Further, the actual activities required and undertaken will vary depending on the nature of the property/portfolio involved and remuneration to the manager should reflect that level of involvement and responsibility. Duties such as tenancy management, collection of rents, control of outgoings and the provision of building services may be the same for a residential property or commercial portfolio, yet the depth, complexity and cost of providing property services to those different types of properties will vary considerably.

In any portfolio, however, the successful property manager will need to be multiskilled. Foremost, as the title implies, the person must be a good manager: well organised; with a professional, steady and consistent approach; and with sound and uniform management systems in place. Because of the importance of controlling income (rents) and outgoings, a good working knowledge of accountancy and IT-based financial systems and reporting is important, as is knowledge of legal matters pertaining to property and building, contracts, risk management and risk mitigation. Additionally, a sound understanding of the property market, building practices and trades, the ability to read plans and a general understanding of how buildings and building services operate are all valuable attributes that the property manager will need to draw upon in carrying out the required duties (Kyle 2004).

Clearly no person could claim to be an expert in all of these fields, but, given that the role of the property manager is to coordinate both activities and people, it is essential at least to 'talk the language', comprehend issues and opportunities and provide sound advice to the owner on how they might be addressed.

The property manager also needs to be aware of his/her knowledge limitations and to be willing to engage specialist help and advice in dealing with complex matters. Of particular note here are detailed contractual and legal matters, building and services opinions and advice and workplace health and safety issues. Some of the most serious failings that emerge in property management activities relate to decisions based on insufficient knowledge, investigations or competencies.

Finally, property management is about people. Many groups and individuals will have an interest in or requirements for a particular building – as an owner, tenant, manager, service contractor or whatever. While some of those interests and requirements will be shared or complimentary, others may well be competing and sometimes in conflict. Therefore, excellent interpersonal skills are perhaps the most important of all characteristics of a successful property manager. This should be based on a knowledgeable, fair and professional approach. It also implies dealing with all stakeholders in a consistent, businesslike, non-emotive, yet friendly manner, recognising the prime responsibility to the principal/owner, but respecting the rights of all other stakeholders.

This is not to suggest that an overly friendly or engaging manner is required. Friendships and personal relationships will often develop, but these need to be earned by both parties and evolved over time. Particularly with tenants and service providers, the property manager is well advised to maintain some professional distance, at least in the first instance. The nature of property use and tenancies is such that there will almost invariably be occasional disagreements. Experience shows it is often difficult to remedy these in a satisfactory and businesslike way if a too casual or relaxed relationship has already developed.

Property management is a mixture of work in the office and fieldwork – inspecting buildings, meeting with owners, tenants, service providers and so on. The property manager needs to be in regular contact with the stakeholders, as well as conducting periodic, physical inspections of the building. It is in those settings that issues and problems can be identified and, hopefully, addressed easily, and where opportunities can be noticed and brought forward. This implies that the property manager needs to spend significant time out with the asset, tenants, service providers and other stakeholders. This approach is sometimes referred to as 'management by walking around'.

Particularly in the case of more complex properties, some of the specific activities of a property manager can include:

Databases, systems, documentation and certifications

Depending on the agreement with the owner, the property manager may be responsible for keeping and maintaining a number of the 'active' databases, records and the like for the building. Under typical agency arrangements, these will remain the property of the owner and be used by the property manager in an appropriate and confidential manner. These may include the following. (Note that, for completeness, the list below is comprehensive. In most cases and depending on the duties required of the property manager, the records held by that person would be substantially reduced.):

- copies of leases and lease schedules, including contact details of tenants
- building and equipment certifications and 'as built' drawings
- details of rates and service charges
- building operations, service protocols and procedures
- budgets and spreadsheets – for both income and expenditure – see Section 13.6
- maintenance details and systems, including the names and contact details of contractors
- asset registers
- a comprehensive diary system, identifying key events and deadlines, with pre-warnings
- details of service agreements and their reviews
- insurance and risk-management programmes or at least copies thereof
- audits, reviews and testing, including fire services and record of statutory compliance confirmations
- copies of leases and lease schedules, including contact details of tenants
- building and equipment certifications and 'as built' drawings
- details of rates and service charges
- building operations, service protocols and procedures
- budgets and spreadsheets – for both income and expenditure – see Section 13.6
- maintenance details and systems, including the names and contact details of contractors
- asset registers
- a comprehensive diary system, identifying key events and deadlines, with pre-warnings
- details of service agreements and their reviews
- insurance and risk-management programmes or at least copies thereof
- audits, reviews and testing, including fire services and record of statutory compliance confirmations.

Tenancy management

Again, depending on the particular property management agreement, these activities may include:

- lease establishment
- providing tenant liaison and information
- budget preparation
- rent invoicing and recovery of debts
- enforcement of lease conditions
- ensuring owners' obligations are met under leases and agreements
- dealing with tenant complaints
- managing rent reviews
- dealing with options and similar
- tenant coordination for joint activities – depending on the property type.

Outgoings/cost control

These activities include establishing a budget for outgoings and the programming, managing and monitoring of those activities; for example, rates and service charges, insurances, maintenance costs – both programmed and emergent – management fees, cleaning, security, waste disposal, and the establishment and monitoring of service contracts.

Property and market information

Providing the owner with agreed property and market information that may include:

- rental levels and market parameters
- actions and likely action of competitors
- movements in the relevant property market
- possible threats, expectations and opportunities.

Building and technical abilities

Ability within reason and as necessary, including other advisors, to:

- instruct, verify and assess issues and problems and supervise service work and contracts
- establish and manage service agreements
- recognise physical and service issues and emerging difficulties
- control and manage any capital funds for upgrades.

Building services

Either directly, through a facilities manager or other advisor, to arrange and manage:

- the opening and closing of the building
- security
- supporting efficiency and sustainability progress
- cleaning (internal and external)
- grounds maintenance
- building and service maintenance
- waste management
- dealing with hazards and hazardous material
- emergency equipment procedures and emergency procedures and testing
- overall adherence to statutory controls and obligations.

Providing adequate reports

Typically, reports will include timely reporting against budget and agreed programmes emphasising financial, tenancy and operational issues, based on established key performance indicators (KPI's) and highlighting exceptions or unusual variations.

Reports should also include matters that, in the professional opinion of the property manager, need to be conveyed, with appropriate recommendations, for the consideration of the owner. These may include, among other things, emerging issues, important market knowledge or upcoming opportunities or threats. These reports will be provided in accordance with the agreement between the owner and the property manager but are normally on a monthly basis. The obligation on the agent/property manager to report and advise, whether requested to or not, is a related consideration here.

13.5 Developing a property asset management plan and system

As previously noted, the property sector is made up of a large variety of sub-sectors and types of properties. These range from quite simple owner-occupied residences or apartments, through to commercial properties, regional shopping centres, factories and so on, some of which are worth hundreds of millions of dollars and are extremely complex to manage and operate effectively.

Particularly for larger, income-producing properties, some form of overall property-management asset plan and system is required. While the basic components of this plan or system may be fairly common to all cases, the unique characteristics of each property requires that plans are developed on a property-by-property basis, accommodating the particular physical, legal and financial features and idiosyncrasies of that particular asset.

The establishment of a comprehensive and effective plan for, say, a large commercial property asset, is a significant task. It is absolutely essential that this be undertaken prior to embarking on the management activities themselves. Without a plan or strategy, there is little chance that the corporate objectives for the asset will be met. Planning is a characteristic of all leading asset and property managers. Fortunately, once a sound initial plan for an asset is put in place, the work to adjust and upgrade it for subsequent financial years is much simpler, with a number of strategic components unlikely to change significantly year-to-year.

Typically, management plans will be developed during March to May of each year so that the new plan is in place and operational for the start of the next financial year in July. (Note however that some organisations operate on a calendar year cycle, in which case, the plans would be developed commencing in September/October each year.)

Like any planning document, this will only be as valuable as the quality and completeness of the data that underpins it. Consequently, the plan must draw together all of the various and diverse components of knowledge that can be secured on the individual property. Therefore they will involve practically all with responsibility for the various components of the building and its operation and will typically be prepared by the asset owner or manager. The property manager will also be deeply involved.

Figure 13.1 identifies the type of information and overall structure that will be required to produce a robust plan. Further comments on each of these components are as follows:

Corporate approach philosophy

This section specifically identifies the demands of the owner. Typically, these will include required return on investment (hurdle rate), risk-management strategies and statutory compliance. They may also include some more specific KPIs or, particularly in

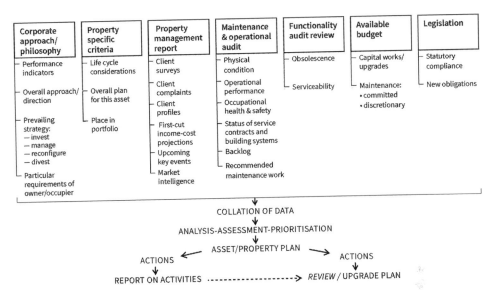

Corporate approach/ philosophy	Property specific criteria	Property management report	Maintenance & operational audit	Functionality audit review	Available budget	Legislation
– Performance indicators – Overall approach/ direction – Prevailing strategy: — invest — manage — reconfigure — divest – Particular requirements of owner/occupier	– Life cycle considerations – Overall plan for this asset – Place in portfolio	– Client surveys – Client complaints – Client profiles – First-cut income-cost projections – Upcoming key events – Market intelligence	– Physical condition – Operational performance – Occupational health & safety – Status of service contracts and building systems – Backlog – Recommended maintenance work	– Obsolescence – Serviceability	– Capital works/ upgrades – Maintenance: • committed • discretionary	– Statutory compliance – New obligations

COLLATION OF DATA

ANALYSIS–ASSESSMENT–PRIORITISATION

ACTIONS ← ASSET/PROPERTY PLAN → ACTIONS

REPORT ON ACTIVITIES - - - - - - - - - - - - - - - - - - -> REVIEW / UPGRADE PLAN

Figure 13.1 Structure of a property asset management plan (for major property).

the case of owner-occupiers, special requirements or conditions that the owner may require the property to deliver.

There are other matters that need to be considered. These include an appreciation of the overall approach and direction of the particular corporation as regards the owner-ship of property, that is, whether they see themselves as long-term owners or investors, or wish to add value, but not hold assets over a long period of time. Furthermore, the asset and property manager would need to be cognisant of the shorter-to-medium term strategies of the owner. These could include whether they are likely to have significant financial reserves to upgrade property; if they simply want to hold it, or value-add to it; or perhaps, because of other issues, they are actually in the process of divesting property assets. All of these issues will make a significant difference to the recommendations on how that particular property is managed into the future.

Finally, in developing such strategies, the plan must be cognisant of any particular service delivery strategies that the owner may require in holding the asset. For example, the approach to asset and property management and maintenance by, say, a state government education department, which has hundreds of assets scattered across a region or state and is required to provide a platform for a particular service – in this case, the education of students over the very long term. Such an organisation will have a different philosophy regarding construction and maintenance from that of, say, an investor in an office building leased out to private-sector tenants.

Property-specific criteria

Once the overall corporate approach and direction is established, the particular property for which the plan is being prepared needs to be further considered. In general terms, it needs to be established how this asset aligns with the overall corporate philosophy discussed above. Second, given that corporate owners will often have a large portfolio of individual properties, this asset needs to be considered, in a financial and risk-management sense, as to how it fits and balances with the rest of the portfolio.

For example, the net income from one secure asset may be required to underpin other, higher risk properties that currently do not have sufficient cash flow. In this case, a significant change in cash flows in the upcoming year may prove quite risky to the entire portfolio. If so, wider issues need to be considered in asset planning.

Finally, the life cycle of the particular asset needs to be considered. Every building goes through different life phases – from when the building is new and requires limited maintenance or capital investment, through to later stages when maintenance costs typically rise, become less predictable, and expensive refurbishments and upgrades may become necessary. At this point in developing the plan, it is important to consider where this property lies in this 'asset journey'. Perhaps astute reinvestment in the asset may extend its effective life and defer additional large expenditures.

Property management report

As part of the development of the asset plan, the advice of the property manager is essential. The property manager has day-to-day contact with tenants, contractors, the building and the wider market – all critical components if functionality, and therefore, rental levels, are to be maintained. The development of the management report considers feedback on how current tenants view the property and any common areas of complaint. Importantly, the property manager prepares income and operational cost estimates for the upcoming year, and these form the basis of some of the main performance criteria.

Functional and audit review

This relates to a wider consideration of the asset and its future by attempting to assess the functionality of the building and its serviceability and the avoidance of obsolescence, compared with other, perhaps newer, competitors in the market. Third-party professional consultants are often used to supply this information.

Available budget

In any enterprise, there are budget limitations. Consequently, there is little purpose in developing elaborate and sophisticated plans for expenditure if that budget is simply not available. The scheduling of capital works and any upgrades also need to be cognisant of these realities.

It should be noted that service and maintenance contracts typically extend from three to five years. Therefore, much of the maintenance and operational budget year-to-year will be pre-committed and not available for discretionary expenditure in any case. Maintenance expenditures are always under pressure; however, care needs to be taken that sufficient well targeted maintenance funding, particularly to meet statutory obligations, is provided. Without necessary expenditures on day-to-day 'fine tuning' maintenance and adjustments, significant asset failures can occur, often within a surprisingly short period of time (Brand 1994).

Maintenance and operational audits

Property assets are physical and in continuous use. They deteriorate over time and their material condition, together with performance and occupational health and safety requirements, needs to be reviewed on a regular basis, preferably by an independent assessor. That person could also review operational performance as compared with industry standards and benchmarks, as well as comment on the suitability and effectiveness of the building services undertaken by day labour and contract staff, considering both costs and effectiveness.

The maintenance of adequate management, administrative records, financial systems and compliance matters should also be reviewed.

Almost invariably, any major asset will have backlog maintenance: works that could be done if funds were available but are currently deferred. That is understandable as financial choices have to be made; however, backlog schedules need to be regularly monitored to ensure that these deferrals do not evolve into unforeseen, major adverse effects.

Those providing this information should also present a schedule of recommended maintenance and upgrade recommendations for the forthcoming year.

Legislation

The complexity of property assets is reflected in the wide range of statutory and other obligations placed upon those who own or manage/operate those assets. An annual review needs to be made to ensure both that existing legislation continues to be complied with – as regards audits, inspections, testing, maintenance regimes and so on – and that any new obligations that have recently emerged are being adequately addressed.

All that information is collated into an asset and property management report for the owner/asset manager, where it is analysed, assessed and prioritised and condensed into a logical form.

It is will almost certain that expenditures, upgrades and operational costs recommended for the upcoming year will substantially exceed available budget. Choices will need to be made. The input reports provided in the development of the plan, particularly those from property management and maintenance and operations audits, will sometimes conflict. It is not a matter of addressing all needs – that is probably impossible – but rather to optimise available funds to meet high-priority demands.

Typically, prioritisation of expenditures will start with statutory compliance and addressing workplace health and safety issues. It needs to be noted however, that no building will ever be 'perfectly safe' and, while obligations to create a safe environment are of the highest order, these matters are best addressed by additional expenditures and adopting good risk-management techniques (see also Environmental Health components in Chapter 15, Section 15.2).

Once statutory and safety issues are suitably accommodated, the prioritisation of works undertaken, and those deferred, need to be considered having regard to the KPIs established for that asset. Typically, the income stream (from rent) needs to be protected and enhanced over time. Consequently, and as a general observation only, priority expenditures should be directed to those areas that address matters of recurrent complaints from tenants or those directly affecting the immediate comfort and serviceability of the asset for tenants and occupants.

As a result of all of that, the property asset and management plan are put into place. It will normally have two linked components (see to Figure 13.1). First, necessary actions and activities are identified, scheduled, prioritised and allocated. Second, the income and operational budget for the asset are established for the upcoming financial year. Those actions will then be subject to ongoing monitoring and reporting, normally on a monthly basis, with cash flows into and out of the asset, as a cost centre, subject to similar review.

With experience, and based on the asset's history, the property manager will normally be able to establish very accurate forward budgets. Nevertheless, in the course of a year, unforeseen expenses will almost always emerge. Emergency maintenance issues can frequently occur, particularly in older buildings, together with unexpected vacancies, and so forth. Consequently, the asset and property management plan for any building needs to be seen as a 'live document', adjusted to meet changing circumstances throughout the financial year. The inclusion of a contingency fund in original budgeting is also good practice, the quantum of which being dependent on the risk profile of that particular property.

13.6 Budget preparation and management

The securing of incoming cash flow and the effective and efficient management of outgoings are critical roles for the asset and property managers in ensuring the performance of the asset overall. The use of reliable systems and standards, including good databases and financial software, is essential. In financial matters, this commences with the development of budgets for both income and anticipated outgoings that provide predictions for the year to come and establish important financial performance benchmarks.

As noted above, management of most property and portfolios is based on a financial year cycle and, under best practice, an asset and property management plan should be assembled in the preceding March to May. An important part of this plan is the preparation of a budget projecting both income and expenditure for the upcoming financial year. A highly simplified income budget is shown in Table 13.1 and presumes a multi-tenanted building where some proportion of outgoings becomes the responsibility of the tenants.

First, the rental at the start of the period, as established in the lease and with realistic estimates of the quantum and timing of any rental increases for the year, is inserted for each tenant. Thereafter, the outgoings for each type of annual expenditure are estimated for the entire building. As relevant, goods and services taxes will need to be included in both rental and outgoing figures (Table 13.1). (Taxations considerations are further discussed in Chapter 14.)

From the total outgoings, allocations are then made in accordance with the obligations on each tenant under the provisions of their respective leases. Any difference between the total amount due and the aggregated amounts for which the tenants are liable for that outgoing becomes the residual responsibility of the owner. It needs to be noted too that some leases are established on a 'gross' basis, meaning that the owner is, in fact, responsible for all outgoings.

For each tenant then, the base-rent responsibilities plus any responsibilities for outgoings are added together and divided by the number of rent periods (e.g. months) and this becomes the total rent invoiced to each tenant.

In a large commercial asset, such as a regional shopping centre, there may be some hundreds of leases to include and adjust. The task is complex and must be done accurately, again reinforcing the need for standardisation and uniformity of leases across the centre.

Table 13.1 Draft income budget (rental and outgoings) – simplified structure

Type of outgoings	Total payable[a]	Tenant A	Tenant B	Tenant C	Owner responsibility
Rates – general services					
Cleaning (common areas)		See note (b)			See note (c)
Electricity (common areas)					
Management fees					
Insurance					
Taxes					
Maintenance (building)					
Maintenance (grounds)					
Security					
Others					
Total outgoing responsibility per annum					
Monthly outgoing responsibility					
+ Monthly rent payable[d]					
Total monthly invoice					

Notes:
a Inclusive of GST
b Obligations as per conditions of subsisting leases
c Owner(s)' responsibilities for outgoing are, in effect, all matters/items/costs not specified as the lessee's responsibility
d As per rental provisions of lease.

Regardless of whether the owner, tenant or some other party is ultimately responsible for those expenditures/outgoings, the owner/property manager will normally pay these expenditures in the first place to ensure they are paid correctly and in a timely manner. Using spreadsheets incorporated in property management software, the quantum, timing and responsibility for the payment of outgoings and other costs will be projected for the upcoming year. Typically, these will be based on the previous year's figures with additional contingencies for inflation and other price rises. The property manager's own knowledge of and experience with that property will also improve the accuracy of these predictions.

This document can now be compared with the previous income schedule to establish net monthly income and benchmarked against expectations to ensure adequate cash flow and cash balances.

Because of the complexity of all of this, it is good practice for the experienced property manager to undertake a final review of all figures to ensure that no unexpected or unjustified aberrations have emerged through any of the calculations.

13.7 Residential property management

Approximately 2.1 million dwellings (or about 27 per cent of all residential property assets) in Australia are rented from private owners for income-producing purposes; another 3.7 per cent of stock is owned by various components of the public and not-for-profit sector, principally for the provision of welfare housing (Australian Government [AIHW] 2019). These combined stocks represent a very significant property investment. The parameters of how this section of the market operates vary considerably from other income-producing property sectors and, therefore, require specific comment. (These matters were also discussed in Chapter 8, Section 8.6.)

While larger commercial assets are typically owned by corporations, trusts and other funds, residential income-producing properties are normally held by individual investors. These assets are usually seen as being affordable (i.e. a relatively low threshold to entry and capital value), have a historically sound record of net capital gain, particularly in the medium to long term, and are perceived as having tax advantages through negative gearing provisions. For small-scale investors, they also have the advantage of being a well understood and manageable commodity. Given that it provides a basic commodity (i.e. shelter), demand will normally be reasonably stable.

While those represent significant advantages, there is also a range of disincentives to residential property investment. Often, there is limited protection of income stream with tenancies normally running for only 6 to 12 months and with significant exposure to bad and doubtful debts. Historically, residential property investment has exhibited a relatively low return on investment and frequently need to have cash flows underwritten from other sources, particularly on highly geared properties. Malicious damage to the asset occasionally occurs and rents are usually paid on a gross basis only, leaving the owner with sometimes considerable exposure to maintenance and other costs. Additionally, while there are tax allowances for maintenance and operating costs, capital gains tax will apply upon sale, reducing the potential for gain, which historically has been required by investors to offset often modest returns on income.

Because of the political sensitivity of matters pertaining to housing, state-based residential tenancy legislation will typically be protective of tenants' rights. These will place specific, but normally not unreasonable, responsibilities on the landlord. These, depending on the particular legislation, require the landlord to maintain the property in a 'habitable state'—regardless of the property's condition at the start of the tenancy, the level of rent sought, or rent concessions that might be provided. Most states have residential tenancy authorities in one form or another to hold and return bond monies as appropriate. They also have enacted special residential tenancy regulations setting out strict procedures for establishing, managing and terminating tenancies that must be followed in all cases. Most legislation also establishes mediation procedures and tribunals for settling disputes without the need for litigation.

It is essential that the owner and property managers involved in this type of property have a detailed knowledge of the legislation and regulations that govern such dealings in that particular state.

From the property manager's point of view, a common sense, practical approach based on good systems and a fair, friendly, but businesslike approach to tenants is far more desirable than time-consuming and sometimes expensive arbitration or litigation.

There is particular value in having good quality, well-maintained properties in the portfolio ('rent roll'). Such properties will typically attract sound and reliable tenants and, once such a tenant is secured, they become an asset, in the broader sense of the word, to the owner and manager alike. For desirable tenants, a slightly conservative approach to rental levels could be considered, thus limiting the cost and loss of rent involved with frequent tenant turnover and, with that, the increased risk of installing a new tenant who may become recalcitrant.

Residential property management provides a reliable, subsidiary business to many real estate offices, assisting with cash flow, presenting opportunities to grow and expand the business and providing referrals to and from the sales component of the agency.

Pivotal to successful residential property management is the selection of good quality incoming tenants. Often these are attracted in the first instance by an 'open house', where the level of rental and bond, together with the proposed rental agreement, normally a pro forma in common industry use, must be provided.

After securing any necessary privacy clearances from the prospective tenants, proof of identity, income, references and a range of other checks should be undertaken. In particular, the comments or observations of agents or landlords who have previously rented property to the prospective tenants are of high value. In various states, a Tenancy Information Centre of Australia (TICA) also operates and identifies individuals with poor tenancy records.

Care must be taken to ensure strict adherence to anti-discrimination legislation at both federal and state levels. Discrimination in these circumstances relates to treating different groups of people unequally because of assumptions about their personal characteristics. All parties involved in any breach, including the owner, property manager or agency office, may be held responsible. In the area of residential tenancies, discrimination can potentially take a number of forms – refusing to rent or sell a property to an individual or group, charging them a different rent, bond or purchase price, imposing different conditions, unfairly evicting individuals or groups or advertising in a particular way that implies the exclusion of certain groups or individuals. For example, it may well be appropriate to advertise that a particular property would 'suit singles or couple', but it may well be illegal to advertise 'no children'.

Regarding property management, however, it should be noted that anti-discrimination does not equate to 'equal opportunity' or 'affirmative action' and, beyond the requirements of the building code, there is no obligation for additional expense to accommodate tenants with specific demands or requirements. (Of course, an owner may wish to carry out such works voluntarily but not by way of obligation.)

Regardless of the particular characteristics of any prospective tenant however, it is quite in order for the owner or agent to undertake normal commercial investigations before making a decision on accepting a particular prospective tenant. These may include the establishing of proof of identity and of that person's ability to pay the rent and to adequately look after the property. References, particularly past landlords, will also be of value here. The owner/property manager may accept, reject or place specific conditions on tenancy agreements, within the parameters of relevant tenancy legislation, based on those commercial considerations. Overall, legislation is justly aimed at providing equal consideration for all, fairness and the objective of eliminating unconscionable conduct in dealings.

Residential tenancies are normally for a shorter term than commercial property; a fixed term of 6 or 12 months with a fixed-gross rental is commonplace. Nevertheless, either from the start, or after that fixed-term tenancy expires, a periodic (e.g. month-to-month) tenancy may be established if both parties agree or is implied by the continued payment and acceptance of rental payments.

Rent rolls (i.e. the rental property portfolio of a particular real estate agent) will often involve some hundreds of properties. Consequently, close adherence to legislation and regulations, good quality operational procedures, IT systems, reliable and strategic links with service and maintenance providers and a programme of regular inspections are all essential for quality, efficient, effective and profitable services. High-quality propriety software packages and residential tenancy management systems are available and in wide use.

As is the case in other property management activities, well-established industry standards, methodologies and procedures, and pro forma documents and agreements are available through regulatory authorities and industry and professional organisations. Quality staff training in legislation, office management and people/negotiating skills is necessary and, together with good diary systems, registers and controls of keys and access rules for prospective tenants, helps provide professional results.

Requirements for dealings with sitting tenants are normally described in detail in the relevant tenancy legislation and clearly explained on relevant government internet sites. Legislation and regulations reinforce the rights of the tenant to the 'quiet, peace and enjoyment of the leased property'. Those legislation and regulations will prescribe requirements for notices of entry, exact definitions for breach and the remedy of any breach that may eventually result in termination of the tenancy and return of the property to the owner. These requirements need to be thoroughly understood and followed precisely by the property manager since even a small error in such dealings can void an action aimed at resolving a particular issue.

13.8 The management of non-income-producing properties

The property sector is obviously part of the wider economy where valuation principles and key performance criteria generally measure asset success in terms of the ability to secure net revenue, manage risk and grow capital value over time. This assumption is quite sound for the ownership and management of the very large proportion of the property sector used for income-producing purposes.

There are, however, a large number of real property assets held for non-income-producing purposes. General observations regarding this asset group were made in Chapter 8, Section 8.5. These include a wide range of government buildings and infrastructure, the assets of church and charitable groups ('not-for-profit organisations') and other non-government organisations (NGOs), publicly owned lands, schools, hospitals, civic buildings, universities and technical and further education (TAFEs), historic and cultural sites, public lands, state forests, national parks and so forth. In the past, this category may well have included power stations, railways, ports, airports, water and sewerage facilities and the like, though, in most parts of Australia, these assets are now typically controlled by government-owned corporations (GOCs), government-owned enterprises (GOEs) or under public-private partnerships (PPPs). Under these arrangements, a more commercial approach is taken to acquisition and operating costs, holding charges, taxation or tax equivalents and revenue generation.

For the balance of those assets, however, there is typically little or no motivation to secure profit or return on investment in the normal sense of the word: the government does not build a school, a hospital or police station with the idea that it is going to generate in a direct financial profit; nor, over time, does it anticipate selling that asset with capital gain. Likewise, a church group would not construct a church, aged care facility, hospital or school with an underlying profit motive.

As a result, the normal commercial methods of management and assessment need to be modified somewhat to reflect the different philosophy, strategic objectives and approach of this type of owner. Nevertheless, in most ways, it should parallel the approach for commercial property. In the first instance, the manager must be aware of, and always respect the basic philosophy, purpose, objectives and approach of the owner organisation and other stakeholders. In this, there will almost always be philosophical, political and corporate parameters. However, it is important that the property manager does not attempt in some way to 'second guess' those agendas. Rather, professional, analytical, property services and advice should be provided to the owners and stakeholders to allow them to make final decisions in full knowledge of all available options and within commercial reality.

The legal parameters of such organisations will normally be similar to that of private owners as regards zoning and other land use controls, workplace health and safety issues et cetera. However, additional legal limitations might also apply as regards those uses and further dealings. For example, a number of these organisations will be established and operate under a specific Act of Parliament or other governance arrangements which may prescribe (and often limit) their property and other activities. Grant of Deed in Trust tenure arrangements are often used in those situations (see also Chapter 7).

Additionally, quite aside from the specific directions from government, trust arrangements involving other individuals, groups or the wider community are frequently encountered. Here care must be taken to establish, in the first instance, the parameters and requirements of any such trust and who, indeed, has final decision-making authority for any significant property dealings.

Other issues that might arise relate to the 'not-for-profit', charitable or tax exempt status that a particular organisation may enjoy, which may be prejudiced by certain commercial dealings. Furthermore, care needs to be exercised where part of the organisation provides services that compete with the private sector. In these cases, under Australian National Competition Policy, complete transparency in trading and, as necessary, tax equivalent payments, need to be put in place to ensure that any arrangements are fully compliant and seen as competitive.

As described in Section 13.4, normal commercial property management has two basic components. The first relates to the securing of revenue/income and the second to the control of outgoings. In the case of many not-for-profit organisations, the revenue generation component may be of much lesser importance than is the case in a commercial environment. However, operationally, cost control and operational matters are equally relevant for commercial or not-for-profit assets. In this case, for the portfolio and each property within it, a business or operational plan is required to identify function, life cycle position and the like. An asset register needs to be established for the portfolio – kept relatively simple, but certainly identifying all property held, their use and condition and 'book value' (i.e. original acquisition cost) and, as appropriate, current value. The management system will need to be based on an asset-by-asset cost centre approach.

This should include identification of any costs incurred, sinking fund[1], maintenance and operations, programmes and certifications, inspections and assessments, capital improvements and checklists to ensure all legal obligations are met.

As with any portfolio, all assets should be kept under regular review to ensure that each is still required, remains functional and fit for purpose and complies with relevant legislation and regulations.

While all corporations will attempt to meet their statutory obligations, there is particular sensitivity, risk and potential embarrassment for typical not-for-profit owners (e.g. government bodies, churches or charities) found in legal breach and for these portfolios, the property or facilities manager needs to be especially vigilant as regards compliance issues.

Another issue arises where government, church and other not-for-profit groups find themselves with assets of various types that are of major civic or heritage interest. There may be unfounded expectations from the public that such properties will be protected and maintained, even when they could be, in fact, of limited further use to the organisation that owns or controls them. These matters are further discussed in Chapter 15, Section 15.3.

It is often the case that not-for-profit organisations are 'asset-rich but cash-poor' with insufficient funds allocated to management or upgrades. This can produce potentially dangerous situations in a financial, physical and legal sense. With the ownership of buildings and assets in the contemporary environment comes a range of obligations and liabilities, especially if the public has immediate or frequent access. It must be reinforced with owners that neither their status, corporate motive nor their limited financial resources provides a defence for breaches or potential litigation.

In summary, therefore, even though the objectives of such owners may well be quite different to those of a commercial portfolio, it is essential that the property and facilities managers undertake their tasks, particularly those relating to building operation, costs and general management, in practically the same way as if the property were commercially owned. If, in strategic decision-making, the owners decide that there are other, non-commercial reasons for particular decisions, then those are matters for the organisation at the most senior levels, not for operational management.

13.9 Property marketing and related dealings

The marketing of real property is outside of the scope of this book and represents an entirely separate field of analysis and skill development – from the actual marketing of various types of property – residential, commercial, retail, industrial, rural and so on – through to advertising and negotiating skills.

From time to time, however, owners and asset or property managers will have the need to engage a professional to secure the best outcome after a decision is made to dispose of an asset.

Rather than a study of marketing, advertising and negotiating per se, this section provides some observations in relation to dealing with marketing agents and the interface between property owners/managers and those service providers. Comments particularly relate to larger income-producing properties rather than residential sales or project marketing. Other sections of this chapter pertaining to the concepts of agency (Section 13.3) and outsourcing (Section 13.11) are relevant.

In deciding to market and dispose of a property, it is essential that the owner and asset manager first clearly understands their motives and priorities, and what they consider to be a satisfactory outcome. Real estate markets are demand driven and to sell into such an environment implies the owner must accept the price effectively set by the market. Nevertheless, there may be a range of other corporate issues that could influence the willingness to sell. Depending on the circumstances of the case, and certainly in confidence, the marketing agent should be made aware of any such issues as early as possible.

To better understand the current market environment, the owner/asset manager may also be well advised at this point to secure an independent valuation, and many corporate and public-sector organisations will require that as a matter of process. The valuer should be advised of the purpose of the valuation and, perhaps, be asked to also provide a range of values within which to set a reserve price or negotiate. The pre-setting of an exact valuation figure as a reserve price, particularly for a publicly listed or government organisation, may frustrate a later, sound negotiated settlement. Again, providing a narrow range of value to allow for reasonable negotiations on price and, perhaps, on some other terms, normally represents good practice.

Also, prior to marketing, the owner or the owner's representatives, such as asset or property manager, should undertake a range of activities to ready the property for sale, ensuring, as much as possible, that vacancies are minimised or eliminated; all agreements, leases, service contracts and the like are in place; all building certifications, tax schedules and other documentation are in order and immediately available and that the building is in good physical condition and state of repair.

When that is in place, the owner or their representative will normally call for marketing submissions from a small number, perhaps two or three, agents considered to have the expertise, knowledge and networks to maximise the opportunity presented. In their submissions, the marketing agents would be asked for general observations, of both the market and the subject property and its marketability, proposed marketing strategy, key personnel involved, likely buyers, advertising proposals and possible costs. The agent may also wish to provide an expected range of achievable value, though, as noted above, the owner will probably by that stage have an independent valuation carried out, which he/she may, or may not, disclose to the agent, depending on the circumstances. It should be noted that for some owners, particularly government and institutions, there may be regulatory requirements as to the manner in which disposals may proceed to ensure public exposure. Care needs to be taken to confirm that no subsisting leases on the property contain 'first right of refusal' options for purchase in favour of the tenant. Such lease provisions, while not common, are sometimes encountered and, in those cases, must be complied with in accordance with the provisions of the lease before open marketing is commenced.

Advertising budgets are often a matter of conjecture. The overall approach to advertising real property sales has changed remarkably in recent years and the simple size of the advertising budget say nothing of its effectiveness. Unnecessary or poorly targeted advertising must be avoided while, at the same time, opportunities to expose the property to as many potential purchasers as possible cannot lost. There are no strict rules, and actual targets and expenditures depend on the circumstances of that case and of that particular property. In the case of public-sector owners or major listed corporations, there will be a latent requirement to ensure that the property has been appropriately exposed to the market place through public advertising. This often means that dealing with properties/owners 'off market' or with a potential purchaser in priority is difficult and sometimes impossible.

These assets are complex and not normally subject to the emotion or subjective decision-making sometimes evident in residential markets. Consequently, the best that 'traditional' forms of advertising, such as press and signage, can do is to simply alert potential purchasers that the property is on the market and direct them to the marketing agent for further details. Many purchasers of commercial property may not reside in the locality and, perhaps, not even in the country. Consequently, the impact of press advertising or signage may be quite limited in those cases.

Over recent years, internet advertising has become dominant and, in the contemporary environment, high-profile placement on the internet, use of the agent's own networks, production of a quality information brochure and/or information sheet, a sign on the property, perhaps some press editorial and relatively modest advertising, well placed in the appropriate press, may well secure cost-effective and quality exposure.

In larger scale, non-residential properties, it is normal to retain only one marketing agent, though of course, there are exceptions. Once the marketing agent is appointed, it is important that a true partnership, high levels of confidentiality and confidence be established on a corporate as well as personal basis. It is in the owner's interests to closely consider and reasonably respond to the advice of the marketing agent who will undoubtedly be more familiar with the day-to-day operations of the market and the needs of prospective purchasers, than the client. Like all other 'agents', the marketing agent must act only in the best interest of, and exclusively for, the principal (i.e. the property owner). While never compromising that, the agent is retained to sell the property and that involves the agent ensuring that the owner has realistic expectations of outcomes. Again, trust and sound communication and information-flows between the owner and his/her agent are critical.

It will be of benefit both to the owner and the agent to recognise and attempt to highlight the key desirable features of the property and, at the same time, to understand and respond positively to the difficulties, issues and imperfections the subject property may have. Given the analytical nature of these markets, such issues will almost certainly be exposed by the prospective purchaser's investigations into the property before settlement (known as 'due diligence').

The marketing agent should also be aware of the competition currently in the marketplace that will form the basis of comparison, together with recent history of relevant market events. This is particularly important as the potential purchaser will not infrequently be drawn from the under-bidders of other recent sales, or other prospects already known to be in the market, but whose requirements remain unfulfilled.

Given the nature of these markets, the marketing approach will often vary considerably from the residential sector. The market here is much more knowledgeable and analytical than the residential sector and there are often corporate limitations to making instantaneous decisions or being in a position to bid at a public auction. Consequently, a tender process is often favoured, although a call of Expressions of Interest (EOI) or a sale by private treaty nominating the asking price, may also be considered, depending on the case.

The marketing agent should and, almost invariably will, provide regular, formal reports on the marketing campaign and feedback from interested parties. Typically, the best responses will come early in the campaign when the property is new onto the market and attract any unsatisfied demand that already exists. If, after an agreed period – normally about a month to six weeks – there is little or no response, there is clearly a fundamental flaw in the existing approach and modification is necessary. Often, that will be a matter of pricing, but there can be a range of other issues, such as target marketing, advertising, property information provided and so forth, which may need to be reconsidered.

The progress of each marketing campaign will be different and may need to adapt over time. Consequently, it is difficult, and perhaps dangerous, to suggest overall or general rules and strategies. Nevertheless, once a firm offer has been received, it is important that this is secured in writing, typically in the form of a contract and complete with the prospective purchaser's signature. That formal offer should include any specific, new conditions. The agent then needs to act as an intermediary and, almost invariably, it is best that the owner uses the agent to distance themselves from face-to-face negotiations. The reason is that, in negotiations, the owner obviously has the most latitude in decision-making and therefore, potentially, the most to give away. The use of the agent as an intermediary provides that additional time to consider offers, contemplate counter offers and so on. Experienced agents are familiar with and normally prefer this approach.

In negotiations, a range of tactics may be used. These might include a statement of pre-conditions or 'not negotiable' issues, setting deadlines or creating delays, putting forward 'what if' scenarios or even presenting ultimatums. Depending on the circumstances, any or all of these may, at times, be suitable or even necessary; however, care needs to be taken with many of these tactics. Failure to follow through on ultimatums, deadlines or pre-conditions will greatly weaken that party's negotiating position.

In summary, it is important to negotiate on the entire proposal, keeping in mind what is particularly important and what final outcomes are really required from the owner/corporation. Other matters – such as tactics, emotion, personalities and personal preferences – need to be kept subservient to overall key goals.

13.10 Facilities management

Like many terms and generic activities undertaken in the property sector, 'facilities management' has a range of possible definitions. In fact, in the case of facilities management, its use extends far beyond simply physical assets. In its widest term, it can relate to the effective provision of infrastructure and logistical support for business and public sector activities across all sectors, including the armed forces. Under this definition, 'facilities management' might include the management of financial resources, human resources, physical resources, both infrastructure and buildings and the management of information technology and the like.

In all of these areas, facilities management will particularly concern itself with the operational capacity of, and planning for, business activities, assisting in risk management and ensuring operational performance and effectiveness of the resources and assets used. Facilities management, in this wider context, is really about coordination and integration of business, people, property and information to secure the best available options for all stakeholders. These skills are required through the various phases of an activity or the life cycle of an asset – from initial planning and development through to the long-term components of the 'asset-in-use'. The facilities manager's role is principally directed into these asset-in-use activities involving building operations, service contracts and efficient and effective use of resources and services such as water and energy.

While having regard to the whole building asset and its services, the facilities manager will frequently be involved in various individual components of the building and its operations (Williams 2006). These activities could include:

• building services and operations
• fit out construction and ongoing management/maintenance and alterations
• furniture, fittings and chattels
• office equipment, consumables, et cetera
• motor vehicles
• production plant and machinery
• work in progress
• raw materials held (stockpiled) and/or
• finished goods/works (inventory).

The typical facets and areas of responsibility within facilities management are shown in Figure 13.2.

In property, the concept of a separate facilities management component is relatively new and its application as a separate activity is confined mainly to large institutional and property-portfolio owners. In many organisations, and, particularly those involving smaller portfolios, facilities management activities are shared among the property owner, the property manager and building operations staff, either in-house or contracted. In any case, there will always be operational overlaps between these groups even where a separate facilities management group is established.

The specifically identified 'facilities management' discipline has emerged in the property sector only in relatively recent years and the involvement and role of this area continues to evolve. Like all other activities, it must be focussed on key commercial and other performance outcomes required of the owner. As noted above, facilities management places particular emphasis on providing better service and higher levels of effectiveness and efficiency from an asset, making it more competitive overall for the building owner, lessees, occupants and customers. These will be largely directed at

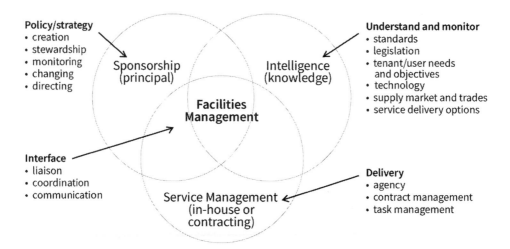

Figure 13.2 Facets of facilities management (Williams 2006 modified).

physical, operational and related management activities. Business and communication skills are particularly important in setting up programmes, presenting them to other staff and tenants and ensuring that tangible outcomes are delivered. In this way, it can be seen that facilities management is, in fact, considerably wider than operations. Its human resource requirements may, therefore, be incorporated as part of the property manager's team (Martin 2006).

In practice, facilities management activities can become an important catalyst for change in the property owner's own corporation or across numerous stakeholders. The acceptance and implementation of environmental protection and sustainability practices provide a contemporary example of this.

In larger organisations, it may well be the facilities manager, and that component of the property management regime, who establishes and manages contracts for services/outsourcing as outlined in Section 13.11. In particular, the skills of the facilities manager would be of considerable value in contract establishment and administration, and in enhancing the quality and relevance of the services provided.

13.11 Outsourcing of building and property services

In the contemporary environment, it is not practical, efficient or effective for the owner of a major property asset or portfolio to attempt to undertake all necessary management and operational activities in-house, that is, by only using the owner's staff.

A key role for asset owners and managers is to establish the right balance between in-house, sometimes called 'day labour' staff, and the services and skills provided under contract, to maximise the operational efficiency of that asset.

The 'outsourcing' of some or all of these activities is defined as the provision of goods and services to an organisation from an independent source, where the consumer organisation has historically provided those goods or services or has the potential or capacity to do so.

The use of 'outsourcing', in one form or another, is common across many business sectors. Sub-contracting in the construction sector and the seeking of a second, specialist opinion or input in professional environments represent two well-known examples (Johnson 1997).

Outsourcing should be seen as a management strategy or option that will be applied in different ways, depending on the circumstances, characteristics and strategic objectives required both for that asset and the portfolio. Overall, it is important to consider outsourcing initiatives on an analytical basis and, if they are to proceed, be applied in a structured and well-managed way (Martin 2006).

There are many reasons to consider outsourcing. Strategically, it may help a company improve its business focus, provide it with access to higher skills and capacities than it could ever secure itself, and act as a risk-management strategy, transferring certain responsibilities to a third party better able to accept and manage that matter. Operationally, outsourcing also allows for new specialisations to be applied, particularly where those activities are required only on an intermittent or 'less-than-full-shift' basis.

Any savings secured will also have a recurring, positive effect and, therefore, enhance capital value. However, it is important to note that while the optimisation of expenditures maybe a reason for considering outsourcing, undertaking this simply to reduce costs may result in fairly disappointing long-term outcomes, all things considered (Johnson 1997).

There is a range of typical parameters and issues encountered in considering or undertaking outsourcing initiatives. These include (Johnson 1997; Angel 2003; Martin 2006):

Test for competitiveness

Before undertaking a significant change in the delivery of services, the owner/ manager needs to be satisfied that there is indeed a significant problem to be addressed. In most cases, this can be established fairly easily by independent assessment and surveys and by benchmarking the operational cost service, quality and condition of the particular asset with similar properties or property groups. Industry associations such as the Property Council of Australia have such comparisons available to members. Care needs to be taken in such investigations that the comparison is, in fact, between assets of similar age, location, design and use parameters.

Before an outsourcing strategy for a service (or services) is commenced, a confirmation of the existence of suitable suppliers in that location needs to be made. This is normally not an issue in major cities but becomes less certain in more remote locations or where highly specialised or unusual skills are required.

People issues

Existing staff involved in servicing the building will almost invariably be best able to identify management and operational issues with that asset and may provide innovative ideas to solve those problems. Management should not presume that existing staff, even those already employed in operations, will be negative about change. Early communication and discussion of issues and options represent good policy.

Being good at new roles and relationships

If outsourcing of some activity is proposed, it is important that the owners and existing managers of the asset become good at their new, and now quite different, role. They need to become informed purchasers with a clear understanding of the services to be provided and the contract establishment and administration involved. For this reason, and particularly in larger organisations, it is good practice to retain some in-house skills and knowledge, not simply regarding the activities themselves, but also in the brokerage and administration of outsourcing contracts.

Clarity and certainty in contracts

As observed earlier, pre-existing industry standards and contracts for service should be used wherever possible. Instead of attempting to describe all activities involved, reference to established standards (e.g. International Organisation for Standardisation [ISO] or Australian Standards [AS]) provide a more comprehensive approach. Such contracts need to identify clear roles, responsibilities and penalties, and provide a fair approach for the owner as well as the contractor. While it is important to secure value for money, contracts that are unfair or unnecessarily onerous on either party will almost invariably result in dispute and unsatisfactory long-term results.

Performance indicators in a particular contract need to be limited in number, specific and outcome or performance based. They need to measure outcomes for which that contractor has responsibility and control.

Ownership and management of intellectual property

Many outsourcing agreements will include the need for the contractor/agent to have access to, and use of, the databases and information of the owner. Major property management companies in their dealings with large corporate owners and assets will typically use their own specialist software and systems. These contain remarkable management information and reporting capabilities which are to the benefit of the owner but also represent commercial comparative advantage to the management company. Consequently, for the protection of both parties, management agreements need to establish the rights to use such systems while, at the same time, ensuring that the building owner's business information remains secure and confidential. Agreements should confirm that any asset-specific information remains in the ownership of the asset owner/principal under the contract and is provided to the contractor for use only in matters at hand. Agency arrangements, as defined in Section 13.3, will often apply and, at the end of the service agreement, it will be normal that asset information be returned to the owner in an acceptable form.

Entry and exit strategies

Contracts of this type will normally extend over three to five years and it is important that, except for wear and tear, the asset and its components not be allowed to run down or deteriorate during that period. Any outsourcing arrangements should include an independent assessment of the building/component condition, including photographic studies as appropriate, at the start of the outsourcing contract and completion. These procedures work for the protection of both the asset owner and the contractor.

Choosing and dealing with contractors

Outsourcing contracts of this type will normally be on the basis of a 'closed tender', that is, a small number, normally three or four, of quality service providers are approached and provided with a scope statement or specification detailing the work to be undertaken. The final decision on the successful tenderer will normally be made on past dealings and performance, competencies and scale, guarantees and price.

Under the management agreements for a particular building, it is common that the property manager be involved in and required to manage those service contracts. If so, it is important that the manager be directly involved in the calling of service contracts, their evaluation and selection.

This ensures appropriate levels of input and responsibility by the property manager at contract establishment and into the future.

Depending on the circumstances, bids from in-house, commercialised groups within the principal's own organisation may be considered, but probity issues and the probable lack of real risk transfer need to be closely considered.

Concentrating on 'big ticket' items

As noted above and, as an overall rule in these matters, quality outcomes and cost-effectiveness are more important than the range of business options available to secure those results. Care needs to be taken to avoid establishing complex and expensive management/outsourcing arrangements where the final impact on overall performance is relatively small.

Outsourcing initiatives, where selected, should be kept as simple as possible and focus on large-scale activities where high costs are involved and, particularly, where higher skill levels, specialist equipment or intermittent, 'less–than–full-shift' activities are involved. These may include some components of maintenance, particularly plant and machinery, lifts, fire services and so on, together with energy management, some components of cleaning and security.

Note

1 A sinking fund is typically money set aside and normally added to incrementally (e.g. annually) which, with interest, will accumulate over an assessed period to meet some known major maintenance or capital replacement requirement into the future.

Chapter 14

Taxation and real property assets

14.1 Introduction

This chapter provides a summary of the Australian taxation regime as it impacts on the holding and management of real property assets, particularly those held for income-producing purposes. The matters are complex and the objective here is simply to establish a context and working understanding of the various components of the taxation system as it applies to that sector.

Such observations also need to be set in a time context. The taxation system, while punctuated by major events such as the introduction of Capital Gains Tax (CGT) in September 1985 and the introduction of a Goods and Services Tax (GST) in July 2000, continues to evolve and is subject to regular amendment.

As for any commercial undertaking, taxation implications for the development, ownership and management of real property must be understood and managed. It is the net income from an asset that needs to be maximised over the short and long term. Therefore, any particular tax advantages that some investment or venture may attract or taxation liabilities that it may incur must be considered by the owner/ financier in the assessment of overall financial benefit.

Tax evasion (i.e. purposely undertaking actions or providing false statements for the purpose of not paying tax due) is illegal and vigorously pursued by the Australian Tax Office (ATO). However, it is legal – and in fact quite expected and anticipated by the ATO – for individuals and other tax-paying entities to arrange their affairs to minimise their exposure to taxation ('tax minimisation').

The taxation system and taxation collections are far more regulated than they were some decades ago, and the self-assessment system in place demands full disclosure and accountability by the taxpayer. Nevertheless, the holding and management of property still provides significant and legitimate tax-minimisation opportunities, which are more attractive than many other forms of wealth holding and wealth generation. At this early point in the study however, it is important to note that structures put in place should not be so complex or contrived as to actually work against clear and simple arrangements that optimise asset performance and efficiency.

14.2 The taxation environment

To fully appreciate and understand the specific implications of tax for real property assets, it is necessary to have a general understanding of the wider taxation environment as well as some of the political, structural and judicial parameters that surround it.

As the economy and society has become more complex, so has the size, role and complexity of government. It is interesting to note that, under the 1901 Australian Constitution, it was prescribed that the Commonwealth would have a very limited and defined role, principally related to foreign affairs and defence and the guarantee of free trade between the states. Although the Constitution has been subject to only very slight variation over more than a century, the whole nature of government and taxation has evolved almost beyond recognition over this period of time.

Much of this changing role of government – and revised fiscal (taxation) policy and collections – stems back to the middle of the last century. During the Second World War, and in defence of the country, much government power (including the ability to collect income tax) was centralised with the federal government in Canberra and, even post-war, much of this centralised economic power remained. The situation was reinforced by a succession of High Court decisions – particularly the earlier Engineers' Case (1920) – which confirmed the primacy of Commonwealth power over the states in areas of joint or overlapping responsibility. These arrangements and their evolution were discussed in more detail in Chapter 6.

As well as the economic power shift, other inherent problems have made the taxation environment difficult to manage. These include:

- the uneasy constitutional relationship between the Commonwealth and the states, often complicated by party politics
- problems in controlling the cash ('black') economy
- the public perceptions of excessive taxation and inequality within the system
  (The overall value for money of tax payments in Australia, say, compared with countries overseas, is a very complex area of analysis. The comparison of abstract levels of taxation in one country compared with another is unhelpful in showing the overall benefit from tax collected. Those collections need to be related to the scope and level of services provided. There is, however, an undoubtedly low-tax morality prevalent in Australia. Many taxpayers, rightly or wrongly, perceive the system to be unfair – sometimes allowing very rich individuals or large corporations to escape with minimal tax. Therefore, there appears to be reluctance by many to be honest in tax declarations and payments. It is interesting to compare this with the situation, say, in the US, where the payment of tax is very much seen as a civic duty and responsibility. In that country, tax evasion is generally perceived as a much more serious crime than in Australia.)
- as the size and complexity of the economy and society rapidly increased and, as noted above, tax morality remained low, the ATO continually finds itself reactive to problems and loopholes. The result has been the evolution of a remarkably complex and, arguably, incomprehensible taxation system based on the Income Tax Assessment Act
- an over-emphasis on pay-as-you-earn (PAYE) taxation – still representing about half of all Commonwealth taxation receipts
- although the above matters have been partially addressed, the taxation system has, at times, been seen as favouring consumption-spending over saving, short-term investment over long-term and speculation over income generation
- because of political imperatives, the taxation regime and structures have regularly been distorted to meet demonstrably uneconomic demands.

The most obvious example of his final observation is the tax treatment of the taxpayer's principal place of residence. Because of the political sensitivity of such matters, the introduction of important and necessary tax changes – including capital gains taxation and certain GST parameters – has expressly excluded the principal place of residence. This leads to the quite remarkable situation where two otherwise identical assets can be subject to two entirely different tax regimes, based on whether or not the property is owner-occupied. Such taxation exemptions can easily promote the diversion of capital into investments that are less than economically optimal. These can also cause market aberrations – for example, the unnecessarily high diversion of private sector investment in Australia into principal places of residences and the subsequent periodic over-heating of those markets.

Furthermore, in these preliminary observations, it needs to be stressed that 'taxation' is an abstract term; it is neither good nor bad. If public and community services are required, as they must in a contemporary society, they have to be paid for. How that tax is collected and those funds distributed are obviously controversial issues determined in the political sphere.

A good tax is one that is simple, easy to collect and with few exceptions. It needs to be clear and understandable in its intent and accepted as generally fair and equitable by taxpayers. A good tax will not fall heavily on production nor exports, nor prove a discouragement to the creation of sustainable jobs and investment. Finally, a good tax needs to promote long-term investment that generates income rather than shorter-term, speculative, more volatile investment that aims principally at the generation of capital gain.

No tax can fully accommodate all of these competing interests, but some are clearly more effective in meeting these macro-economic objectives than others. It would be argued by many economists that property taxes reasonably meet many of these criteria and yet they remain under-represented in the overall 'tax-mix' within the country.

14.3 Ad valorem taxation systems

Despite significant shifts towards taxation based on net income and capital gains from property, taxes on ownership and use remain significant and represent around 7 per cent of all taxation receipts collected in Australia (Australian Government 2012). These taxes are typically applied at state and local government levels and include land tax, general council rates, stamp duties, and a range of largely other levies and charges applied variously across jurisdictions and under a range of state-based legislation. General rates, levies and (where applicable) service charges for water, sewerage, garbage collection and so on, represent by far the largest income source available to local authorities across Australia.

Property taxes represent one of the oldest forms of revenue collection by governments, with a history dating back several thousands of years. As well as general and service rates levied by all Australian local authorities, states and territories governments also levy land tax on certain types of property. Land Tax is, in effect, an asset or wealth tax normally applied at a rate of 2 or 3 per cent of unimproved value (UV) or site value (SV), depending on the legislation involved. Typically, the taxpayer's principal place of residence is exempt and there are thresholds before such taxes become payable. This means that the taxation base here is very narrow, but its impact on those affected is significant (Australian Government 2010).

These taxes are normally *ad valorem*, that is, the rate of tax payable is on the basis of the value of the asset, normally applied as a percentage or expressed as 'cents-in-the-dollar'. In many ways, these forms of taxation sit well with the definitions of 'a good tax' outlined above. They are 'reasonably' progressive and simple to collect – based as they are on registered ownership and property identified by survey and title – and, more importantly, these taxes are generally well-recognised, understood and accepted by tax payers (McCluskey & Franzsen 2005; API 2007). Service rates components are typically applied on an equitable 'user pays', 'payment based on quantity/level of usage' basis.

While the fundamentals of ad valorem taxation systems are fairly uniform across Australia, the fact that they are state and local authority based means that the method of application varies to some extent. In most states, a Site Value (SV) approach is used for urban lands, that is, the land is valued without structural improvements but, in this case, ground improvements such as filling, clearing, levelling, and retaining walls are considered part of the land, and, therefore, included in the land value for taxation purposes. In most states, rural land is assessed on an unimproved basis – that is, for the purpose of the valuation, the surrounding developments/infrastructure is as it exists as at the date of valuation, but the subject property is considered as in its natural state. This variation was aimed at ensuring that rural land-owners/producers were, in fact, not penalised/additionally taxed for the groundworks and improvements previously applied to their land.

Individual state legislation sets out the frequency of valuations–in many jurisdictions, annually, provided that there has been apparent sales evidence of changes in value. The methodology for such mass appraisals normally involves the analysis of key sales in localities and sub-markets to determine variations from the benchmark of previous assessments. Computer-based models are then used to apply the new standard.

14.4 Major taxation shifts 1985–2004

As noted above, taxation systems are dynamic and organic. They continually evolve to meet changing situations, though unfortunately, in practice, that sometimes degenerates into closing loopholes and applying remedies to what, in Australia, is an increasingly inefficient system.

While that system continues, there have been, since the mid-1980s, several key events that have changed the nature of taxation with particular effect on real property assets.

In 1985, Paul Keating, as Treasurer, introduced sweeping changes to the taxation of assets whereby indexed capital gains on the sale of most assets were taxable. To balance against this new tax, a wide regime of new tax allowances, including building depreciation allowances, was introduced (see below).

As with many reform agendas, these changes, which were introduced by the Hawke–Keating Labor government, were in fact maintained and further developed under the conservative Howard–Costello government, with a succession of refinements continuing to the present day.

A second major initiative was developed under the Howard–Costello government in 1999 – the Ralph Report – which recommended a range of business tax initiatives that would effectively have reduced company tax and simplified taxation systems for business and investment. While a number of these recommendations have been approved and introduced in part, the entire Ralph Report was never fully implemented – undoubtedly to the detriment of Australian business.

The third and much more successfully implemented initiative was the introduction, on 1 July 2000, of a comprehensive Goods and Services Tax (GST) regime in Australia (see below).

It will be noted that, during this period, there has been a change in taxation emphasis as real property was brought much more into the 'taxation mainstream'. This has, in effect, seen the moving of the property tax base from one of taxation based on ownership towards much more specific taxation on the income and capital gain derived from those assets.

The former (i.e. taxes on ownership) is mainly state-based taxation: land tax, local authority general rates and stamp duty. That is not to say that these are not still important and considerable in magnitude, clearly they are. The point here, however, is that their relative importance for property owners and managers has diminished compared with the Commonwealth-based taxation, represented by income tax, CGT and GST.

Introduction of capital gains taxation and depreciation offsets

In September 1985, the then Hawke–Keating government introduced CGT on profits made on sale of assets. It is important to note that it is not a property tax per se, but applies to all capital assets, real property, chattels, motor vehicles, plant and equipment and shares – all assets where a net capital gain had been realised. A significant exception is the taxpayer's principal place of residence.

CGT is payable on transfer, for example, on sale of the property; but not through transmission by death, though such a 'tax event' recalibrates liability for later CGT liability unless prior principal-place-of-residence status is claimed. That status normally is defined as continuous occupation for the six months prior to that tax event.

Since the introduction of CGT, there have been a number of iterations of its assessment. In all cases, tax liability has been assessed after the entry and exit costs of buying and selling the property are deducted. In the case of income-producing properties, holding charges, such as interest and operational costs, cannot be taken into account in assessing net capital gains liability. The reasoning is that these expenses are already deductible by the taxpayer as an expense incurred in earning income.

The tax was not retrospective, that is, it cannot be imposed on a taxpayer relative to an equitable interest in an asset where that interest was held by that taxpayer prior to 20 September 1985.

The assessment of taxation payable is on net capital gain but the method of assessing the tax payable had been subject to subsequent amendment and variation. For example, as from September 1999, the assessment criteria for properties held over 12 months established that a component (namely, 50 per cent) of the net capital gain was to be untaxed and the balance of the gain was to be transferred and added to other taxable income of the owner(s) in proportion to their share of ownership. Capital losses can normally only be claimed against capital gains.

As well as the obvious implications for transaction planning and decision-making, the introduction of CGT had two significant effects on the property market:

- Because of the tax exempt status of the principal place of residence, large capital inflows progressively moved into the residential market, particularly in higher-value sectors. The increase in demand, combined with other factors, has typically encouraged price rises well in excess of rises in many other non-residential sectors.
- Even though all of this was introduced three decades ago, there remains some reluctance by owners of properties purchased prior to September 1985 to place these properties on the market, as the sale and re-investment will lose their CGT-exempt status – given that CGT was not retrospective to properties held prior to September 1985.

Negative gearing

Negative gearing refers to the ability to transfer losses from one source of income to positive taxable income in another, so reducing tax liability from that second source.

This concept is by no means new and, even before 1985, and certainly subsequently, the transfer of such losses is common, particularly in the early years of investment into highly geared real property where high-interest payments will often result in an operational loss.

The ATO, in accepting negative gearing arrangements, will need to be reasonably assured that the venture currently incurring a loss has the expectations of generating profits in later years (i.e. that it is truly a commercial undertaking and not simply a contrivance for tax purposes).

As an overall observation, it needs to be recognised that, to secure access to negative gearing provisions, a loss on that undertaking for that tax period is a pre-requisite. Despite the fact that the loss will be recognised and at least partly compensated through the reduction of taxable income, it still represents a trading deficit, which must be supported through cash from the taxpayer's other sources.

As with negative gearing, the tax deductibility of operating costs, such as maintenance, insurances, council rates, management charges, et cetera, had always been allowable. However, the increased focus on tax on income and capital gains from property post-1985 brought such matters into clearer focus and placed more importance on them. The access to these allowances and the potential to transfer such losses remains controversial. Political debate continues given its potential impact on investment levels and prices for residential property, particularly in major Australian cities.

Maintenance and building depreciation allowances

To act as a balance (offset) against the introduction of CGT, a range of depreciation allowances were introduced for plant and equipment and, for the first time, against the capital costs of buildings constructed post-September 1985.

The rationale behind this was that plant and equipment, and indeed, to a lesser extent, the buildings used for income-producing purposes, suffered wear and tear and depreciation as a result of their use to produce income (rent). It was reasonable, therefore, that such depreciation of the owner's asset be recognised. Once that depreciation was established for a particular item/building in a depreciation schedule, the annual depreciation allowance could be deducted as an operating expense, so reducing taxable income.

Items defined as 'maintenance' are 100 per cent deductible in the financial period during which that expenditure took place.

The definition of 'maintenance' for taxation purposes is (in summary):

- the replacement or renewal of a worn out part of something, but not the whole
- the restoration to original condition or function without changing character
- restoration of efficiency in function, but not necessarily the exact repetition in form or substance
- rectification of an incipient fault in the building or the control of health risks.

Depreciation allowances are offsets against capital expenditures (i.e. plant and equipment and, since 1985, buildings).

Capital expenditures include:

- the renewal of material different from the original
- work that has effectively been an improvement
- work that has increased the value of the asset from the original or when purchased
- expenditures to reduce the likelihood of further repairs.

As noted above, for real property assets for taxation purposes, capital works fall into two basic categories (Renton 2009), namely:

Plant and equipment

These are, in effect, the functional items within a building and also defined as 'tools of trade'. As the term 'plant and equipment' would imply, such things as air-conditioning equipment, lift cars and motors, fire services, chattels and so on are included. The depreciation rate varies from item to item, but is normally taken as 'straight line, non-indexed depreciation over the effective anticipated life of the asset plus 20 per cent'.

This approach, based on self-assessment, has replaced extensive schedules identifying plant and equipment and the anticipated depreciation rates previously published by the ATO. In a practical sense, however, these depreciation rates typically vary from about 20 per cent (i.e. written off over five years) through to rates of 8–10 per cent (write-off over 10–12 years) depending on the type of plant or equipment involved.

As well as depreciation allowances for active plant, depreciation may also be attracted on 'passive plant' items such as lift wells, swimming pools and so on.

For both plant and building depreciation allowances, a depreciation schedule is established when the item is first acquired, with depreciation deductibility for each year of the anticipated life. Under taxation changes from July 2017, depreciation allowances can no longer be claimed on second hand assets.

It should also be noted that, from time to time, government may seek to adjust the rate of depreciation allowable for business purchases of plant and equipment. These accelerated depreciation provisions are typically only available over a relatively short period of time (a 'window') but are often effective both in supporting business and in stimulating spending and business investment.

Building

As part of the major taxation changes described above, building depreciation allowances were also progressively brought in for all income-producing properties constructed after September 1985. Despite some variations, the depreciation allowance on buildings has normally been at the rate of 2.5 per cent (i.e. written off over 40 years). Clearly, such percentage allowances are not as attractive as full, direct and immediate deductions against income, as are maintenance expenses, but they are nonetheless highly significant, particularly for large capital projects.

The term 'building' in this context refers in the first instance to the overall cost of building, less the costs of the land, land preparation, external works and plant, developer's profit and holding charges. The date of commencement of allowances is effectively the date of practical completion; once that date is established for a particular

> **Maintenance**
> If an expense on an income-producing building can be defined as 'maintenance', it becomes an operating expense and therefore can be claimed 100 per cent as a deduction against taxable income in the same period as the expenditure was incurred.
>
> **Capital expenditure (Capex)**
> - Plant and equipment (tools of trade)
>   Defined as: 'a functional item' depreciation: typically written off over the 'effective life of the asset plus 20 per cent'.
> - Building depreciation allowance
>   For income-producing buildings constructed after early 1985
>   Defined as: 'the setting for business' (construction cost only; straight line depreciation)
>   Depreciated through a depreciation schedule for that improvement; rate applied in individual cases will vary depending on date of construction... (In most cases, 2.5 per cent i.e. written off over 40 years).

Figure 14.1 Summary of building expenses (inc. maintenance) and capital depreciation allowances.

development, it remains relevant for the 'write-off life' of the building, regardless of ownership. Additional depreciation schedules can be established as further capital items are procured – either as plant and equipment or building (e.g. additions or refurbishment) (Figure 14.1).

Goods and Services Tax (GST)

A comprehensive Goods and Services Tax (GST) was introduced by the Commonwealth government in July 2000.

Proponents of a GST would claim that it is a simpler system of tax, without the complexity of wholesale and other taxes, is widely applied and, therefore, fairer, and in part, addresses the problems of the cash economy and encourages savings over consumption. However, its application in Australia, owing to political process, has been more complex than first envisaged. In summary, the provisions of the GST are as follows.

The tax was set into legislation on 8 July 1999 and introduced from 1 July 2000. It is a 10 per cent tax on all domestically produced goods and services, effectively paid for by consumers, but collected along the supply chain. The tax is levied on outputs but reclaimed on inputs (tax credits).

It applies to all goods and services at point of sale with the exception of food, excluding certain prepared/restaurant meals, health services, education, certain sanitary products, child care, exports, water and sewerage, sales of going-concern businesses, certain government dealings, religious items and incoming duty-free goods.

In return for the introduction of this tax, a number of other taxes were to be withdrawn, including most sales and wholesale taxes, and personal income tax rates were reduced. In practice, not all of these proposed reductions or withdrawals were realised, particularly at state level.

The introduction was remarkably well managed, being monitored both by the Australian Competition and Consumer Commission (ACCC) and by consumer groups.

There was an increased obligation on business to register under an Australian Business Number (ABN) to keep an adequate audit trail and to report through Business Activity Statements (BAS). Under general provisions of the legislation, suppliers do not have to charge GST if they do not wish, but they are obliged to pay it. GST is not to be shown separately from final list prices although, noting 'GST included' can be used.

The GST has had significant impacts on the cash economy at the business-to-business level, given the need to quote an ABN of a supplier before tax deductibility for expenses is allowed. However, it has been less effective on individual services supplied to consumers.

In summary, the application and impact of GST on the property sector is as follows:

Leases commencing before 8 July 1999 were GST-free until 1 July 2005 or until a rent-review period after 8 July 1999 under the lease, provided that the lessee is entitled to full input tax credits. If the lessee is not entitled to full input tax credits for the lease, the lease was GST-free if commenced before 2 December 1998. The lease was GST-free until 1 July 2005 or until a rent-review period after 2 December 1998 under the lease.

GST is payable on the full value of any transaction, lease and so forth, including the net impact of incentives.

- Farms/grazing properties
  Land that has been used as a farm for at least five years is exempt from GST if sold to a purchaser who uses the land as a farm.

The margin scheme under goods and services taxation

An important GST option for the property and, particularly, the development sector is what is known as a 'Margin Scheme' allowable under GST legislation. GST applies to the supply of land in situations that include, but not limited to:

- the sale of new residential property
- the sale of commercial or industrial property
- the sale of vacant land in the course of the developer's enterprise but exclude (1) the sale of residential property that is not new, or for commercial purposes and (2) the sale of a property as part of a going-concern business.

The choice of whether to accept liability for GST in the normal way (i.e. 10 per cent added to the price at the point of final sale, less input tax credits), or to apply the Margin Scheme, is at the option of the taxpayer.

Where real property is sold under the margin scheme, instead of the vendor paying GST on the full sale price and claiming tax credits for input costs, the developer only has to pay GST on the 'margin'. Depending on when the property was acquired, the margin is either the difference between the sale price and the acquisition price of the property, or the difference between the sale price of the property and the market value of the property as at 1 July 2000.

The simple formulation of this is as follows:

$$\text{GST (Margin Scheme)} = 1/10 \times (\text{price paid by vendor less price to be paid by the purchaser})$$

$$\text{Normal GST} = 1/10 \times \text{price to be paid by purchaser}$$

(Note that prices to be used in any of those calculations will be GST exclusive.)

The Margin Scheme is widely used in the assessment of GST liability in a range of property subdivision and strata-title development markets, particularly in the residential sector. While the use of the scheme was subject to litigation in 2008, its continued use was confirmed by the Court, though certain anomalies regarding 'going-concern' provisions were tightened.

Whether or not the use of the Margin Scheme is in the best interest of the developer/ taxpayer and the project overall will depend on a large number of factors and must be considered on a case-by-case basis. Note that the Margin Scheme is not available to the sale of land previously acquired under normal GST provisions.

Some likely benefits of the Margin Scheme include a lower final price and reduced exposure to stamp duty, interest and CGT. It may also provide the basis for the continued use of the Margin Scheme in downstream transactions. However, the Margin Scheme may be disadvantageous because the purchaser would be unable to claim a GST input tax credit. This again emphasises the importance of securing quality professional advice in such matters and considering them on a case-specific basis.

14.5 Implications for asset ownership and asset/property management

Because of the complexities of taxation legislation and operations, strategies will be put in place, largely by the property owner's taxation and accounting specialists to effectively minimise tax payable. As part of this, there is a need for asset and property managers to recognise key parameters here and to undertake the following activities that will assist in minimising taxation exposure and risks related thereto:

- Understand and remain current on the overall parameters of the taxation system as it applies to real property.
- Understand the importance of maintaining depreciation schedules and securing those schedules, and the history of the ownership and development of a particular asset. It should be noted, however, that under changes in taxation regulations in July 2017, depreciation on second-hand assets, such as those on-sold as plant and equipment included in a property sale will no longer attract depreciation allowances for the incoming owner.
- Understand the importance of the apportionment of sales on transfer. This apportionment will have implications for possible exposure to GST and, potentially, to CGT.
- The importance of quality records.
    The contemporary tax system is based on self-assessment, but audits and investigations can extend back as far as seven years, should they be required by the ATO. Consequently, good, clear records, including reasoning and photo essays, file notes and the like, to justify certain actions, activities and costs involved, may prove to be particularly valuable.
- Margin Scheme decisions.
    Where a development is considering the use of a Margin Scheme on a particular project, it will be important to make that decision early in the process and based on professional opinion relating to the particular circumstances of that case.

- The importance of classification/allocations of costs

  Because of the different taxation treatments of maintenance, plant and equipment, capital depreciation and building depreciation, there are clearly benefits, within the parameters of the law, in being able to identify activities as either maintenance or plant and equipment or building works. Again, clear file notes and justifications are important.

- The need for prioritisation.

  Once potential works have been classified as above, there will be an advantage, all other things being equal, in carrying out maintenance items first, as these expenditures can be recognised and, to the extent of tax liability, effectively recovered at the next taxation return. Once those amounts are recovered, those funds can then potentially be applied to capital items. Obviously, the reverse is not the case as it will take much longer, through depreciation allowances, to recover capital expenditures.

  There will be sound reasons why this prioritisation is not always possible, for example, workplace health and safety, pressing tenant demands for capital upgrades and so on, but, wherever possible, there will be, on the face of it, tax advantages in expending maintenance funds before capital works.

14.6 Ongoing change

As observed throughout this chapter, the Australian taxation system at all levels remains subject to frequent change and ongoing evolution. Many of those changes effect real property investment and use – matters of concern where such long-term assets rely on high levels of certainty as regards taxation liability.

Nevertheless, a number of major and difficult improvements have been achieved – including CGS, comprehensive depreciation provisions and GST.

Given the complexities of the social, economic and political environment within Australia, it would be naive to believe that the taxation system can be significantly simplified and that tax policy will become more settled, at least in the short to medium term.

As regards real property, its development and use, major challenges lie in meeting rapidly increasing demands for urban infrastructure (particularly transportation and water). Those very large, long-term liabilities can no longer be met from government's consolidated revenue and more innovative and holistic solutions are urgently required (O'Donnell 2017; Stein 2019).

Current examples of rapid change across many states and the Commonwealth relate to absentee (i.e. foreign) ownership. For non-residents, there have been major (adverse) changes in stamp duty and land tax payable, reduction in CGT concessions (and withholding provisions), vacancy fees for unused, residential assets and a number of policy changes.

Whether or not these limitations/disincentives on investment into Australia represent sound economic policy may be debatable. It does highlight however the continued willingness of government at all levels to continue to adjust and use taxation policy to address a range of issues. On that basis, further (Keynesian-based) change can be anticipated into the future as new political and economic challenges emerge.

Chapter 15

Contemporary issues

Environmental health, heritage and sustainability

15.1 Introduction

Real property is a dynamic asset, that is, despite the perception of 'bricks and mortar' permanency, the function and, therefore, value of any of these assets change over time, impacted by a range of external influences and factors. These may relate to shifts in consumer demand and preference, the changes in the immediate and wider locality or from economic, technical, legal and environmental change. As communities and societies become more sophisticated and complex, the number, range and potential impacts on property assets also increase.

As discussed in Chapter 5, managing risks and securing competitive advantage from opportunities requires, in the first instance, a detailed understanding and analysis of those threats and opportunities. Thereafter, it involves establishing strategies and actions that will reduce the risk profile or enhance the ability to secure maximum benefit, depending on the particular case.

This chapter considers three such influences and their potential impacts for real property: environmental health, heritage issues and sustainability, in the broad meaning of that word. These areas have various positive and negative implications for asset owners. If misunderstood or badly managed, they can be highly detrimental to the functionality and value of the asset, raising its risk profile considerably. The same factors, however, if properly assessed and effectively managed, can, in fact, lower risk and increase capital value compared with competing assets where the changing circumstances are not well recognised and addressed.

15.2 Environmental health

The majority of people spend a great proportion of their lives enclosed within the built environment. Depending on design, layout, locality and a range of other factors, this can be a successful and pleasant experience, enhancing liveability and productivity, and protecting health. That, of course, is not always the case. It is only through research in recent years that the importance of the built environment has been recognised in protecting and enhancing physical health and general well-being.

Glaeser (2011) observes that the real triumph of urbanisation and cities in the last 150 years was perhaps not simply about the architecture or even the economic activity generated but, critically, it was about human health and well-being. He notes that with the rise of contemporary cities, the populous, for the first time in human history, could be provided with truly safe drinking water and sanitation systems, both of which had dramatic effects in reducing death rates and increasing overall health and well-being.

Overcrowding and poor air quality, sanitation and physical surrounds have well-recognised, adverse effects on physical, physiological and emotional well-being and, in a business environment, productivity. In the more recent past, the widespread use of asbestos and chlorofluorocarbons (CFCs) in buildings exemplifies the magnitude of the problems that can emerge in these areas. For reasons outlined below, such matters have become increasingly important issues for all property owners and particularly those providing these assets to third parties, given the duty of care now well recognised.

Environmental health in the built environment includes a range of physical, chemical, biological, social and psychological factors that impact on both physical health and quality of life. It also refers to the theory and practice of assessing, correcting, controlling and preventing those factors in the environment that can, potentially, adversely affect the health of present and future generations (US Department of Health and Human Services 1998).

Environmental health in this context, therefore, does not simply negate potential damage or injury but, in a much more positive aspect, aims to promote and enhance health and well-being. Clearly, these matters relate to the concepts of sustainability discussed in Section 15.5. They are also linked to building design, environmental influences and factors including air quality, humidity and temperature.

Environmental health is now of fundamental concern to employers (as lessees) and building occupants alike, and, by implication, to asset owners and managers. It is increasingly becoming a matter of competitive advantage/ disadvantage and a point of difference between various properties on offer.

The nature of contemporary office work and those involved with it now places a greater emphasis on the physical environment, occupational safety and the provision of a healthy and safe environment than was the case in the past. This has legal as well as economic implications for owners and managers. On the face of it, it is the primary responsibility of any employer to provide a safe working environment. However, there is, underpinning that, a responsibility and 'duty of care' that falls not only to the tenant (i.e. employer), but also the building owner and manager – that the building will be safe for the use for which it is offered (e.g. office, retail, residential or other). This can be an onerous and serious obligation, since fires, outbreaks of airborne diseases and other emergency situations can occur within the building.

Additionally, the management of these risks is often difficult. A building can never be made completely safe. Sometimes problems and risks are difficult to identify and manage in advance of an incident. Sometimes too, adverse effects of air-conditioning, chemicals in use and so on will vary widely depending on the susceptibility to those issues for the individual involved. The level of risk will also vary with the size, characteristics and composition of the population using a particular asset. Such volatility, both in cause and effect, makes the task of management particularly challenging.

For all of these difficulties, however, environmental health, like practically all other risks, can be managed if it is understood and addressed in an efficient and cost-effective way.

Environmental health risks can emerge from a huge number of sources. However, analysis would clearly show that the vast majority of issues emerge from four key areas: indoor air quality (IAQ); the quality, state of repair and cleanliness of surfaces and finishes; compliance with regulations, standards and acceptable industry practice; and finally, effective and provable systems and operations.

Problems with environmental health can typically be addressed at one or more levels – in design (i.e. eliminating the problem at the source in new construction, redesign or refit), management and management systems, and/ or operations and operational procedures.

15.3 Air quality

As noted above, many of the environmental health issues in the management of environmental health in buildings relate to IAQ and the design and operations of air-conditioning systems (May 2006).

These problems typically fall into two main categories:

- Building-Related Illnesses (BRIs): These are caused by specific organisms and contaminants in the air known to adversely affect human health. They include a range of infections and airborne disease, including the notorious Legionnaires' disease and a range of others. These organisms are typically airborne but require warm water in which to breed, thus making air-conditioning systems ideal habitats unless properly operated and maintained. Also included in this group are a range of toxins, carcinogens, chemical emissions, pollens, petrol, tobacco smoke, and other fumes or airborne particles that could cause either illness or allergic reactions.
- Sick Building Syndrome (SBS): As well as the specific and identifiable illnesses identified as BRI's, it is also now recognised that some buildings appear to induce higher incidences of absenteeism, illness and occupant complaints, even though scientific or environmental tests cannot find specific illnesses or toxins within them.

When these issues began to emerge some decades ago, they were thought to be perhaps psychosomatic, or caused by other matters related to contemporary work, and not necessarily caused by the built environment. Even then, they were recognised by the owners and managers of buildings under the generic title 'SBS'. It was clear that they would have to be addressed because of the adverse impact on leasing capacity and income generation in the longer term.

Eventually it was established that many of these issues did emanate from a certain category of building and, in particular, the air-conditioning systems within them.

For a number of reasons, there was a large increase in the number and scale of commercial office buildings in Australia from the late 1960s to the 1980s. While some were of high quality, many were of a lesser 'investment' standard. This occurred during the period of the first oil crisis and, as a result, many buildings were constructed with under-capacity air-conditioning, often relying on large proportions of recycled air, ostensibly to save energy. Subsequently, the heat load on these buildings increased with rising office density and a huge increase in the use of heat-generating office equipment, particularly, computers. The inevitable result was that buildings typically operated at a temperature significantly above the benchmark of 24°C, and often with poor air circulation. Occupants repeatedly complained that the buildings felt stuffy. The buildings became 'tight', as variable air volume (VAV) systems and plant struggled to meet the additional loads.

Even though these problems were identified, retrofitting and rectification were invariably expensive and, in some instances, physically almost impossible.

Owner and manager responses

For reasons outlined above, environmental health has now emerged as a high priority issue for which both owners and managers must take responsibility. Responses are at a number of levels.

Where possible, for example, in new construction or major refurbishment, the best solutions normally lie in 'designing in' features and components of the new/upgraded asset which, within budget parameters, promote a healthy and attractive physical space. Whatever the asset, however, proactive property management and high-quality maintenance and building operation will be fundamental to success.

Good design represents the basis of an efficient and effective building but, if those fundamentals are not in place, it can be extremely difficult for any level of management to compensate. Redesign or retrofitting will always be an expensive and disruptive option. In addressing environmental health, good design can in-build a range of passive and active features that encourage good health and increased productivity, liveability and comfort. Some of these design features include: correct building orientation; building footprint and shape; the optimum location of service core; the avoidance of deep floor plans; overall image, aesthetics and scale; shade and use of vegetation; simple logical layouts; provision of necessary services; use of natural ventilation (where possible); use of appropriate colours; use of natural light, including appropriate glass tinting, external shading and size of apertures and the use of atriums, light wells and so on; use of artificial light, including diffused light, ensuring light consistency and the avoidance of glare; consideration of floor plan and layout that supports business operations; generous floor-to-ceiling heights; and the convenience of, and ability to control, services by occupants.

Particular care in air-conditioning design with emphasis on flexibility, quality systems, control, zoning, filtering, location of inputs and the accommodation of seasonal variations and changing directions of sunlight is required. Sucher (2003) notes particularly that it is the 'human scale' and finer features in the design and operations of buildings and the wider built environment (i.e. public and civic spaces) that most influence the level of convenience, liveability, familiarity, safety, predictability and comfort of a place. These attributes are arguably more important to the human experience of the built environment and, therefore, to well-being and health than simply good design and innovation for its own sake (Week 2002). These matters were discussed in more detail in Chapter 4.

High-quality management and maintenance systems are critical to promoting environmental health. There are, for many building components, statutory requirements and standards that must be understood and require strict adherence. These particularly apply to air-conditioning, machinery and electrical systems. This is not simply to reflect good property practice or avoid strong legal penalties; they are matters of human health where injury or even death can result from poor practice.

Environmental health needs to be instilled as a critical issue for all property management and maintenance/operational staff and good communication, information-sharing and training are important components of that. Management systems must include up-to-date and accessible practice and procedure manuals. Practical facilities management also requires adherence to training, practice drills, maintenance schedules and inspections (both formal and 'management by walking around'); regular inspections by experts; and ensuring the maintenance of all equipment in a clean, serviceable and efficient condition.

In complex undertakings, such as the management and use of major assets, mishaps and problems will invariably occur. Good management, however, learns by looking continually for root causes and trends, and by managing risk issues at the source.

Within the building, works (particularly fit-out) and other activities by tenants need to be properly managed to ensure that health and safety considerations are not compromised. Covenants within commercial leases will normally address such issues, but regular inspections should be undertaken to ensure compliance. The safe and efficient use of all building and office equipment should be encouraged. Particular care needs to be taken with the use of chemicals (for cleaning and equipment) and with the storage and handling of dangerous materials (May 2006).

As regards management and maintenance of air-conditioning equipment, the fundamental considerations are cleanliness (e.g. of the air systems, vents filters and ducting), the maintenance of consistent temperatures across all parts of the building (the standard being from 22°C to 24°C), and the control of humidity and movement of air through the building. In most locations and across most seasons in Australia, the temperature and the quality of air being drawn into a building is quite acceptable. Often, using only the inexpensive fan component of an air-conditioning system to blow air through otherwise stuffy buildings may be an efficient and effective way to maintain comfort levels.

Managing environmental health, like many other components of property management, is based on successful dealings with individuals, whether they are the lessees, occupants or visitors. Good relationships here depend on the manager being reasonably proactive, asking for tenant input, involvement and opinion and providing information and explanations of issues and activities around the building. Being responsive to reasonable complaints, following up outstanding matters and facilitating reasonable tenant/occupant requests all represent good practice and support sound relationships between all stakeholders in the building. Many such matters are the responsibility of the employer (the lessee) and their dealings with their employees (e.g. the provision of quality office furniture, quality fit-out and provision of sufficient space for activities). There is a range of other matters, however, that requires the combined effort of both the tenant/employer and the owner/manager to deliver desirable, safe, habitable and logical accommodation. The fostering of a cooperative approach provides long-term benefits to the success of the asset as a whole.

As noted above, there are a number of potential areas of high physical risk within buildings and these need to be the basis of regular inspections and monitoring. The building construction and safety codes and standards from various Australian jurisdictions and organisations identify key issues including:

- emergency exits (e.g. compliance, design, identification, accessibility, and lighting)
- fire services (e.g. compliance, type, location, servicing, maintenance and checks, pressure testing, watch for leakages, evacuation procedures, practices and records)
- electricity (e.g. certifications and inspections, loading of switchboards, efficiency and safety of motors, adequate signage, safe use of power cables and leads)

- air-conditioning (e.g. statutory compliance, particularly as regards water quality; air testing, cleanliness, particularly of filters; checks on zoning and fit-out, plant maintenance and efficiency testing)
- aisles and walkways (e.g. free from obstruction, rectification of slippery or broken surfaces, adequate lighting)
- stairways and landings (e.g. design, width, hand rails, landing size, pitch, tripping hazards, broken or damaged surfaces, cleanliness, accessibility, lighting)
- floors (e.g. remedy wet, greasy, slippery or broken surfaces)
- hydraulics (e.g. testing of pumps and systems, checks for clogged drains and water ponding/seepage/leakages)
- lifts (e.g. statutory compliance, scheduled maintenance and regular condition and operational investigations and reports)
- general (e.g. correct handover procedures, maintenance of handbooks, records and 'as built' drawings, training and records of training and drills, including training of new staff and occupants, statutory compliance, maintenance of quality records, reporting and fault/ failure analysis).

15.4 Heritage

As communities mature, become more educated and are exposed to rapid growth and change, they typically become more interested in their own past experiences and history. Sometimes these interests relate to nostalgia, emotion or an attempt to align to a perceived more stable and simpler past era (Mumford 1961). Often, they have a base in culture, tradition, sense of identity, ethnicity and a general interest in both distant and more recent history.

Australia represents an interesting case study in these matters. In the first instance, its Aboriginal history represents one of the most ancient in the world and it is only in recent years that this tradition has been widely recognised and appreciated – given, however, their nomadic existence, little of that relates to the built form. Meanwhile, European settlement in Australia is, by the benchmark of other cultures, very recent – little more than 230 years old. Until contemporary times, places and buildings, machinery and other items associated with the past were often seen as old-fashioned, dated and anachronistic. Many were destroyed and replaced by more modern built forms and practices.

In recent decades however, as Australia attempted to better define its own identity, links to history have gained greater importance. Sometimes that interest pertained to some historical event of national or regional importance, sometimes to a particular activity, industry or undertaking. In other cases, interest focussed on a particular iconic place or building, a precinct, or entire town that presented some important historic feature, style or design. Often, the place or building was important not simply for its physical characteristics, but for its location at which some particular, important event occurred.

Historical interest increased as the community, and particularly interest groups, observed the physical loss of, or significant change to, these places. The move to protect heritage buildings and precincts reached a watershed in the early 1970s where, led by building unions, 'green bans' were applied to large areas of The Rocks and Woolloomooloo in Sydney, thus saving them from demolition and redevelopment. Other civic, community and union action followed to protect similar buildings and sites of heritage value across many parts of the country. In the face of overt public opinion, the New South Wales government, followed by the federal and other state governments, enacted heritage protection legislation that has been further developed and extended since that time.

Few would argue with the value proposition that heritage is important to the community and regional and national identity. With globalisation and the rapid sharing of knowledge, iconic sites and the presentation of threatened cultures and communities have become matters of international relevance and priority (Labadi & Long 2010). Consequently, it was seen that important heritage places within the built environment needed to be protected and carefully managed for current and future generations. Many sites are in public ownership where restoration and management could reflect community objectives while, hopefully, securing asset functionality. Difficulties often arise, however, where assets are in private ownership and the ability to restore a building to an appropriate standard may be problematic compared with the economic returns derived.

Definitions, legislation and guidelines

Heritage can be said to constitute

> the valuable features of our environment which we seek to conserve from development or decay. It recognises the existence of underlying deep-rooted but undefined values. It evokes a lofty sense of obligation to one's ancestors and to one's descendants and it secures the high ground of principle for conservationists.
>
> (Davison & McConville 1991)

Bell (1990) put it more succinctly, defining heritage as 'our collective memory – the physical relics of our history, our sense of place, time and community'.

Regarding the built environment, historic or heritage sites and buildings are those which have intrinsic to them some broad cultural, political, economic or social history relevant to the nation, state or community. They exemplify to an observer, the history, culture and heritage of that location (National Trust for Historic Preservation 2012).

It is important not to confuse the concept of heritage with simple fashion or taste which in themselves will prove transitory. Moreover, heritage is not confined to major or iconic buildings, a certain type of design, attractive features or styles, nor to restored properties. It will include the humble along with the great, and those more recent in time than 'historical' in the normal sense of the word.

Essential criteria here are whether the site in question is of a physical character and/ or social representation of an era or a period, or if its state of preservation and authenticity are worthy of special consideration, or protection, in its current or restored form. It might be observed that many of the criteria identified are, to some extent, subjective and certainly open to various interpretations when applied to a particular site. This lack of certainty often sits uneasily with the economic realities of property ownership and investment.

Individual properties and sites clearly have spatial dimensions; the location of a particular site, its ambience, surrounding development and streetscape may be very important in assessing both its heritage value and a case for its preservation. On this basis, even though one part of a property or building may be of heritage significance, the whole of that property/holding may well be identified for the purpose of heritage control/ management. During the long history of the asset, there will almost certainly have been considerable modification, additions and upgrades of the original building. Consequently, it is often difficult to accurately identify and assess the underlying heritage value/components of that asset, site or precinct in an overall sense.

Because of the unique nature of individual sites and their heritage characteristics, each must be considered on the merits of that particular case. It is generally true to say that exemplification (i.e. maintenance of a small section of the property in its original state) or the retention of a simple, restored façade, as some reminder of previous buildings and developments, fails to preserve heritage values. The basis of this observation is that, if the heritage characteristics of a property are in fact worthy of preservation, it is highly likely that a small area or a well-preserved façade will be little more than a caricature of the matter of importance.

Legislative control and influence on heritage buildings, or buildings with heritage potential, emerge from federal, state and local government throughout Australia. The number of sites identified as having heritage significance continues to grow.

Although the Commonwealth Government would seem to have limited, specific powers over land, except in its own small territories, it is noted that the Australian Heritage Commission has the ability to exert substantial influence through a register of the National Estate and, in significant cases, the ability to secure World Heritage status through the International Council on Monuments and Sites (ICOMOS).

Of more significance overall is the heritage legislation enacted in each of the states and territories of Australia over the last three decades. As discussed in Chapter 6, state legislation is of particular importance because of the direct legislative control that state governments exert over land, tenure and land uses. While there are important differences between legislative parameters in each state, all have some form of heritage protection legislation, typically setting up a heritage register to identify places considered of heritage importance, preservation and/or management. Most establish some form of regulatory body (such as a Heritage Council) to assist in the management of that act and, particularly, to approve changes to the physical structure, use, additions, demolitions, et cetera, of a heritage-listed property.

Practically all of these pieces of legislation refer to an accepted national standard for the conservation of places of cultural significance, known as the 'Burra Charter', named after a conference held in Burra, South Australia in 1979, and has been amended as the need arose. The charter recommended a two-stage process for dealing with a heritage property. In the first step, a methodology was established for the identification and qualification of the heritage value of a particular site, building, use/activity or the like, known as a 'Statement of Cultural Significance'. Second, guidelines for the development of a Conservation Plan were established to preserve and manage the items of significance identified in the first stage.

It is important to note that the Burra Charter does not specifically address such issues as the economic impact or the financial feasibility of development that might be allowed under the Conservation Plan. Typically, the Conservation Plan will provide little or no guidance on future uses or changes that might be applicable for that site, leaving specific issues for future application by the owner to regulatory authorities. While this is understandable given the purpose of the charter, it will often lead to a disconnect between the demands of legislation and the reasonable demands for economic return from such assets by owners.

Approvals and regulations pertaining to particular state heritage legislation are normally triggered only by some change in use, or a development application of one type or another. For the most part, therefore, heritage legislation is not proactive. Unless an owner seeks some change in use or development approval, the legislation does not require the maintenance or specific presentation of existing heritage values or features, though owners are typically prohibited from destroying or altering them without approval. Therefore, heritage legislation does not guarantee protection per se, let alone the enhancement of most heritage sites.

Under the provisions of most of these acts, any person, including members of the general public, may seek to have a particular property included on a state heritage register at any time. While, of course, a Heritage Council will undertake all necessary investigations to assess the validity of these applications, these provisions must adversely influence investor confidence, given that inclusion on a state register will bring a range of additional development controls and approvals into force.

Finally, many larger local authorities or those whose areas include significant heritage precincts will often have their own heritage register and planning by-laws, heritage advisory committees and, sometimes, identification under ubiquitous terms such as 'properties of interest'. The power to regulate and enforce local-authority requirements varies with jurisdiction, but at times this involvement can create more uncertainty than assistance.

Overall, the multiple layers of control, as well as a number of heritage registers, which do not necessarily coincide, often provide a confusing and uncertain environment for property owners, managers and tenants alike.

Implications of heritage controls on property and its management

The actual and potential effects of heritage control on property assets, their development, use and value are diverse. For the large majority of properties, the chance of heritage listing is very remote and, therefore, the issue does not arise. For other properties, the recognition provided by a listing with the National Trust, or a state Heritage Register, may well prove beneficial. Examples include certain hotels, tourist accommodation and attractions, and historic precincts. Such buildings/precincts will often attract a certain type of tenant or use, with those users willing to pay proportionally higher rent for that address, image or ambience.

For many other properties, a heritage listing or even its potential brings complexity and uncertainty. The raising of finance and the analysis of property value become problematic because of the difficulty in ascertaining whether certain future uses will be permitted, the level of restoration work required and ongoing special management obligations. Regardless of heritage requirements, the refurbishment of such buildings typically brings a range of challenges including structural integrity, water ingress, ability to secure specialist materials and trade skills, net-to-gross inefficiencies, provisions of contemporary services and tenant demands, and issues in meeting current building standards, particularly fire regulations – all of which potentially add to development and construction risk. A generous approach to contingency allowances within development costs is therefore prudent.

Even assets held within the public sector can bring added difficulties. Viable and sympathetic uses are often difficult to identify and involve additional restoration and maintenance costs compared with alternative property accommodation. Therefore, for those public sector asset managers, the challenges were similar to those for their private-sector counterparts. The investment to secure usable space through renovations on heritage buildings could not normally be justified financially and, as a result, the heritage building was often left unrestored unless special funding becomes available or specific political direction emerged.

In either the public or private sector, acceptable uses may be difficult to identify. Activities that encourage public visitation or use – meeting places, museums, restaurants, galleries and the like – would appear to have potential. However, in many cases there is only a finite demand for such uses and, in any case, the need for renovation, additions for parking, climate control and the provision of back-of-house facilities including kitchens, toilets, loading facilities and storage areas, often make adaptations difficult or intrusive to the heritage characteristics.

For privately owned heritage buildings, potential issues arise between public and private interests. The general community may well enjoy the streetscape or precinct provided by heritage buildings, but, where the individual assets are in private ownership, there is little or no public contribution to restore or maintain that vista. More specifically, relevant legislation may prescribe a standard of restoration, perhaps using Burra Charter criteria, that may well involve costs to the owner that cannot be recovered through rent or later sale. Where this gap is too great to be accommodated in the development, the project will simply not proceed and it is, therefore, unlikely that the building will be restored.

The withdrawal of this potential reduces functionality and the property progressively deteriorates – the antithesis of the objectives of the community, regulators and the property owner.

In most jurisdictions, land-value reductions and/or concessions are available from general rates and land tax to recognise any potentially negative impact of heritage controls. Such benefits are of relatively small scale and unlikely to significantly influence investment decisions. In many town planning schemes and to assist recognising the impediment that heritage listing may present, Transferable Development Rights (TDR's) were made available which facilitated the potential development that would otherwise have been available to the heritage site to be transferred to another site, normally in the same vicinity. While those rights may well have been of value, the ability to fully benefit often proved more limited in practice.

Proposals have been advanced in the past, suggesting that additional benefits, perhaps through increased depreciation allowances, or some relief from capital gains taxation, would improve the value proposition of investment in heritage properties. These have generally not met with support from taxation agencies.

15.5 Sustainability

As a result of the frequent and pervasive use of the term 'sustainability' in contemporary language, the concept has become somewhat clichéd and its real meaning and gravity are at risk of becoming eroded. The complexity and interrelations of its components are likewise often not fully appreciated. Sometimes, because of their magnitude, it is easy to consider such issues as 'someone else's problem'.

Furthermore, in recent times, the word has been employed by a range of individuals and groups from across the public and private sectors willing to use its currency to prosecute any number of political, corporate, marketing or other agendas. All of this confuses and dissipates the real message of 'sustainability' which, in the widest and correct use of the word, represents arguably the greatest challenge of this generation.

Definitions and parameters

While a range of meanings and descriptions can be put forward, a definition of 'sustainability' (based on that established by the World Bank) is:

> … the strategies and methodologies that optimise current function and use of physical, economic and community resources, while ensuring that options for the use of those resources into the future are not unnecessarily prejudiced or compromised.

Related to this definition is the concept of 'triple bottom line' (TBL). Under this mantra, business and other corporate activities take place within a framework and reporting regime that aligns with sustainability aspirations. In its broadest terms, TBL is defined as an integrated philosophy requiring a company or other venture to address its economic, social and environmental values, strategies and processes, and report on each measure. These components are commonly referred to as 'the three Ps' – people, planet and profit (Slaper & Hall 2011).

It can be noted from these definitions that, while there is a close association between sustainability and environmental protection and management, they are clearly not the same. Sustainability does not imply conservation or preservation per se. It relates to all forms of resources: physical/natural environments, economic/business, human resources, community, social structures and activities. Rather than being passive or negative, it is about optimisation, efficiency and innovation – ways to manage development and consumption, not destroy growth and the benefits that come with it.

Within a semi-planned capitalist economy such as Australia's and other OECD countries, a balance needs to be struck between the legitimate, immediate microeconomic rights and self-interest of the individual owner and the 'macro' interests of the community, region and country. That balance lies at the core of successful sustainability initiatives and government has a range of policy, innovations, knowledge transfer, incentives and taxation options available to achieve that.

Some background

For some of the reasons outlined above, many of the initiatives and activities undertaken under the general concept of 'sustainability' have only tenuous links to the core issues and challenges of the topic. Typically, these activities hold to 'tag lines' such as 'environmentally friendly', 'good for the planet' or, perhaps, the most meaningless of all – 'sustainable'.

An important starting point must place the debate in a historical, economic, environmental and social context.

There is no doubt of the compounding nature and effects of change and development over recent decades. Typically, in times of great advances and upheavals, those directly involved find it difficult to grasp the totality of the events under way.

The Industrial Revolution represents a case in point. Even though it commenced in the late 1700s, the term 'industrial revolution' and an analysis of all its causes and effects did not emerge until the 1870s. Even the father of economics, Adam Smith, who lived through some of the early stages of this era, failed to recognise the advances in technology, development and changes in industry structures that it represented. Similarly, in the contemporary environment, commentators use words such as 'sustainability', 'globalisation', 'information revolution' and a range of others, to attempt, probably inadequately, to explain current events (Brand 2010).

In reality, the contemporary, economic, physical and social/demographic environment represents the aggregated effect of fundamental, technological and other drivers that have emerged over the last century or more. These include the wide use and adaptation of electricity, the dominance of the internal combustion engine, dramatic and compounding breakthroughs in medical science, with major impacts on demographics, exponential growth in the access to data and knowledge and the aggregation of political and financial power, notably in the US. Many of these impacts have been positive, but some have had negative, ongoing subsequent effects.

Encapsulating much of this, however, is the dominant theme of urbanisation. All of the factors identified above have, for a range of reasons, resulted in an ever-increasing population that is concentrating more and more in urban footprints. These are predominately located in littoral and estuarine areas around the world, but particularly so in the tropics and subtropics and around the Asia–Pacific rim. Significantly, from around 2007, and for the first time in history, more people lived in an urban rather than rural environment. That trend will continue for some decades to come (Brugmann 2009).

These matters were further discussed in Chapters 1 and 4. This is the true urban context of sustainability. Clearly, this context is confined not only to the physical environment, but also to community and business activities fundamental to those concentrations (ETN Communications 2007; Montgomery 2013). Urban cities have historically been able to concentrate finance, technology, innovation, community action and political will to generate solutions to fundamental problems. Furthermore, while not underestimating the challenges ahead, concentrated richer populations in urban areas may prove, over time, an easier environment in which to address sustainability issues than widely scattered, poorer rural regions (Glaeser 2011). The 2015 Paris Climate Change Agreement indicates the ability of the world community to eventually coordinate a response to the climate component of the sustainability debate. It may also bring investment in renewable, rather than non-renewable energy technologies, into the economic and investment mainstream.

The physical environment represents the basic manifestation of the sustainability issue. This takes two partly related forms: global climate change and, second, loss of biodiversity. It is critical, however, to observe that these are larger manifestations – the physical end result of a diverse range of issues and drivers emanating principally from the actions of industry, the urban environment and primary production. The final outcomes – climate change and loss of biodiversity – are global, immense and, so complex; they are unresolvable as totalities (i.e. as single issues). Rather, they have to be deconstructed into their components and their upstream contributors. As noted above, one of the critical contributors is the urban-built environment.

In considering the issues of environmental change and sustainability, the role of economic growth and levels of income are particularly important. In this, there are contradictions. As noted above, as incomes rise, so do levels of consumption and, therefore, the potential to increase ecological footprints, particularly regarding the consumption of energy. This is the experience in most developed countries over decades, and now occurring across the developing world (Rubin 2009). Ironically, however, it is those wealthy countries that have the government funding, governance and legal control, research and education to take affirmative action to address environmental degradation and challenges to sustainability. It is countries like the US, the UK, Canada and Australia, notwithstanding their quite appalling records of consumption of non-renewable resources, which, in fact, have the most sophisticated commitment, laws and practice in environmental protection and sustainability.

Several other external forces including the role of technology and innovation, and the sometimes countervailing forces of existing fixed-capital investments, need also to be noted here.

The insightful work of Joseph Schumpeter (1883–1950) noted that history over the past two centuries is dominated by what he called 'creative destruction', that is, in the face of changing demand, new, disruptive technologies can quickly sweep through economies, fundamentally changing practically everything (McCraw 2009). In the past, the steam engine, the internal combustion engine and the widespread use of electricity present clear examples of these intermittent waves of breakthrough technology. The widespread use of information and communication technology (ICT) provides a contemporary example.

Consequently, and repeated through human history, when a challenge has arisen, such as sustainability in the current era, new practices, particularly innovation and new technologies, have risen quickly to meet the challenge. Without underestimating the sustainability challenge for the planet, there are now emerging models such as 'ecologically sustainable development' (ESD) and 'environmentally sustainable practices' (ESP) that are capable of addressing these issues over time. There would appear to be a strong and urgent case to rethink existing models of economic activity and objectives in view of the impact of the current, growth-driven approach to wealth creation (McKibben 2007; Jackson 2009; Gilding 2011). The ramifications of that for economic policy and behaviour, and by implication, the development and use of the built form are yet to become clear. They will, however, almost certainly widen the interpretation of economics into a civic and society-based model with changes in business practices, lifestyles and in energy and transportation systems (Morgan 2013). The COVID 19 pandemic will provide a further major catalyst to such changes in the resetting of economic and civic priorities.

The problem is, however, that a fundamental change comes at a cost – the obsolescence and redundancy of the huge investment in existing fixed capital. This could include petrol-driven cars, coal-fired power stations and so on, which are typically highly specialised and cannot easily be adapted. Disruptive technologies, especially in their early stages, are, therefore, not particularly welcomed and are frequently challenged – not just by large corporate interests but, indeed, by continued consumer preferences and habit, and also by governments, many of whom find political safety in avoiding radical change.

Given that scenario, fundamental change will not be easy within a capitalist system. While there exist strong profit motives to introduce innovation and new technologies to solve contemporary problems, there is also considerable inertia against that change by the owners of the existing entrenched technologies and fixed-capital assets challenged by change. Nevertheless, as Schumpeter observed, there comes a point where the case for that new technology and its benefits become overwhelming and, despite those strong countervailing forces, the old technologies will collapse and be replaced, often much more quickly than might have otherwise seemed possible. This is known, obviously enough as a 'tipping point' (Gilding 2011).

In the case of sustainability, the impacts of peak oil (i.e. maximum rate of economic extraction) and changes in energy prices, new carbon taxes in one form or another, changing consumer demands, the rise of ethical investment, the impact of rapid technological advancement, continued challenges to old technologies through sector regulations and, finally, meaningful international agreement will together represent the fundamental challenge to a 'status quo' future.

The built environment

The built environment is highly significant in the sustainability debate. In a physical sense, buildings consume about 32 per cent of all energy consumed (International Energy Agency 2015). While impressive, that figure could reasonably be expected, given the amount of time that humans spend in a range of activities inside the built environment. These statistics also reflect the rapid urbanisation of communities and societies over recent decades (Jowsey 2011).

As 'machines', however, buildings have been less than efficient in their traditional form. Often they were built with little regard to efficient passive design, correct building orientations or energy efficiency.

Less obvious, but arguably, equally important, is the role the built environment plays in the overall response to sustainability and a Triple Bottom Line ('TBL') approach to business. Buildings surround people's lives. They provide a platform for business, and a 'place' for all manner of community activities and interactions. The influence of the built environment on the sustainability debate is, therefore, much greater than an inactive role or simply providing 'passive space'. The success, integration, liveability and, broadly defined, sustainability of individuals, businesses, communities and the surrounding physical environment are all impacted, either positively or negatively, by the effectiveness of the built environment that is occupied (Appleby 2011). As Winston Churchill observed in 1943: '... we shape our buildings, and afterwards our buildings shape us'.

A key component in all of this is time. Land and various types of building development remain operational far beyond commercial horizons or indeed human life spans. Development and buildings are, in effect, predictions of the future and predictions that, to a varying or greater extent, will prove wrong. Few could have predicted the nature of change that has occurred over the last few decades. Exact predictions of the future would appear to be problematic at best. On the face of it, this creates an enormous challenge to fixed assets, such as property, which do not accommodate change well.

Nevertheless, buildings can and do adapt to change in demands over time. Strategies to 'in-build' flexibility and adaptability from the outset would appear critical if obsolescence is to be avoided (Brand 1994).

Over the last decade, advances in sustainable practice associated with property and development have been very significant compared with that in other sectors such as manufacturing and primary production. There is a raft of different initiatives, some led by government, but many initiated within the private sector, which represent major advances (Yudelson 2008). These include such region-wide schemes as overall land use and infrastructure plans, the development of master planned communities, transport oriented developments, the facilitation of infill developments, the establishment of integrated urban villages, and higher levels of investment in public transport. Likewise, innovative design based on new rating schemes and government regulations are now producing a quite different built product and urban development than in the past.

A wide range of contemporary information relevant to the above is available on websites and elsewhere. Many, particularly those provided by government, universities and other institutions and professional and industry groups are of very high quality and value. However, as noted in Section 1.6 of Chapter 1, care must always be taken in confirming the quality, independence and currency of any information in this or other similar fields, given the importance of subsequent decisions that rely on that information.

While there are a range of initiatives to produce more sustainable new buildings and assist refurbishment projects, annual increases in new commercial building stock, for instance, has averaged only 2.1 per cent over the ten years from 2000 to 2010 (PCA 2015). Urban development and built assets have a very long life cycle and, important as the sustainable qualities of the new developments are, the upgrading of existing stock to contemporary standards is likely to be protracted (Sassi 2006).

Legislative action is important but, by its nature, legislation tends to be reactive, typically following behind public opinion and industry standards. It will, understandably, only set regulatory minimums rather than encourage or reward innovation and excellence.

Another important observation is that the property market is segmented and diverse. For example, a large-scale owner/occupier of commercial space, or the owners of high-value residential property, may well place a high priority on specifying, and paying for, inbuilt, sustainable features. The same cannot be said for more cost-competitive sectors of the same markets.

While all understand the environmental and long-term economic benefits of sustainable practices, a first home buyer will have little available resources to purchase additional sustainable features. Likewise, a typical small office tenant may see little day-to-day benefit or reward in an ongoing commitment to sustainable practices. Again, government regulation can assist these sections of the market, but there may well be political sensitivities to actions that add to the overall cost and create adverse effects on affordability.

Nelson et al. (2008) note a frequent, generalised criticism – that increased regulation and design emphasis on sustainable and/or environmental features adds considerably to cost and time delays in new construction. However, that research goes on to largely contradict those impressions. Approval authorities may be slow and inefficient, but that represents a separate issue. Increasingly, it would seem, those regulations simply reinforce sustainable design and operations demanded by the market.

While there is strong political and community interest in the area of sustainability, it is economic drivers that will provide the catalyst for major change. For example, the cost of housing and, therefore, demand, is a function not simply of building costs, but also energy, transportation and a range of other living or operational costs. As those costs continue to rise over forthcoming years, the market will favour product where overall costs are more predictable and comparatively lower while maintaining comfort and convenience. These factors, combined with ongoing government regulation and incentives, will ensure the market evolves into quite new and different forms, especially when the impacts of rapidly increasing private transportation costs are recognised.

In the case of commercial buildings, the stated preference of major owners and tenants for higher environmental rating for buildings will be critical. As well as compliance with statutory requirements, any developer would now wish to 'future proof' their asset by either incorporating features of a suitable sustainability/energy rating, or at least providing the ability to facilitate upgrades and changes to meet rising standards and market expectations. Lower quality older buildings without that ability will increasingly find themselves marginalised to the extent of demonstrably higher vacancies, higher capitalisation rates and lower capital values. As with most strategies in high-value long-term assets, flexibility, adaptability and pre-planning for change are of irreplaceable value.

Market drivers

For any economic or financial change to occur, there needs to be some particular, strong market driver or stimulus that changes existing behaviours and encourages a new approach to investment, management and operations. There would be little debate as to the benefits of a more sustainable approach to business and investment, particularly in the built environment. Nevertheless, a general or community feeling that there is 'merit good' in the adoption of sustainable philosophies does not, by necessity, translate into changes in investment and management patterns.

These observations reflect back to those made in Chapters 5, 7 and 8, which considered the characteristics of various property sectors. Therein it was observed that the majority of property in Australia was privately owned. A conservative approach dominates given the typical investment profile, whereby the owner invests large amounts of capital 'up front' hoping the market would reward that investment over time in the form of rent and capital gain. Additionally, the built environment and property markets are typically slow to react, given that so much of the investment is held over a very long time in existing stocks. Given those market characteristics, developers and investors need to have a logical, evidence-based case before they would be willing to implement significant change to product (Freeman et al. 2000).

When addressing investment in contemporary features such as sustainability, the developer needs to be confident that the additional investment in sustainable features will indeed be reflected in higher rents and later capital values into the future.

At this point, with limited sales evidence available in these novel areas, such a causal relation may be difficult to establish. Expressions of community and political interest, and the stated wish of tenants to be supplied with more sustainable product, may not translate into a willingness to pay higher rent for such services. These observations are particularly true for smaller tenants where even strict adherence to sustainable management operations and use may not reflect in significant financial savings for the business, given the marginal cost savings that may be involved.

As noted above, governments will also be sensitive to the impact of raising the standards required of existing buildings. This is due to the potential cost imposed and the avoidance of what might be seen as 'de facto' retrospective legislation. Pragmatically governments are aware of the financial and other impacts that such changes would have on their own extensive portfolios. Consequently, legislation will typically be aimed at setting regulatory minimums.

Government incentives and information-sharing, while of value, may have fairly limited long-term impact on behaviour.

In the commercial market, it will, as usual, be the market leaders that drive change. In the case of major commercial buildings, for example, it is already the stated preference of major owners and tenants to pursue high sustainability/energy ratings for the buildings they occupy. Part of this is because of corporate positioning and identity. Moreover, in these very large-scale assets, the savings achieved by accepting sustainable practices are considerable, and as energy, transport and other costs continue to increase, the cost advantages of sustainable practices also increase over time.

Led by these major asset owners and tenants, particularly all levels of government and large corporations, the adoption of sustainable practices will become commonplace, with legislation, regulations and rating schemes responding accordingly over time. Even the owners of smaller-scale assets will be likely to embrace these changes, knowing that, into the future, assets without those attributes and suitable ratings will simply not be considered by a range of public sector and major corporate tenants.

Valuation issues

For owners and investors, these additional costs need to be reflected in increases in market value.

Rapid changes in the property market make value recognition difficult. In the first instance, there are still fairly limited, though growing, sales and leasing evidence whereby comparative analysis could identify the valuation impact of new sustainability features. No doubt, however, that situation will rectify itself even in the short-to-medium term, as more and more buildings with sustainability ratings are developed, providing necessary valuation evidence. Perhaps in corollary, the situation has now arrived were properties will not be considered by prospective public sector or corporate purchasers or tenants unless they have achieved a certain sustainability standard.

For valuers, there can be difficulties in actually identifying the existence of many sustainable features, given that many of them are incorporated in passive design, and/or plant and equipment, which may not be easy to ascertain. The use of appropriate rating schemes for particular buildings will assist here; however, broader valuer education is also required.

In assessing capitalised values, recognition needs to be given not only to the shorter-term increases in rental that may be identified, but also the wider benefit, which may also emerge in a lowering of the risk profile of the asset into the future – 'future proofing' – as discussed above. Again, depending on market evidence and the professional opinion of the valuer, it may be that such an investment gives the property a higher level of certainty into the future, consequently reducing somewhat the acceptable rate of return demanded in the market and, therefore, having a positive capitalised effect on market value (Kilbert 2005).

The final impact, in terms of lower operating costs, higher capital value and environmental protection, will occur only with the concerted efforts of designers, managers and owners, together with tenants and their employees in the day-to-day operations of the building. Education and information-sharing, together with innovative and contemporary methods of communication and engagement of all groups, is vital if sustainability is to be delivered in practice as well as theory.

Rating schemes

Performance and comparison of physical assets such as buildings require some accepted form of measurement. Given the complex and multi-faceted contemporary issues involved (including environmental health, heritage and sustainability more generally), it is unrealistic to expect that a single metric or assessment tool could fully rate 'sustainable performance'.

Consequently, over recent years, a number of rating schemes/methodologies and assessment bodies have emerged in countries such as the UK, the US and Australia, with many of the Australian schemes based on the former countries mentioned. Some rating schemes specifically concentrate on environmental impacts, and/or levels of energy and services consumed by an individual building. Others also address wider, and often more qualitative, sustainability issues. Schemes most commonly encountered include:

- BREEAM (Building Research Establishment Environmental Assessment Method) – UK Building Research Establishment. This methodology measures the overall environmental performance of any building and provides an evidence-based scoring system that includes such issues as overall management, health and well-being, energy, transport, water usage, materials and waste, land use and ecology, and pollution levels produced.
- LEED (Leadership in Energy and Environmental Design) – US Green Energy Council. LEED is a voluntary certification system for buildings, providing measures based, effectively, on a TBL approach. It includes components assessing site suitability, water efficiency, energy and atmosphere, materials and resources, indoor environmental quality, locational linkages, awareness and education, innovation in design and regional priorities and issues.
- Green Star – Green Building Council of Australia (GBCA). Green Star is a rating scheme for new and retrofitted commercial buildings and includes the assessment of building attributes such as energy, water efficiency, indoor environmental quality and materials used (i.e. principally related to physical building attributes). While a voluntary scheme, it is increasingly used as an industry standard.
- NABERS (National Australian Building Environment Rating Scheme) – New South Wales Office of Environment and Heritage. This is an Australian national scheme, authorised originally by the New South Wales Office of Environment and Heritage, and provided an assessment and rating of existing buildings; it is increasingly used for statutory disclosures for commercial buildings.
- EnviroDevelopment – Urban Development Institute of Australia (UDIA) The UDIA has also now established an innovative industry-based standards system, EnviroDevelopment, which recognises high levels of sustainable performance for land developments (including subdivisions) in ecosystems, waste, energy, materials usage, water levels and community development.

Rating schemes are important, given their high level of recognition and the increasing demand for their disclosure in any subsequent dealings.

It is important to note, however, that the simple establishment of a rating level for a building does not of itself guarantee suitable performance, and a partnership between owners, tenants and occupants will be necessary to ensure that, in operations and use, the benefits of the established performance potential will be maximised.

For all of that, the wider use and application of rating schemes will provide the additional benefit of better differentiating the various classes of buildings. In turn, this should enable the environmental and wider sustainability performance of the asset to be recognised more easily in valuation assessments and in the wider property marketplace. It remains unrealistic, however, to believe that a single rating scheme system can, or indeed should, ever encapsulate a measure of all of the characteristics of a building, reflecting its sustainability profile and, even in the longer term, more than one measure will be required.

15.6 Environmental banking and carbon capture

The impact of global warming on habitability, use and economic prosperity is now well recognised and obvious. This has international as well as national and regional implications. Despite ongoing political debate as to appropriate responses, international agreements are now in place, which legally commit practically all countries, including Australia, to specific action to stabilise and thereafter to reduce to levels of carbon dioxide emissions into the atmosphere which were seen as a primary cause of these significant changes.

The most important, contemporary international agreement in this regard was the Kyoto Protocol to the United Nations Framework Convention on Climate Change which, from 2005, legally committed the 128 signatories, including Australia, to stabilise and then progressively reduce greenhouse gas (GHG) emissions (notably carbon dioxide) from their respective countries.

Australia has particular challenges in addressing obligations to reduce such emissions given its reliance on fossil fuels (particularly in coal-fired power generation), animal husbandry practices and dense urban forms and congestion. At the same time, potential impacts on low-lying developed coastal lands, adverse effects on national assets such as the Great Barrier Reef and the growing frequency of destructive weather events render Australia particularly vulnerable to these challenges.

As with other OECD countries, Australia's response has involved a number of initiatives – some representing coordinated strategies, others more stand alone. Obvious priorities have been the growing move away from fossil fuels and towards renewable energy sources. Other initiatives act to reduce energy usage and waste in economic activity and processes across various sectors and in attempting to modify consumer behaviour. As noted above, changes in the design, management and use of buildings have played a leading role in that evolution.

By way of penalty rather than innovation, a 'carbon tax' leveed on heavy polluters has been successfully applied in some other countries as a type of 'user pays' (here, 'polluter pays') stimulus to reduce such emissions. An advance approach on that envisages a 'carbon market' wherein credits could be bought and sold between emitters and those involved in activates that actually reduce or 'capture' emissions (see below).

While on several occasions Australia has come close to establishing such market-oriented frameworks, political differences and misunderstandings have failed to bring such initiatives to fruition. Another tranche of initiatives are compensatory in nature – that is, proposals to provide off-sets (or 'carbon sinks' as they are sometimes called) to emissions from large industry sectors. On the face of it, these present significant positive options. Australia remains heavily dependent on fossil fuels for power generation and transportation, and it would be naive and economically damaging to radically change that situation in other than the medium to long term. Therefore, some trade-off/penalty arrangement will remain necessary for a considerable time into the future.

Many of these emerging initiatives require significant land resources and thus providing new opportunities and income sources for land holders. These include the storage of carbon in soils ('soils sequestration'), the storage of carbon in underground geological formations ('geo-sequestration') and the storage of carbons in forests and other vegetation ('plant sequestration').

Soil sequestration can occur through changed farming practices particularly involving decreases in tillage, better crop rotation, improved water conservation and the encouragement of diverse crop plantings. In many cases, these simply represent good contemporary farming practices and should be pursued in any case. However, while such activities may indeed add to carbon absorption, it seems almost impossible to accurately measure those carbon absorption rates or 'net' economic returns from such activities. Consequently, the ability to establish a trading regime for such activities appears unlikely (Sanderman et al. 2009).

Geosequestration schemes typically envisage the actual capture of emissions from the sources of greenhouse gas emissions from, for example, a coal-fired power station and their redirection into underground geological formations. These concepts, also known as carbon capture and storage (CCS), are often associated with 'clean coal' initiatives; however despite government investment and research over a long period of time, the technical and economic feasibility remain unproven. Additionally, there continues to be strong negative local reaction to such proposals, given their potential to affect subterranean aquifers.

The best potential of these types of schemes lies in plant sequestration. These can involve the establishment of new, reserved forest plantations, 'afforestation' (i.e. the return of previous forest lands that in the past have been lost to other uses) or 'reforestation' (i.e. the restoration of degraded/ marginal agricultural or grazing land or forest) (Nijnik 2010).

Through any of these routes, there is potential for a new, long-term income that is likely to be more consistent and reliable than other farm incomes. The property owner's returns can vary from a passive 'rental' return for the use of that land through to, depending on their level of involvement in development, production and management, income, royalties or profit share from timber production and, of course, the sale of carbon off-sets. To be established, security must be provided through tenure arrangements that ensure the forestry usage is protected and guaranteed for a very long term (normally 40 years or more). Further, there must be in place the ability to transfer those 'carbon rights' and, as appropriate, the protection of the right to receive that ongoing income for the land owner and successors in title.

Such initiatives are common in Europe where forestry areas continue to expand, across both private and public ownership (Breuer 2018). As well as providing the traditional sources of timber resources, these forests are increasingly used for a range of water catchment, erosion protection and environmental protection and trade-off purposes. In Australia, there is a long history of the development and use of forestry resources. The country already has almost 2 million hectares of commercial timber plantations (ABARE 2019).

About half of this consists of earlier plantings of softwoods, principally exotic pine species in plantations owned and managed by various state agencies. Later, as suitable plantation hardwood species became available, new plantations were established, mainly using private-sector capital and expertise. International forestry corporations and investors were typically involved.

There are large tracts of land in Australia, much currently marginal agricultural land, which could be converted for such purposes. Rapidly growing hardwood species have been developed, which would be well suited to reforestation projects though their viability also depends on suitable climatic conditions and the availability of water. Good prospects for forestry expansion existed particularly in Queensland, where 39 per cent of Australia's forests are already located, together with the Northern Territory, Western Australia and New South Wales also having further potential in this regard (ABARE 2018).

Such plantations are progressively developed, planted and harvested in a continuous cycle and, therefore provide an ongoing 'carbon storage'. The plantation system in fact enhances the process as maximum carbon absorption occurs when the young trees are vigorously growing. Post-harvest, the absorbed carbon remains 'locked' in the finish timber products and re-plantings immediately commence.

There remain, however, three fundamental matters to be addressed:

- The Commonwealth government is yet to provide, with any level of certainty, requirements on industry in establishing carbon off-set arrangements or a market price or trading regime for those rights.
- Most states have now enacted property-based legislation to identify and protect off-set areas but unfortunately there is diversity in the approach taken. Some are based on a profit-a-prendre agreement, which recognises rights to future income derived from the land or crop. Others establish a more specific, contractual property arrangement, akin to an easement or a land lease. The latter appears to provide a much more certain environment; however national uniformity in approach would appear essential, given that such trade-offs would not be confined by state boundaries. In time too, such off-sets may be tradable on an international market where a consistent, simple approach and framework would represent a key requirement to attract large-scale investor interest.
- There are currently very low levels of knowledge and expertise in these areas, not simply with industry or potential investors but, surprisingly, among property professionals and other experts and advisors who will have an essential role in helping to develop and support those opportunities into the future (Blake 2016).

While having fairly obvious, long-term potential, the financial viability of such projects, including land tenure arrangements, establishment, management and production costs and timing cannot be determined until government policy is clear and established. Understandably, no large-scale investment can be anticipated until such policy is in place and operational.

Bibliography

Chapter 1: A contemporary approach

Alder, A 2017, *Irresistible: The Rise of Addictive Technology and the Business of Keeping us Hooked*, Penguin, New York, US.

Australian Bureau of Statistics (ABS) 2015a, *Australian System of National Accounts, 2014–15*, Cat. No. 5204.0, 2014–15, Australian Capital Territory, Australia.

Australian Bureau of Statistics (ABS) 2015b, *Regional Population Growth, Australia, 2013–14*, Cat. No. 3218.0, 2013–14, Australian Capital Territory, Australia.

Australian Government (ABARE) 2019, Snapshot of Australian Agriculture, viewed 11 December 2019, www.agriculture.gov.au/abares/publications/insights/snapshot-of-australian-agriculture.

Blastland, M & Dilnot, A 2009, *The Numbers Game*, Gotham Books, London, England.

Brugmann, J 2009, *Welcome to the Urban Revolution: How Cities Are Changing the World*, University of Queensland Press, Queensland, Australia.

Brynjolfsson, E & McAfee, A 2014, *The Second Machine Age: Work, Progress and Prosperity in a Time of Brilliant Technologies*, Norton & Company Inc., New York, US.

Cairo, A 2019, *How Charts Lie: Getting Smarter about Visual Information*, Norton & Company Inc., New York, US.

Cole, J 2009, *The Great American University*, Public Affairs Books, Perseus, New York, US.

Cooper, D & Schindler, P 2001, *Business Research Methods*, 7th edn, McGraw-Hill/Irwin, New York, US.

Department of Infrastructure and Transport 2015, *State of Australian Cities 2014–15*, Major Cities Unit, Australian Government, Australian Capital Territory, Australia.

Ferguson, N 2011, *Civilisation*, Penguin Books, London, England.

Fiorilla, P, Kapas, M & Liang, Y 2012, *A Bird's Eye View of Global Real Estate Markets: 2012 update*, Prudential Real Estate Investors, New Jersey, US.

Florida, R 2002, *The Rise of the Creative Class: And How It's Transforming Work, Leisure, Community and Everyday Life*, Basic Books, New York, US.

Fung, K 2010, *Numbers Rule Your World*, McGraw-Hill, New York, US.

Giradet, H 2004, *Cities, People, Planet*, Wiley, West Sussex, England.

Glaeser, E 2011, *Triumph of the City*, Penguin, New York, US.

Hair, J Jr, Money, A, Samouel, P & Page, M 2007, *Research Methods for Business*, Wiley, West Sussex, England.

Harding, A, Scott, A, Laske, S & Burtscher, C 2007, *Bright Satanic Mills*, Ashgate, London, England.

Himmelfarb, G 2005, *The Roads to Modernity, the British, French and American Enlightenments*, Vintage Books, New York, US.

International Energy Agency 2015, *About Us – Frequently Asked Questions: Energy Efficiency,* viewed 10 December 2016, https://www.iea.org/reports/energy-technology-perspectives-2015.

Kaplan, M & Kaplan, E 2006, *Chances Are...: Adventures in Probability,* Viking, London, England.

Kealey, T 2008, *Sex, Science and Profits,* William Heinemann, London, England.

Komninos, N 2002, *Intelligent Cities,* Spon Press, Taylor & Francis, New York, US.

Kotkin, J 2006, *The City: A Global History,* Random House, New York, US.

Kotkin, J 2010, *The Next Hundred Million: America 2050,* Penguin Press, London, England.

Landry, C 2006, *The Art of City Making,* Earthscan, Virginia, US.

Maylor, H & Blackmon, K 2005, *Researching Business and Management,* Palgrave Macmillian, London, England.

Mazur, J 2016, *Fluke: The Maths and Myths of Coincidences,* Oneworld Publications, London, England.

Mitchell, W 1996, *City of Bits (Space, Place and the Infobahn),* MIT Press, Massachusetts, US.

Montgomery, J 2007, *The New Wealth of Cities,* Ashgate, Surrey, England.

Newport, C 2019, *Digital Minimisation,* Portfolio, Penguin, New York, US.

O'Connor W & Lines WJ 2008, *Overloading Australia,* Envirobooks, New South Wales, Australia.

Pearce, F 2010, *The Coming Population Crash: and Our Planet's Surprising Future,* Beacon Press, Massachusetts, US.

Phys.org 2019, *News Website,* viewed 11 December. 2019, https://phys.org/news/2018-05-percent-world-population-urban-areas.html.

Population Reference Bureau 2015, *Human Population Urbanisation,* Washington, US.

Ratcliffe, J, Stubbs, M & Keeping, M 2009, *Urban Planning and Real Estate Development: The Natural and Built Environment Series,* 3rd edn, Routledge, London, England.

Reader, J 2004, *Cities,* Random House, London, England.

Ropeik, D 2010, *How Risky Is It, Really?: Why Our Fears Don't Always Match the Facts,* McGraw-Hill, New York, US.

Saunders, M, Lewis, P & Thornhill, A 2009, *Research Methods for Business Students,* 5th edn, Pearson Education, Essex, England.

Short, J 1996, *The Urban Order,* Blackwell Publishers, Massachusetts, US.

Smith, D 2011, *Population Crisis,* Allen and Unwin, New South Wales, Australia.

Stein, S 2019, *Capital City... Gentrification and the Real Estate State,* Verso. London. UK.

Taleb, N 2010, *The Black Swan: The Impact of the Highly Improbable,* Random House, New York, US.

The World Bank 2014, *Rural Population: Australia,* extracted from the World Development Indicators, The World Bank, Washington, District of Columbia, US.

United Nations Department of Economic and Social Affairs 2017, New York, US.

United Nations, Department of Economic and Social Affairs 2019, *World Population Prospects 2019,* viewed 11 December 2019, https://population.un.org/wpp/Download/Standard/Population/.

Walsh, T 2018, *2062 – The World that A.I. Made.* La Trobe University Press, Melbourne, Australia.

Chapter 2: Economic foundations

Alexander, M 2002, *The Kondratiev Cycle: A generational interpretation,* Universe, Nebraska, US.

Akerlof, G & Shiller, R 2009, *The Animal Spirit,* Princeton University Press, New Jersey, US.

Australian Government (Austrade) 2019, *Why Australia,* viewed 20 December 2019, www.austrade.gov.au/International/Invest/Why-Australia/robust-economy.

Baumol, W & Blinder, A 2006, *Microeconomics, Principles and Policy,* 10th edn, Thomson, Ohio, US.

Bishop, M 2009, *Essential Economics: An A–Z guide,* 2nd edn, Bloomberg Press, New York, US.

Bowles, S 2004, *Microeconomics: Behavior, Institutions, and Evolution,* Princeton University Press, New Jersey, US.

Conway, E 2009, *50 Economics Ideas You Really Need To Know*, Quercus, London, England.

Dasgupta, P 2007, *Economics: A Very Short Introduction*, Oxford University Press, Oxfordshire, England.

Ekelund, R Jr & Hebert, R 1990, *A History of Economic Theory and Method*, 3rd edn, McGraw-Hill, New York, US.

Feigenbaum, S & Hafer, R 2012, *Principles of Macroeconomics: The Way We Live*, Worth, New York, US.

Ferguson, N 2011, *Civilisation the West and the Rest*, Allen Lane, London, England.

Florida, R 2000, *The Rise of the Creative Class and How It Is Transforming Work, Leisure, Community and Everyday Life*, Basic Books, New York, US.

Forstater, M 2007, *Economics*, ABC Books, New South Wales, Australia.

Gans, J, King, S & Mankiw, N 2005, *Principles of Microeconomics*, 3rd edn, Thomson, Victoria, Australia.

Groenewegen, P (ed.) 1998a, *Alfred Marshall: Critical Responses*, vol. 1, Routledge, London, England.

Groenewegen, P (ed.) 1998b, *Alfred Marshall: Critical Responses*, vol. 2, Routledge London, England.

Haakonssen, K 2006, *The Cambridge Companion to Adam Smith*, Cambridge University Press, New York, US.

Häring, N & Storbeck, O 2009, *Economics 2.0: What the best minds in economics can teach us about business and life*, Palgrave Macmillan, New York, US.

Hayek, F 1944, *The Road to Serfdom*, republished Routledge Classics, 2001, London, England.

Himmelfarb, G 2005, *The Roads to Modernity: The British, French and American enlightenments*, Vintage Books, New York, US.

International Monetary Fund 2015, *World Economic Outlook Database 2015*, Washington, District of Columbia, US.

Jackson, J & McIver, R 2007, *Economic Principles*, 2nd edn, McGraw-Hill/ Irwin, New York, US.

Jowsey, E 2011, *Real Estate Economics*, Palgrave Macmillan, London, England.

Keynes, J 1936, *The General Theory of Employment, Interest and Money*, Macmillan, London, England (reprinted 2007).

Leadbeater, C 2000, *Living on Thin Air: The New Economy*, Penguin, London, England.

Lewis, M 2018, *The Fifth Risk… Undoing Democracy*, Allen Lane, London, England.

Maslow, A 1943, 'A Theory of Human Motivation', *Psychological Review* 50(4), 370–396.

Maslow, A 1998, *Maslow on Management*, Wiley, New York, US.

Maslow, A 2000, *The Maslow Business Reader* (Deborah C Stephens ed.), Wiley, New York, US.

Marx, K & Engels, F 1988, *Economic and Philosophic Manuscripts of 1844 and the Communist Manifests* (translated by Martin Milligan), Prometheus Books, New York, US.

McCraw, T 2009, *Prophet of Innovation: Joseph Schumpeter and Creative Destruction*, Harvard University Press, Massachusetts, US.

McLeod, S 2014, *Maslow's Hierarchy of Needs*, viewed 17 October 2015, www.simplypsychology.org/maslow.html.

Medearis, J 2001, *Joseph Schumpeter's Two Theories of Democracy*, Harvard University Press, Massachusetts, US.

Mills, J 2003, *A Critical History of Economics*, Palgrave Macmillan, London, England.

O'Rourke, P 2007, *On the Wealth of Nations*, Allen & Unwin, New South Wales, Australia.

O'Sullivan, A, Sheffrin, S & Perez, S 2008, *Economics: Principles, Applications and Tools*, Pearson Education, New Jersey, US.

Pyhrr, S & Born, W 2006, *Theory and Practice of Real Estate Cycle Analysis*, paper presented at the ARES Annual Meeting, University of Florida, Florida, US.

Psalidopoulos, M (ed.) 2000, *The Canon in the History of Economics: Critical Essays*, Routledge, London, England.

Reed, R & Wu, H 2009, 'Understanding Property Cycles in a Residential Market', *Property Management*, 28(1), 33–46.

Robbins, L 1998, *A History of Economic Thought: The LSE lectures* (Steven G Medema and Warren J Samuels eds.), Princeton University Press, New Jersey, US.

Roncaglia, A 2005, *The Wealth of Ideas: A History of Economic Thought*, Cambridge University Press, Cambridgeshire, England.

Roubini, N 2010, *Crisis Economics: A Crash Course in the Future of Finance*, Allen Lane, New York, US.

Schumpeter, J 1976, *Capitalism, Socialism and Democracy*, 5th edn, George Allen & Unwin, London, England.

Schumpeter, J 1997, *Ten Great Economists: From Marx to Keynes* (first published 1952, George Allen & Unwin, London) reissued 1997, Routledge, London, England.

Scott, P & Judge, G 2000, 'Cycles and Steps in British Commercial Property Values', *Journal of Applied Economics*, 32(10), 1287–1297.

Screpanti, E & Zamagni, S 2005, *An Outline of the History of Economic Thought*, Oxford University Press, New York, US.

Shearmur, J 1996, *Hayek and After: Hayekian Liberalism as a Research Program*, Routledge, London, England.

Skidelsky, R 2009, *Keynes: The Return of the Master*, Penguin Books, London, England.

Skousen, M 2007, *The Big Three in Economics: Adam Smith, Karl Marx and John Maynard Keynes*, Sharpe, New York, US.

Smith, A 1776, *An Inquiry Into the Nature and Causes of the Wealth of Nations* (E. Cannan ed.), vol. 1, The University of Chicago Press, 1977 edition, Chicago, US.

Stanford, J 2008, *Economics for Everyone*, Pluto Press, New York, US.

Vaggi, G & Groenewegen, P 2003, *A Concise History of Economic Thought: From mercantilism to monetarism*, Palgrave Macmillan, London, England.

Vines, D, Maciejowski J & Meade J 2013, *Demand Management: Stagflation*. Routledge. London. UK.

Wapshott, N 2011, *Keynes Hayek: The Clash That Defined Modern Economics*, WW Morton and Co., New York, US.

Wheaton, W, Torto, R & Evans, P 1997, 'The Cyclic Behaviour of the Greater London Office Market', *Journal of Real Estate Finance and Economics*, 15, 77–92.

Wheelan, C 2010, *Naked Economics: Undressing the Dismal Science*, WW Norton, New York, US.

Chapter 3: The financial sector

Australian Bureau of Statistics (ABS) 2019, *Managed Funds Australia June 2019*, Australian Government, Australian Capital Territory, Australia.

Australian Government Treasury 2016, *The Strength of the Australian Financial System* Australian Government, Australian Capital Territory, Australia.

Australian Institute of Health and Welfare (AIHW) 2019, viewed 19 December 2019, www.aihw.gov.au/reports/australias-welfare/home-ownership-and-housing-tenure.

Australian Productivity Commission (APC) 2018, *Inquiry Report in the Australian Financial System*, viewed 20 December 2019, www.pc.gov.au/inquiries/completed/financial-system/report/financial-system-overview.pdf.

Australian Superannuation Funds Association (ASFA) 2019, *Superannuation Statistics*, ASFA, New South Wales, Australia.

Australian Trade Commission 2011, Report: *Australia's Banking Industry*, Australian Government, Australian Capital Territory, Australia.

Australian Trade Commission 2012, Report: *Strong and Sophisticated Financial Services Sector*, Austrade Australian Government, Australian Capital Territory.

Australian Trade and Investment Commission (ATIC) 2017, *Australian Managed Funds 2017 Update*, Australian Government, Australian Capital Territory, Australia.

Brown, G & Matysiak, G 2000, *Real Estate Investment: A Capital Market Approach,* Pearson Education, London, England.

Brueggeman, W & Fisher, J 2008, *Real Estate Finance and Investments,* McGraw-Hill/Irwin, Massachusetts, US.

Das, S 2011, *Extreme Money, the Master of the Universe and the Cult of Risk,* Penguin Books, Victoria, Australia.

Das, S 2015, *A Banquet of Consequences: How We Consume our Own Future,* Viking Books, Victoria, Australia.

Deloitte 2015, *Dynamics of the Australian Superannuation System – The Next 20 Years,* viewed 2 December 2019, www2.deloitte.com/content/dam/Deloitte/au/Documents/financial-services/deloitte-au-fs-dynamics-australian-superannuation-nov-2015.pdf.

Deloitte, 2018, *Banking Industry Outlook 2018,* Deloitte, Sydney.

Doeleman, D & Rogers, R 1986, *Real Estate Financial Management,* 2nd edn, Realtors National Marketing Institute, Illinois, US.

Ferguson, N 2008, *The Ascent of Money: A Financial History of the World,* Allen Lane, Victoria, Australia.

Ferguson, N 2013, *The Great Degeneration: How Institutions Decay and Economics Die,* Allen Lane, New York, US.

Fox, J 2009, *The Myth of the Rational Market: A History of Risk, Reward, and Delusion on Wall Street,* HarperCollins, New York, US.

Hayek, F 1944, *The Road to Serfdom,* republished by Routledge Classics 2001, London, England.

Hefferan, M 2006, *An Examination of the Interface Between Commercial Property Assets and Contemporary Knowledge-intensive Firms – Demands, responses and priorities,* PhD Thesis, Queensland University of Technology, Queensland, Australia.

Hefferan, M 2011, *Housing Choice: Demand preferences in the Sunshine Coast,* University of the Sunshine Coast, Queensland, Australia.

Hollander, S 2008, *The Economics of Karl Marx Analysis and Application: Historical Perspectives on Modern Economies,* Cambridge University Press, Cambridge, England.

IG, 2018, Top Ten – most traded currencies in the world, viewed 2 December 2019, www.ig.com/au/trading-strategies/the-top-ten-most-traded-currencies-in-the-world-180904.

International Monetary Fund (IMF) 2018, *Banking and Financial System Report,* IMF, Washington, District of Columbia, US.

Koukoulas, S 2015, *Myth-Busting Economics,* John Wiley & Sons Australia, Queensland, Australia.

KPMG 2015, *Supertrends: The Trends Shaping Australia's Superannuation Industry,* New South Wales, Australia.

Lewis, M 2018, *The Fifth Risk… Undoing Democracy,* Allen Lane, London, England.

Malkiel, B & Ellis, C 2013, *The Elements of Investing,* Wiley, New Jersey, US.

Newell, G & Sieracki, K (eds.) 2010, *Global Trends in Real Estate Finance,* Blackwell, Iowa, US.

Organisation for Economic Cooperation and Development (OECD) 2005, *Glossary of Statistical Terms: Financial Sector,* viewed 10 December2015, http://stats.oecd.org/glossary/detail. asp?ID=6815.

Phillips, K 2008, *Bad Money: Reckless Finance, Failed Politics, and the Global Crisis of American Capitalism,* Viking, New York, US.

Renton, N 2004, *Renton's Understanding Investment Property: The Essential Reference for Ordinary Savers,* BAS Publishing, Victoria, Australia.

Reserve Bank of Australia 2015, *Assets of Financial Institutions,* RBA, Australian Capital Territory, Australia.

Reserve Bank of Australia 2019, *Snapshot – Components of the Australian Economy,* viewed 8 December 2019, www.rba.gov.au/education/resources/snapshots/economy-composition-snapshot/.

Rowland, P 2010, *Australian Property Investment and Financing,* Thomson Reuters (Professional), New South Wales, Australia.

Saul, J 2009, *The Collapse of Globalisation and the Reinvention of the World,* Penguin Books, New York, US.

Savings.com 2019, *By the Numbers,* viewed 8 December 2019, www.savings.com.au/home-loans/by-the-numbers-australian-home-ownership-tenancy-statistics.

Shiller, R 2012, *Finance and the Good Society,* Princeton University Press, New Jersey, US.

Statista 2018, *Leading Stock exchanges in the Asia Pacific 2018,* viewed 2 December 2019, www.statista.com/statistics/265236/domestic-market-capitalization-in-the-asia-pacific-region/.

Turner, G 2009, *No Way to Run an Economy: Why the System Failed and How to Put It Right,* Pluto Press, London, England.

US Department of Commerce 2019, *Select USA Financial Services Spotlight Report,* viewed 2 December 2019, www.selectusa.gov/financial-services-industry-united-states.

Valdez, S & Wood, J 2003, *An Introduction to Global Financial Markets,* 4th edn, Palgrave Macmillan, London, England.

Wapshott, N 2011, *Keynes Hayek: The Clash That Defined Modern Economics,* Norton, New York, US.

World Population Review 2019, *GDP ranked by country,* viewed 2 December 2019, http://worldpopulationreview.com/countries/countries-by-gdp/.

Chapter 4: The use of land resources – history and trends

Bower, T 2009, *Oil – Money, Politics and Power in the 21st Century,* Grand Central Publishing, New York, US.

Brugmann, J 2009, *Welcome to the Urban Revolution: How Cities Are Changing the World,* University of Queensland Press, Queensland, Australia.

Bryson, B 2015, *The Road to Little Dribbling: More Notes from a Small Island*, Random House, London, England.

Department of Infrastructure and Transport 2015, *State of Australian Cities* 2014 –15, Major Cities Unit, Australian Government, Australian Capital Territory, Australia.

Ehrenhalt, A 2013, *The Great Inversion and the Future of the American City,* Vintage Books, New York, US.

Ellis, B 2009, *The Capitalist Delusion*, Penguin Books (Australia), Victoria, Australia.

Eversole R 2017, 'Economies with People in Them: Regional Futures through the Lens of Contemporary Regional Development Theory', *Australasian Journal of Regional Studies* 23(3), 305–320.

Ferguson, N 2008, *The Ascent of Money: A Financial History of the World,* Allen Lane, Victoria, Australia.

Ferguson, N 2011, *Civilisation,* Penguin Books, London, England.

Florida, R 2002, *The Rise of the Creative Class: And How It's Transforming Work, Leisure, Community, and Everyday Life,* Basic Books, New York, US.

Florida, R 2008, *Who's Your City?* Basic Books, New York, US.

George H 1879, reprint 2016, *Progress and Poverty – An Enquiry into the Cause of Increase of Want with the Increase in Wealth: A Remedy,* Cathedral Classics, Durham, England.

Gittens, R 2010, *The Happy Economist: Happiness for the Hard-Headed,* Allen and Unwin, New South Wales, Australia.

Glaeser, E 2011, *Triumph of the City: How Our Greatest Invention Makes Us Richer, Smarter, Greener, Healthier, and Happier,* Penguin, New York, US.

Glover, D 2015, *An Economy Is Not a Society: Winners and Losers in the New Australia,* Redback 7, Victoria, Australia.

Horan, T 2000, *Digital Places: Building Our City of Bits,* Urban Land Institute, Washington, US.

Jacobs, J 1961, *The Death and Life of Great American Cities*, Reprinted 1993, Random House, New York, US.

Katz, A 2005, *Our Lot: How Real Estate Came to Own US*, Bloomsbury, New York, US.

Kealey, T 2008, *Sex, Science and Profits*, William Heinemann, London, England.

Keen, S 2011, *Debunking Economics – Revised and Expanded Edition: The Naked Emperor Dethroned*, Zane Books, London, England.

Kelly, J & Donegan, P 2015, *City Limits: Why Australian Cities Are Broken and How We Can Fix Them*, Melbourne University Press, Victoria, Australia.

Knox, P (ed.) 2014, *Atlas of Cities*, Princeton University Press, New Jersey, US.

Kotkin, J 2006, *The City: A Global History*, Random House, New York, US.

Kotkin, J 2010, *The Next Hundred Million: American 5050*, Penguin Press, London, England.

Leadbeater, C 2000, *Living on Thin Air: The New Economy*, Penguin Books, London, England.

Libert, B 2010, *Social Nation: How to Harness the Power of Social Media*, John Wiley and Sons, New Jersey, US.

Mallach, A & Bradman, L 2013, *Regenerating Americas Legacy Cities Report*, Lincoln Institute of Land Policy, Massachusetts, US.

Moretti, E 2013, *The New Economy of Jobs*, Mariner Books, New York, US.

Montgomery, C 2013, *Happy City: Transforming Our Lives through Urban Design*, Farrar, Straus and Giroux, New York, US.

Montgomery, J 2007, *The New Wealth of Cities*, Ashgate, Surrey, England.

Morrison, E & Hutcheson S 2019, *Strategic Doing: Ten Skills of Agile Leadership*, John Wiley and Sons, New Jersey, US.

Mumford, L 1961, *The City in History: Its Origins, Its Transformations and Its Prospects*, Harvest Book, New York, US.

Neal, P 2003, *Urban Villages and the Making of Communities*, Spar Press, London, England.

O'Connor, M & Lines, W 2008, *Overloading Australia*, Envirobooks, Canterbury, New South Wales, Australia.

O'Donnell, E T 2017, *Henry George and the Crisis of Inequality*, Columbia University Press, New York, US.

Pearce, F 2010, *The Coming Population Crash: And Our Planet's Surprising Future*, Beacon Press, Massachusetts, US.

Polèse, M 2009, *The Wealth and Poverty of Regions: Why Cities Matter*, The University of Chicago Press, Illinois, US.

Porter, M 1998, *Clusters and the New Economics of Competition*, Harvard Business Review (Nov–Dec Issue), Massachusetts, US.

Quiggin, J 2010, *Zombie Economics: How Dead Ideas Will Work amongst Us*, Princeton University Press, New Jersey, US.

Reader, J 2004, *Cities*, Random House, London, England.

Rubin, J 2009, *Why Your World Is About to Get a Whole Lot Smaller: Oil and the End of Globalization*, Random House, New York, US.

Saar, M & Palang, H 2009, 'The Dimensions of Place Meanings', *Living Reviews in Landscape Research*, 3.3, 1–24.

Short, J 1996, *The Urban Order*, Blackwell Publishers, Massachusetts, US.

Smith, D 2011, *Population Crisis*, Allen and Unwin, New South Wales, Australia.

Stein, S 2019, *Capital City… Gentrification and the Real Estate State*, Verso, London, UK.

Storper, M 2013, *Keys to the City: How Economics, Institutions, Social Interaction and Ethics Shape Development*, Princeton University Press, New Jersey, US.

Sucher, D 2003, *City Comforts*, City Comforts Inc., Washington, District of Columbia, US.

Tietz, J 1999, *The Story of Architecture of the 20th Century*, Konemann, Cologne, Germany.

Tuan, Y 1977, *Space and Place: The perspective of experience*, University of Minnesota Press, Minnesota, US.

Wasik, J 2009, *The Cul-de-sac Syndrome: Turning around the Unsustainable American Dream*, Blooms-bury Press, New York, US.

Wardner, P 2013, *Reassessing the Value Added by Centres Providing Non-retail Employment in Mas-terplanned Communities in South-east Queensland*, PhD thesis, University of the Sunshine Coast, Queensland, Australia.

Weber, K 2012, *Last Call at the Oasis: The global water crisis and were we go from here*, Pegasus Books, New York, US.

Chapter 5: Real Property as an asset

Australian Bureau of Statistics (ABS) 2012, *Fifty Years of Labour Force: Now and then*, Year Book Australia, Cat. No. 1301.0, ABS, Australian Capital Territory, Australia.

Brand, S 1994, *How Buildings Learn – What Happens to Them after They're Built*, Viking Penguin, New York, US.

Churchill, Winston (ed.) 2013, *Never Give In!: Winston Churchill's Speeches*, Bloomsbury Revelations, London, England.

Fiorilla, P, Kapas, M & Liang, Y 2012, *A Bird's Eye View of Global Real Estate Markets: 2012 Update*, Prudential Real Estate Investors, New Jersey, US.

Gleeson, B & Steele, W (eds.) 2010, *A Climate for Growth: Planning South East Queensland*, University of Queensland Press, Queensland, Australia.

GlobalData 2018, Global Construction Outlook to 2022, GlobalData UK, viewed 20 December 2019, https://store.globaldata.com/report/gdcn0010go--global-construction-outlook-to-2022-q3-2018-update/.

Haight, T & Singer, D 2005, *The Real Estate Investment Handbook*, Wiley, New York, US.

Infrastructure Australia 2010, *State of Australian Cities 2010*, Major Cities Unit, Commonwealth of Australia, Australian Capital Territory, Australia.

Jowsey, E 2011, *Real Estate Economics*, Palgrave Macmillan, London, England.

Kolbe, P & Greer, G 2006, *Investment Analysis for Real Estate Decisions*, 6th edn, Dearborn Real Estate Education, Illinois, US.

O'Sullivan, A 2009, *Urban Economics*, 7th edn, McGraw-Hill/Irwin, New York, US.

Pawley, M 1998, *Terminal Architecture*, Reaktion Books, London, England.

Property Council of Australia (PCA) 2015, *Office Market Report*, PCA, New South Wales, Australia.

Rowland, P 2010, *Australian Property Investment and Financing*, Thomson Reuters (Professional), New South Wales, Australia.

Roubini, N 2010, *Crisis Economics: A Crash Course in the Future of Finance*, Allen Lane, New York, US.

Stein, S 2019, *Capital City… Gentrification and the Real Estate State*, Verso, London, UK.

Week, D 2002, *The Culture Driven Workplace, Using Your Company's Knowledge to Design the Office*, Royal Australian Institute of Architects, Australian Capital Territory, Australia.

Chapter 6: Legal and government parameters

Australian Government 2010, *Australia's Future Tax System*, Australian Government, Australian Capital Territory, Australia, viewed 8 December 2014, http://taxreview.treasury.gov.au.

Australian Government 2012, *Budget 2012–13*, Australian Government, viewed 6 November 2014, www.budget.gov.au.329.

Australian Property Institute (API) & Property Institute of New Zealand 2006, *Professional Prac-tice: Your Guide to Being a Member of an Industry-Leading Professional, Property Institute*, Australian Property Institute, Australian Capital Territory, Australia.

Australian Property Institute (API) 2007, *Valuation Principles and Practice*, Australian Property Institute, Australian Capital Territory, Australia.

Bradbrook, A, MacCallum, S, Moore, A & Grattan, S 2011, *Australian Real Property Law*, 5th edn, Thomson Reuters (Professional), New South Wales, Australia.

Chambers, R 2008, *An Introduction to Property Law in Australia*, Thomson Reuters (Professional), New South Wales, Australia.

Duncan, W 2008, *Commercial Leases in Australia*, 5th edn, Thomson Reuters (Professional), New South Wales, Australia.

Hepburn, S 2012, *Real Property Law*, The Law Book Company, Sydney, Australia.

Hinchy, R 2008, *The Australian Legal System: History, Institutions and Method*, Pearson Education, New South Wales, Australia.

Leo, K, Hoggett, J, Sweeting, J & Radford, J 2009, *Company Accounting*, Wiley, Queensland, Australia.

Lonergan, W 2003, *The Valuation of Businesses, Shares and Other Equity*, 4th edn, Allen & Unwin, New South Wales, Australia.

McCluskey, W & Franzsen, R 2005, *Land Value Taxation: An Applied Analysis*, Ashgate, London, England.

Meek, M 2008, *Australian Legal System*, Thomson Reuters (Professional), New South Wales, Australia.

O'Donnell, E 2017, *Henry George and the Crisis of Inequality*. Columbia University Press, New York, US.

Parkinson, P 2010, *Tradition and Change in Australian Law*, Thomson Reuters (Professional), New South Wales, Australia.

Renton, N 2009, *Renton's Understanding Taxation for Investors: A Simple Guide for Families with Shares and Property*, BAS Publishing, Victoria, Australia.

Richardson, K 2018, *Property Law*, 2nd edn, Thomson Reuters (Professional), New South Wales, Australia.

Tooher, J & Dwyer, B 2008, *Introduction to Property Law*, LexisNexis (Australia), New South Wales, Australia.

Wallace, A, Weir, M & McCrimmon, L 2015, *Real Property Law in Queensland*, 4th edn, The Law Book Company, Thomson Reuters (Professional), New South Wales, Australia.

Chapter 7: Real property – tenure and dealings

Australian Property Institute (API) 2007, *Valuation Principles and Practice*, Australian Property Institute, Australian Capital Territory, Australia.

Bradbrook, A, MacCallum, S, Moore, A & Grattan, S 2011, *Australian Real Property Law*, 5th edn, Thomson Reuters (Professional), New South Wales, Australia.

Brennan, F 1998, 'Legislating Liberty: A Bill of Rights for Australia', *QUT Law Review* [S1] 14, 246–249. QUT Brisbane, Australia.

Chambers, R 2008, *An Introduction to Property Law in Australia*, Thomson Reuters (Professional), New South Wales, Australia.

Duncan, W 2008, *Commercial Leases in Australia*, 5th edn, Thomson Reuters (Professional), New South Wales, Australia.

Hyam, A 2004, *The Law Affecting Valuation of Land in Australia*, 3rd edn, The Federation Press, New South Wales, Australia.

Meek, M 2008, *Australian Legal System*, Thomson Reuters (Professional), New South Wales, Australia.

Parkinson, P 2010, *Tradition and Change in Australian law*, Thomson Reuters (Professional), New South Wales, Australia.

Price, R & Griggs, L 2008, *Property Law in Principle*, Thomson Reuters (Professional), New South Wales, Australia.

Reynolds, H 1996, *Aboriginal Sovereignty... Reflections on Race, State and Nation*, Allen and Unwin, Sydney.

Richardson, K 2018, *Property Law*, 2nd edn, Thomson Reuters (Professional), New South Wales, Australia.

Tooher, J & Dwyer, B 2008, *Introduction to Property Law*, LexisNexis (Australia), New South Wales, Australia.

Whipple, R 1986, *Commercial Rent Reviews Law and Valuation Practice*, Thomson Reuters (Professional), New South Wales, Australia.

Chapter 8: Property sectors - urban

Australian Bureau of Statistics (ABS) 2019, *Household Income and Wealth 2017–18*, Cat. No. 6523.0, ABS, Australian Capital Territory, Australia.

Australian Bureau of Statistics (ABS) 2018, *Housing Occupancy and Costs 2017–18*, Cat. No. 4130.0, ABS, Australian Capital Territory, Australia.

Australian Bureau of Statistics (ABS) 2015a, *Labour Force, Australia*, Cat. No. 6202.0, ABS, Australian Capital Territory, Australia.

Australian Bureau of Statistics (ABS) 2015b, *Household and Family Projects, Australia, 2011 to 2036*, Cat. No. 3236.0, ABS, Australian Capital Territory, Australia.

Australian Bureau of Statistics (ABS) 2015c, *Australian National Accounts: State Accounts, 2014–15*, Cat. No. 5220.0, ABS, Australian Capital Territory, Australia.

Australian Bureau of Statistics (ABS) 2015d, *Consumer Price Index, Australia*, Cat. No. 6401.0, ABS, Australian Capital Territory, Australia.

Australian Bureau of Statistics (ABS) 2013a, *Building Activity*, Cat. No. 8752.0, ABS, Australian Capital Territory, Australia.

Australian Bureau of Statistics (ABS) 2013b, *FA Data for Info Consultancies*, Australia, ABS, Australian Capital Territory, Australia.

Australian Bureau of Statistics (ABS) 2011, *2011 Census QuickStats*, ABS, Australian Capital Territory, Australia.

Australian Bureau of Statistics (ABS) 2010, *Building Approvals*, Cat. No. 8731.0, ABS, Australian Capital Territory, Australia.

Australian Bureau of Statistics (ABS) 2009, *Australian Social Trends*, Cat. No. 4102.0, ABS, Australian Capital Territory, Australia.

Australian Bureau of Statistics (ABS) 2001, *Building Approvals, Australia*, Cat. No. 8731.0, ABS, Australian Capital Territory, Australia.

Australian Competition and Consumer Commission (ACCC) 2018, *Report on the Australian Petroleum Market*, ACCC, Canberra, Australia.

Australian Government (Australian Institute of Health and Welfare AIHW) 2018, Older Australia at a glance, viewed 19 December 2019, www.aihw.gov.au/reports/older-people/older-australia-at-a-glance/contents/health-aged-care-service-use/aged-care-assessments.

Australian Government (Australian Institute of Health and Welfare AIHW) 2018, *Housing Assistance in Australia*, viewed 19 December 2019, https://www.aihw.gov.au/reports/housing-assistance/housing-assistance-in-australia-2018/contents/social-housing-dwellings.

Australian Government (Australian Institute of Health and Welfare AIHW) 2019a, *Home Ownership and Housing Tenure*, viewed 19 December 2019, www.aihw.gov.au/reports/australias-welfare/home-ownership-and-housing-tenure.

Australian Government (Australian Institute of Health and Welfare AIHW) 2019b, *Deaths in Australia*, viewed 19 December 2019, www.aihw.gov.au/reports/life-expectancy-death/deaths-in-australia/contents/life-expectancy.

Australian Government (Australian Institute of Health and Welfare AIHW) 2019c, *Home ownership and Housing Tenure*, viewed 19 December 2019, https://www.aihw.gov.au/reports/australias-welfare/home-ownership-and-housing-tenure.

Australian Government (Department of Education, Skills and Employment) 2019, *Child Care in Australia Report 2018*, Australian Capital Territory, Australia.

Australian Government (Department of Foreign Affairs and Trade, DFAT) 2017, *The Importance of Services Trade to Australia*, viewed 20 December 2019, https://dfat.gov.au/trade/services-and-digital-trade/Pages/the-importance-of-services-trade-to-australia.aspx.

Australian Government Parliamentary Library 2016, *Employment in Australia*, viewed 20 December 2019, https://www.aph.gov.au/About_Parliament/Parliamentary_Departments/Parliamentary_Library/pubs/BriefingBook45p/EmploymentAustralia.

Australian Productivity Commission (APC) 2018, *Inquiry Report in the Australian Financial System*, viewed 20 December 2019, www.pc.gov.au/inquiries/completed/financial-system/report/financial-system-overview.pdf.

Australian Property Institute (API) 1996, *Specialist Valuations in Australia and New Zealand-theory and practice*. API, Canberra Australia.

Australian Property Institute (API) 2007, *Glossary of Property Terms*, Australian Property Institute, Queensland, Australia.

Ball, M, Lizieri, C & MacGregor, B 1998, *The Economics of Commercial Property Markets*, Routledge, New York, US.

Baker Consulting 2018, *Reported in Shopping Centre Council of Australia – Key Facts 2018*, viewed 19 December 2019, www.scca.org.au/industry-information/key-facts/.

Bankwest Limited 2019, *Focus on Childcare*, viewed 10 December 2019, www.bankwest.com.au/content/dam/bankwest/documents/business/insights/focus-on-childcare-report-2019.pdf.

Belbin, M 2010, *Size Matters: How Many Make the Ideal Team?* viewed 15 March 2011, https://www.belbin.com/resources/blogs/size-does-matter/

Borg, A 2012, 'Architecture Australia, January 2012', *Architecture Australia* 101(1), Architecture Media.

Brynjolfsson. E & McAfee A 2014, *The Second Machine Age: Work, Progress and Prosperity in a Time of Brilliant Technologies*, Norton & Company Inc., New York, USA.

Burgess Rawson 2018, *Childcare Investment Report 2018*, viewed 11 December 2019, https://lp.burgessrawson.com.au/childcare.

CBRE 2018, *Logistics and Retail Converge amid Australia's Omni-Channel Revolution*, viewed 20 December 2019, https://www.cbre.com.au/about/media-center/logistics-and-retail-converge-amid-australias-omni-channel-revolution

CommSec. 2018, *Economic Insights* – Housing Size Trend Report, viewed 10 November 2019, www.commsec.com.au/content/dam/EN/ResearchNews/2018Reports/November/ECO_Insights_191118_CommSec-Home-Size.pdf.

Department of Education, Employment and Workplace Relations (DEEWR) 2011, *Australian Jobs 2011*, Australian Government, Australian Capital Territory, Australia.

Ernst & Young 2015, *2015 PCA Office Market Report*, Sydney, NSW, Property Council of Australia.

Florida, R 2002, *The Rise of the Creative Class: And How It's Transforming Work, Leisure, Community and Everyday Life*, Basic Books, New York, US.

Florida, R 2004, *Cities and the Creative Class*, Routledge, New York, US.

Glover D 2015, *An Economy Is Not a Society: Winners and Losers in the New Australia*, Black Inc. Melbourne, Australia.

Green Building Council of Australia & Ernst and Young 2016, *Mid-Tier Commercial Office Buildings in Australia*, Sydney, Australia.

Haight, T & Singer, D 2005, *The Real Estate Investment Handbook*, Wiley, New York, US.

Hefferan, M 2006, *An Examination of the Interface Between Commercial Property Assets and Contemporary Knowledge-intensive Firms – Demands, Responses and Priorities*, PhD Thesis, Queensland University of Technology, Queensland, Australia.

Hefferan, M 2011, *Housing Choice: Demand Preferences in the Sunshine Coast,* University of the Sunshine Coast, Queensland, Australia.

Housing Industry Association (HIA), 2018 *Dwelling Unit Commencements Data 1955–2018,* Housing Industry Association, Canberra, Australia.

Hugo, G 2001, *A Century of Population Change in Australia,* Australian Capital Territory, Australia, Australian Bureau of Statistics (ABS).

ID Community 2016, *Community Profile Australia,* viewed 20 December 2019, https://profile.id.com.au/australia/household-size.

Johnson, D & Johnson, F 1997, *Joining Together: Group Theory and Group Skills,* 6th edn, Allyn & Bacon, Massachusetts, US.

Kelly, A 2012, *IBIS World Industry Report 411: Commercial and Industrial Building Construction in Australia,* Ibis, Victoria, Australia.

Kelly, A 2015, *IBIS World Industry Report E3021: Commercial and Industrial Building Construction in Australia,* Ibis, Victoria, Australia.

Kolbe, P & Greer, G 2006, *Investment Analysis for Real Estate Decisions,* 6th edn, Dearborn Real Estate Education, Illinois, US.

Kotkin, J 2006, *The City: A Global History,* Random House, New York, US.

Kotkin, J 2010, *The Next Hundred Million: America 2050,* Penguin Press, London, England, New York, US.

Koukoulas, S 2015, *Myth-Busting Economics,* John Wiley & Sons Australia, Queensland, Australia.

Kusher, C 2015, *Home Value Growth Across Australia's Capital Cities Has Slowed over Time,* CoreLogic RP Data, Report, CoreLogic, Sydney, Australia.

Leer, A 1999, *Welcome to the Wired World: Key Strategic Agendas for Commerce in the Digital Age,* Financial Times/Prentice Hall, London, England.

Libert, B 2010, *Social Nation: How to Harness the Power of Social Media to Attract Customers, Motivate Employees, and Grow Your Business,* Wiley, New Jersey, US.

Marine Industries Association of Australia 2010, *Size and Characteristics of the Australian Marina Sector,* Industry Report, MIAA, Sydney, Australia.

McCraw, T 2009, *Prophet of Innovation: Joseph Schumpeter and Creative Destruction,* Harvard University Press, Massachusetts, US.

McDonald, J & McMillen, D 2010, *Urban Economics and Real Estate: Theory and policy,* Wiley, New York, US.

National Australian Bank (NAB) 2015, *Online Retail Sales Inbox – Update,* Victoria, Australia.

Nguyen, J 2007, *Definitions, Shopping Centre Directories,* Property Council of Australia, Sydney, Australia.

O'Sullivan, A 2009, *Urban Economics,* 7th edn, McGraw-Hill/Irwin, New York, US.

Parliament of Australia 2018, *Child Care and Early Childhood Education,* viewed 9 December 2019, www.aph.gov.au/About_Parliament/Parliamentary_Departments/Parliamentary_Library/pubs/BriefingBook46p/Childcare.

Pech, J, Nelms, K, Yuen, K & Bolton, T 2009, *Retail Trade Industry Profile,* Australian Fair Pay Commission, Australian Government, Australian Capital Territory, Australia.

Pawley, M 1998, *Terminal Architecture,* Reaktion Books, London, UK.

Profile id 2019, *Australia, Community Profile – Household Size,* viewed 12 December 2019, https://profile.id.com.au/australia/household-size.

Property Council of Australia (PCA) 2015, *Office Market Report,* PCA, New South Wales, Australia.

Property Council of Australia (PCA) 2019, *Office Space – Australia,* unpublished data viewed Brisbane 16 October 2019.

Property Council of Australia (PCA) 2017, *Fuel Stations of the Future,* on-line article, viewed 6 December 2019, www.propertycouncil.com.au/Web/News/Articles/News_listing/Web/Content/News/National/2017/Fuel_stations_of_the_future.aspx.

Putnam, R 2000, *Bowling Alone: The Collapse and Revival of American Community*, Simon & Schuster, New York, US.

Pyhrr, S, Cooper, J, Wofford, L, Kapplin, S & Lapides, P 1989, *Real Estate Investment*, Wiley, New York, US.

Raymond, R 2019, *Insights. Reshaping the Industrial Real Estate Sector: 2019 Outlook*, viewed 20 December 2019, https://lpc.com.au/insights/reshaping-the-industrial-real-estate-sector-2019-outlook.

Reed, R & Wu, H 2009, 'Understanding Property Cycles in a Residential Market', *Property Management*, 28(1), 33–46.

Reserve Bank of Australia (RBA) 2016, *Indicator Lending Rates – F5*, Australian Capital Territory, Australia.

Reserve Bank of Australia (RBA) 2019, *Snapshot – Composition of the Australian Economy*, viewed 20 December 2019, www.rba.gov.au/snapshots/economy-composition-snapshot/.

Roy Morgan Research 2019, *Research Summary. More Australians Intend to Retire Despite Inadequate Savings Levels*, viewed 20 December 2019, www.roymorgan.com/findings/7949-retirement-intention-201904260129.

Shopping Centre Council of Australia (SCCA) 2007, *Shopping Centre Facts*, viewed 7 February 2011, www.scca.org.au/HTML%20Pages/Research.htm.

Smith, M 2008, *What Is a Group?*, viewed 15 March 2011, www.infed.org/groupwork/what_is_a_group.htm.

Statista 2019, Retail e-commerce sales 2014–2019, viewed 20 December 2019, www.statista.com/statistics/379133/e-commerce-share-of-retail-sales-in-australia/.

Trading Economics 2019, *Australian Retail Sales MoM*, viewed 20 December 2019, https://tradingeconomics.com/australia/retail-sales-annual.

Urbis JHD 2007, *Australian Shopping Centre Industry: Information update*, March, report for Shopping Centre Council of Australia, Urbis JHD, New South Wales, Australia.

Urbis JHD 2015, *Australian Shopping Centre Industry: Scale and Performance Measures, August*, report for Shopping Centre Council of Australia, Urbis JHD, New South Wales, Australia.

Week, D 2002, *The Culture Driven Workplace: Using Your Company's Knowledge to Design the Office*, Royal Australian Institute of Architects, Australian Capital Territory, Australia.

Chapter 9: Property sectors – rural

Australian Government (Department of Agriculture and Water Resources) 2018a. *ABARE Agricultural Commodities Report September 2019*, Commonwealth Government, Canberra.

Australian Government (Department of Agriculture and Water Resources) 2018b. *ABARE Beef Cattle Facts Sheet, November 2019*, viewed 10 December 2019, https://www.agriculture.gov.au/abares/research-topics/surveys/beef.

Australian Government (Department of Agriculture and Water Resources) 2018c. *ABARE Dairy Industry Facts Sheet, September 2019*, viewed 10 December 2019, www.agriculture.gov.au/abares/research-topics/surveys/dairy.

Australian Government (Department of Agriculture and Water Resources) 2018d. *Agricultural Lending Data 2016–17*. Commonwealth Government, Canberra.

Australian Government (ABARE) 2019, *Farm Debt 2014–15 to 2016–17*, viewed 17 December 2019, https://www.agriculture.gov.au/abares/research-topics/surveys/debt.

Australian Government (ABARE) 2018, *Agricultural Lending Data 2016–17*. Department of Agriculture and Water Resources, Canberra, Australia.

Australian Government (ABARE), 2019, *2016 Snapshot of Australian agricultural Workforce*, viewed 17 December 2019, www.agriculture.gov.au/abares/publications/insights/snapshot-of-australias-agricultural-workforce.

Baxter, J & Cohen, R 2009, *Rural Valuation: The Australian guide to complex rural valuation practice*, Australian Property Institute, Australian Capital Territory, Australia.

Chan G 2018, *Rusted Off, Why Country Australia is Fed Up*. Vintage Australia North Sydney Australia.

Eversole R 2017, 'Economies with People in Them: Regional Futures through the Lens of Contemporary Regional Development Theory', *Australasian Journal of Regional Studies* 23(3), 305–320.

Grain Research and Development Corporation (GRDC) 2018, *Research Development and Extension Plan 2018 – 2023*, GRDC, Barton, Australian Capital Territory, Australia.

Hort Innovation 2019, *Australian Horticulture Statistics Handbook 2018/9*, Hort Innovation, North Sydney, NSW, Australia.

Lewis, M 2018, *The Fifth Risk… Undoing Democracy*, Allen Lane, London, England.

Moretti, E 2013, *The New Economy of Jobs*, Mariner Books, New York, US.

Morrison D 2019, *The Wine Gourd – Exploring the Worlds Wine Data*, viewed 20 December 2019, http://winegourd.blogspot.com/2019/.

National Farmers' Federation 2017, *Farm 2017* and *Farm Facts*, National Farmers' Federation, Canberra, Australia.

O'Rourke, P J 2018, *None of my Business*, Grove Press. London, England.

Polèse, M 2009, *The Wealth and Poverty of Regions: Why Cities Matter*, The University of Chicago Press, Illinois, US.

Regional Australia Institute (RAI) 2019, *Regional Population Growth – Are We Ready?* RAI, Canberra, Australia.

Scarratt, D & Osborne, S 2014, *Property Valuations: The Five Methods*, Routledge, Oxfordshire, England.

The World Bank 2014, *Rural Population: Australia,* extracted from the World Development Indicators, The World Bank, Washington, District of Columbia, US.

Wine Australia 2019, *Australian Wine: Production, Sales and inventory 2017–18,* Wine Australia, Adelaide South Australia.

Chapter 10: Valuation theory and applications

Australian Property Institute (API) 1996, *Specialist Valuations in Australia and New Zealand: Theory and Practice,* Australian Institute of Valuers and Land Economists, Australian Capital Territory, Australia.

Australian Property Institute (API) 2007a, *Glossary of Property Terms*, Australian Property Institute, Australian Capital Territory, Australia.

Australian Property Institute (API) 2007b, *Valuation Principles and Practice*, Australian Property Institute, Australian Capital Territory, Australia.

Australian Property Institute (API) 2007c, *The Valuation of Real Estate: Australian Edition of the Appraisal of Real Estate*, 12th edn (Richard Reed ed.), Australian Property Institute, Australian Capital Territory, Australia.

Australian Property Institute and Property Institution of New Zealand 2012, *Australian and New Zealand Valuation and Property Standards,* API, Australian Capital Territory, Australia.

Ball, M, Lizieri, C & MacGregor, B 1998, *The Economics of Commercial Property Markets,* Routledge, New York, US.

Baxter, J & Cohen, R 2009, *Rural Valuation: The Australian guide to complex rural valuation practice,* Australian Property Institute, Australian Capital Territory, Australia.

Commonwealth of Australia 1901, *Australian Constitution*, Parliament of Australia, Australian Capital Territory, Australia.

Enever, N & Isaac, D 2002, *The Valuation of Property Investments*, Estates Gazette, London, England.

Eves, C 2004, *The Use of Income Valuation Methods to Value Rural Property*, International Real Estate Research Symposium, 13–14 April 2014, conference paper.

Ferguson, N 2013, *The Great Degeneration: How Institutions Decay and Economies Die*, Allen Lane, New York, US.

Fung, K 2010, *Numbers Rule Your World*, McGraw-Hill, New York.

Haight, T & Singer, D 2005, *The Real Estate Investment Handbook*, Wiley, New York.

Hartley, S 2009, *Project Management Principles & Strategies*, Pearson Education, New South Wales, Australia.

Harvey, J & Jowsey, E 2004, *Urban Land Economics*, Palgrave Macmillan, London, England.

Hefferan, M 2015, *Changing Dynamics – Pastoral and Remote Property Assessment in Australia*, 21st Pacific Rim Real Estate Society Conference, Kuala Lumpur, Malaysia, 18–21 January 2015, conference paper.

Hoesli, M & MacGregor, B 2000, *Property Investment: Principles and Practice of Portfolio Management*, Pearson Education, London, England.

Hyam, A 2004, *The Law Affecting Valuation of Land in Australia*, 3rd edn, The Federation Press, New South Wales, Australia.

International Valuation Standards Committee (IVSC), 2013, *International Valuation Standards,* 8th edn, International Valuation Standards Council, London, England.

Isaac, D & O'Leary, J 2012, *Property Valuation Principles*, 2nd edn, Palgrave Macmillan, London, England.

Kolbe, P & Greer, G 2006, *Investment Analysis for Real Estate Decisions*, 6th edn, Dearborn Real Estate Education, Illinois, US.

Newell, G & Sieracki, K (eds.) 2010, *Global Trends in Real Estate Finance*, Blackwell, Iowa, US.

Petrole, J 2007, *Financial and Investment Analysis for Commercial Real Estate,* Dearborn Real Estate Education, Illinois, US.

Property Council of Australia (PCA) 2019, *Glossary of Terms*, viewed 15 December 2019, https://www.propertycouncil.com.au/Web/Events___Services/Research_Services/Glossary_of_Terms.aspx

Reich, R B 2018, *The Common Good*, Borzoi Books, New York, US.

Reinhart, C. & Rogoff, K 2009, *This Time Is Different*, Princeton University Press, New Jersey, US.

Rowland, P 2010, *Australian Property Investment and Financing*, Thomson Reuters (Professional), New South Wales, Australia.

Royal Institute of Chartered Surveyors (RICS) 2014, *RICS Valuation – Professional Standards (the 'Red Book')*, RICS, London, England.

Scarratt, D & Osborne, S 2014, *Property Valuations: The Five Methods,* Routledge, Oxfordshire, England.

Siggelkow, N. & Terwiesch, C 2019, *Connected Strategy*, Harvard Business Review Press, Massachusetts, US.

Whipple, R 1986, *Commercial Rent Reviews Law and Valuation Practice,* Thomson Reuters (Professional), New South Wales, Australia.

Whipple, R 2006, *Property Valuation and Analysis*, 2nd edn, Thomson Reuters (Professional), New South Wales, Australia.

Wyatt, P 2007, *Property Valuation in an Economic Context*, Blackwell, Massachusetts, US.

Chapter II: Compulsory Acquisition

Brown, D. 2009, *Land Acquisition,* 6th edn, LexisNexis, Butterworths, Australia.

Carr R K 1942, *The Supreme Court and Judicial Review*, Farrar and Binehart New York, US.

Hepburn S 2012, *Real Property Law*, Law Book Company Pyrmont, Australia.

Hyam, A, 2014, *The Law Affecting Valuation of Land in Australia*, 5th edn, The Federation Press, Sydney.

Reich R B, 2018, *The Common Good*, Borzoi Press, New York, US.

Chapter 12: Land subdivision and property development

Dobie, C 2007, *A Handbook of Project Management: A complete guide for beginners to professionals,* Allen & Unwin, New South Wales, Australia.

Ehrenhalt, A 2013, *The Great Inversion and the Future of the American City,* Vintage Books, New York, US.

Jowsey, E 2011, *Real Estate Economics,* Palgrave Macmillan, London, England.

Kelly, J & Donegan, P 2015, *City Limits: Why Australian Cities Are Broken and How We Fix Them,* Melbourne University Press, Victoria, Australia.

Malkiel, B & Ellis, C 2010, *The Elements of Investing,* Wiley, New Jersey, US.

Mendleson, R 2009, *The 10 Mistakes Businesses Make and How to Avoid Them,* New Holland, New South Wales, Australia.

Miles, M, Berens, G, Eppli, M, Weiss, M & Urban Land Institute 2000, *Real Estate Development: Principles and process,* Urban Land Institute, Washington, District of Columbia, US.

Peca, S 2009, *Real Estate Development and Investment: A Comprehensive Approach,* Wiley, New Jersey, US.

Reed, R & Wu, H 2009, 'Understanding Property Cycles in a Residential Market', *Property Management,* 28(1), 33–46.

Sutton, G 2005, *Corporate Canaries: Avoiding Business Disasters with a Coal Miner's Secrets,* Nelson Business, Tennessee, US.

Wardner, P 2013, *Reassessing the Value Added by Centres Providing Non-retail Employment in Masterplanned Communities in South-east Queensland,* PhD thesis, University of the Sunshine Coast, Queensland, Australia.

Chapter 13: Property management and facilities management

Australian Bureau of Statistics (ABS) 2011, *2011 Census Quick Stats,* ABS, Australian Capital Territory, Australia.

Australian Government (Australian Institute of Health and Welfare AIHW) 2019, viewed 19 December 2019, www.aihw.gov.au/reports/australias-welfare/home-ownership-and-housing-tenure.

Angel, P 2003, *Facility Management Contracting Guidelines,* Facility Management Association of Australia, Victoria, Australia.

Brand, S 1994, *How Buildings Learn: What Happens after They're Built,* Viking Penguin, New York, US.

Duncan, W 2006, *Real Estate Agency Law in Queensland,* 4th edn, Thomson Reuters (Professional), New South Wales, Australia.

Duncan, W 2008, *Commercial Leases in Australia,* 5th edn, Thomson Reuters (Professional), New South Wales, Australia.

Edwards, V & Ellison, L 2003, *Corporate Property Management: Aligning Real Estate with Business Strategy,* Blackwell, Oxfordshire, England.

Johnson, M 1997, *Outsourcing in Brief,* Butterworth-Heinemann, Oxfordshire, England.

Kyle, R 2004, *Property Management,* 7th edn, Dearborn Financial, Illinois, US.

Martin, D 2006, *The A–Z of Facilities and Property Management,* Thorogood, London, England.

Pyhrr, S, Cooper, J, Wofford, L, Kapplin, S & Lapides, P 1989, *Real Estate Investment,* Wiley, New Jersey, US.

Scarrett, D 1995, *Property Asset Management,* 2nd edn, Spon, London, England.

Williams, B 2006, *Facilities Economics in the European Union,* International Facilities and Property Information, London, England.

Chapter 14: Taxation and real property assets

Australian Government 2010, *Australia's Future Tax System*, Australian Government, Australian Capital Territory, Australia, viewed 8 December 2014, http://taxreview.treasury.gov.au.

Australian Government 2012, *Budget 2012–13,* Australian Government, viewed 6 November 2014, www.budget.gov.au.

Australian Property Institute (API) & Property Institute of New Zealand 2006, *Professional Practice: Your Guide to Being a Member of an Industry-Leading Professional*, Property Institute, Australian Property Institute, Australian Capital Territory, Australia.

Australian Property Institute (API) 2007, *Valuation Principles and Practice*, Australian Property Institute, Australian Capital Territory, Australia.

Leo, K, Hoggett, J, Sweeting, J & Radford, J 2009, *Company Accounting*, Wiley, Queensland, Australia.

Lonergan, W 2003, *The Valuation of Businesses, Shares and Other Equity*, 4th edn, Allen & Unwin, New South Wales, Australia.

McCluskey, W & Franzsen, R 2005, *Land Value Taxation: An applied analysis*, Ashgate, London, England.

O'Donnell, E T 2017, *Henry George and the Crisis of Inequality*, Columbia University Press, New York, US.

Renton, N 2009, *Renton's Understanding Taxation for Investors: A Simple Guide for Families with Shares and Property*, BAS Publishing, Victoria, Australia.

Stein, S 2019, *Capital City… Gentrification and the Real Estate State*, Verso, London, England.

Chapter 15: Contemporary issues – environmental health, heritage and sustainability

Appleby, P 2011, *Integrated Sustainable Design of Buildings,* Earthscan, London, England.

Australian Government (Department of Agriculture and Water Resources) 2019, *ABARE, Forests*, November 2019, viewed 10 December 2019, https://www.agriculture.gov.au/abares/research-topics/forests.

Bell, P 1990, *Heritage Conservation in Australia*, published paper, ICOM CD3 Conference, Heritage Futures for Queensland, Queensland, Australia.

Blake, A G 2016, Carbon sequestration: Evaluating the impact on rural land and valuation approach. PhD thesis, Queensland University of Technology, Brisbane, Australia.

Brand, S 1994, *How Buildings Learn: What Happens to Them after They're Built*, Viking, New York, US.

Brand, S 2010, *Whole Earth Discipline: Why Dense Cities, Nuclear Power, Transgenic Cities Restored Wildlands and Geoengineering Are Necessary*, Viking Penguin, New York, US.

Breven, M 2018, *The European Union and Forest*, European Union Forest Factsheet, EU, Brussels, Belgium.

Brugmann, J 2009, *Welcome to the Urban Revolution: How Cities Are Changing the World*, University of Queensland Press, Queensland, Australia.

Churchill, Winston (ed.) 2013, *Never Give In!: Winston Churchill's Speeches*, Bloomsbury Revelations.

Davison, G & McConville, C 1991, *A Heritage Handbook*, Allen & Unwin, New South Wales, Australia.

ETN Communications 2007, *Sustainable Nation: Managing Australia's future*, Design Masters Press, New South Wales, Australia.

Freeman, R Pierce, J & Dodd, R 2000, *Environmentalism and the New Logic of Business*, Oxford University Press, New York, US.

Gilding, P 2011, *The Great Disruption*, Bloomsbury, London, England.

Giradet, H 2004, *Cities, People, Planet*, Wiley, West Sussex, England.

Glaeser, E 2011, *Triumph of the City*, Penguin, New York, US.

Gleeson, B 2010, *Lifeboat Cities*, University of New South Wales Press, Sydney, New South Wales, Australia.

Hamin, E, Geigis, P & Silka, L (eds.) 2007, *Preserving and Enhancing Communities: A Guide for Citizens, Planners and Policymakers*, University of Massachusetts Press, Massachusetts, US.

Hargroves, K & Smith, M (eds.) 2005, *The Natural Advantage of Nations*, Earthscan, London, England.

Hyde, R, Watson, S, Cheshire, W & Thomson, M 2007, *The Environment Brief: Pathways for Green Design,* Taylor & Francis, Oxfordshire, England.

International Chamber for the Conservation of Movements and Sites 1988, *The Australian ICOMOS Charter for the Conservation of Places of Cultural Significance (The Burra Charter)*, National Trust for Historic Preservation, ICOMOS, New South Wales, Australia.

International Energy Agency 2015, *About Us – Frequently Asked Questions: Energy Efficiency*, viewed 10 December 2016, www.iea.org/reports/energy-technology-perspectives-2015.

Jackson, T 2009, *Prosperity without Growth: Economics for a Finite Planet*, Earthscan, Virginia, US.

Jowsey, E 2011, *Real Estate Economics*, Palgrave Macmillan, London, England.

Kilbert, C 2005, *Sustainable Construction: Green Building Design and Delivery*, Wiley, New Jersey, US.

Labadi, S & Long, C (eds.) 2010, *Heritage and Globalisation*, Routledge, London, England.

May, J 2006, *My Office is Killing Me! …the Sick Building Survival Guide*, Johns Hopkins University Press, Maryland, US.

McCraw, T 2009, *Prophet of Innovation: Joseph Schumpeter and Creative Destruction*, Harvard University Press, Massachusetts, US.

McKibben, B 2007, *Deep Economy: The Wealth of Communities and the Durable Future*, Holt, New York, US.

Michaels, P 2004, *Meltdown: The Predictable Distortion of Global Warming by Scientists, Politicians and the Media*, Cato Institute, Washington, District of Columbia, US.

Miller, G 2007, *Living in the Environment*, 15th edn, Thomson Brooks/Cole, Kentucky, US.

Montgomery, C 2013, *Happy Cities: Transforming Our Lives through Urban Design*, Farrar, Straus and Giroux, New York, US.

Morgan, T 2013, *Life after Growth,* Harriman House, Hampshire, England.

Mumford, L 1961, *The City in History: Its Origins, Its Transformations and Its Prospects,* Harvest Book, New York, US.

National Trust for Historic Preservation 2012, *A Brief History of the National Trust*, viewed March 2012, www.preservationnation.org/ who-we-are/history.html.

Nelson, A, Randolph, J, McElfish, J, Schilling, J, Logan, J & LLC Newport Partners 2008, *Environmental Regulations and Housing Costs*, Island Press, Washington, District of Columbia, US.

Nijnic, M 2010, 'Carbon Capture and Storage in Forests', *Issues in Environmental Science and Technology* 29.

Noon, K & Ward, J 2007, *Green Wealth,* Squareone, New York, US.

Office of the Chief Scientist 2019, *Resource and Energy Quarterly*, viewed 21 December 2019, https://publications.industry.gov.au/publications/resourcesandenergyquarterlydecember2019/documents/Resources-and-Energy-Quarterly-December-2019.pdf.

Owen, D 2009, *Green Metropolis: Why Living Smaller, Living Closer, and Driving Less Are the Keys to Sustainability*, New York, US.

Property Council of Australia (PCA) 2015, *Office Market Report*, PCA, New South Wales, Australia.

Property Council of Australia (PCA) 2010, *A Guide to Office Building Quality*, Guidelines Document, Property Council of Australia, Sydney, Australia.

Rubin, J 2009, *Why Your World Is About to Get a Whole Lot Smaller: Oil and the End of Globalization*, Random House, New York, US.

Sanderman, J., Farquharson, R & Baldock, J 2009, *Flagship Sustainable Agriculture Soils – Carbon Sequestration Potential – A Review for Australian Agriculture*. CSIRO, Adelaide, South Australia.

Santamouris, M (ed.) 2006, *Environmental Design of Urban Buildings: An Integrated Approach*, Earthscan, Sterling, Virginia, US.

Sassi, P 2006, *Strategies for Sustainable Architecture*, Taylor & Francis, New York, US.

Slaper, T & Hall, T 2011, 'The Triple Bottom Line: What is it and how does it work?', *Indiana Business Review*, 86(1), 4–8.

Sucher, D 2003, *City Comforts*, City Comforts Inc, Washington, US.

US Department of Health and Human and Services 1998, *An Ensemble of Definitions of Environmental Health*, Washington, District of Columbia, US.

Venning, J & Higgins, J (eds.) 2001, *Towards Sustainability*, University of New South Wales Press, New South Wales, Australia.

Week, D 2002, *The Culture Driven Workplace: Using Your Company's Knowledge to Design the Office*, Royal Australian Institute of Architects, Australian Capital Territory, Australia.

Yudelson, J 2008, *The Green Building Revolution*, Island Press, Washington, District of Columbia, US.

Index

Note: **Bold** page numbers refer to tables; *italic* page numbers refer to figures and page numbers